January 2008

LAYING THE HOE

A CENTURY OF IRON MANUFACTURING IN STAFFORD COUNTY, VIRGINIA

WITH GENEALOGICAL NOTES ON OVER 300 FAMILIES

Jerrilynn Eby

HERITAGE BOOKS
2007

HERITAGE BOOKS
AN IMPRINT OF HERITAGE BOOKS, INC.

Books, CDs, and more—Worldwide

For our listing of thousands of titles see our website
at
www.HeritageBooks.com

Published 2007 by
HERITAGE BOOKS, INC.
Publishing Division
65 East Main Street
Westminster, Maryland 21157-5026

Other books by the author:

Laying the Hoe: A Century of Iron Manufacturing in Stafford County, Virginia

Men of Mark in Stafford County, Virginia: A Listing of County Officials, 1664-1991, with Accompanying Biographical Annotations

Men of Mark: Officials of Stafford County, Virginia

Stafford County, Virginia Officials, 1664-1991: Taken from the 1783 Records in the Stafford County Courthouse

They Called Stafford Home: The Development of Stafford County, Virginia, from 1600 until 1865

International Standard Book Number: 978-1-58549-863-5

Table of Contents

List of Illustrations

Introduction

Cooking pots, firebacks, shovels, horseshoes, nails, andirons. Iron was an essential commodity during the colonial period and a preponderance of metal items available to settlers were made of iron. The Chesapeake region, which included Virginia, Maryland, Delaware, and Pennsylvania, offered early industrialists an abundance of high-quality iron ore in easily worked surface deposits. The vast forests provided a seemingly inexhaustible supply of timber for charcoal used as fuel for smelting. The region was further blessed with strong streams for water power and suitable furnace sites were usually located on or near navigable water. The combination of these important features assured the development of a robust iron industry in the Chesapeake region.

Iron making was the first industry to be established in the New World. In 1607/08 a bloomery was built at Jamestown and produced small quantities of iron from local ore. The first actual furnace was built in 1622 at Falling Creek just below what is now the city of Richmond, Virginia. This was far from a successful venture. Indians destroyed the facility during its first season and killed most of the workers. Nearly a century passed before serious smelting was again attempted. By the early 1720s iron making was firmly established and quickly became a thriving industry. The growth of the Chesapeake iron industry was such that, by the outbreak of the American Revolution, there were more functional blast furnaces in America than in Britain.

The focus of this book is on Accokeek Furnace and Rappahannock Forge, two large 18th century iron works located in Stafford County, Virginia. Little research has heretofore been done on either of these works despite the fact that they had a major impact on the social, political, and economic life of eastern Virginia. It is impossible to study Accokeek and Rappahannock without looking at the Chesapeake iron industry as a whole, for these two businesses were merely individual components of a much larger entity. Both of these operations, however, were unique in their own ways. For some 25 years Accokeek, which was in business during the first half of the 18th century, was the headquarters of the Principio Iron Company, colonial America's largest iron producer. Rappahannock Forge, which operated during the second half of the century, was one of the New World's largest industrial complexes and, in addition to iron, produced a wide variety of manufactured goods.

An 18th century iron works was comprised of many individual parts and covered thousands of acres. In addition to the furnace, the operation also included mines, charcoal pits, extensive tracts of forests, raw material dumps, waste dumps, ore processing areas, residential and commercial areas, a grist mill, roads, ship and boat landings, a plantation, a store, and blacksmith, wheelwright, and cartwright shops. A large number of people, both skilled and unskilled, were required to operate the furnace and its support facilities. An iron works, then, was a complex social entity, largely self-sufficient, yet in constant contact with adjoining counties, neighboring colonies, and with the British Isles. The furnace employees expected staple and luxury goods to be available for barter and purchase from the company store. Pig iron, bar iron, castings and other manufactured goods had to be shipped away from the furnace, either for sale or for further processing at a forge. These two essential functions were fulfilled by the complex mercantile system that evolved to support the tobacco industry. The same ships that carried tobacco to Britain and mainland Europe, picked up rum, sugar, and molasses from the West Indies, and delivered manufactured goods and essential supplies to the iron works. Pig and bar iron were sent to England and Scotland as ballast on board these same tobacco ships. Thus, the iron and tobacco industries were married, each dependent upon the other. By means of this mercantile system, the colonial iron producers met their obligation to supply the Mother Country with iron, made enough to meet the needs of the colonists, and grew and prospered.

Early in the 18th century, a handful of English ironmasters brought their technological expertise to the New World and undertook or oversaw the building of several early bloomeries and iron furnaces in the Chesapeake region. Very few of these highly skilled men could be persuaded to leave the safety and comfort of English soil and those who did come commanded exorbitant wages. Because there were far fewer ironmasters than prospective iron companies, the masters traveled from furnace to furnace, sharing their expertise over a wide geographical area. Within a generation, the colonies were able to draw on locally-born talent and were no longer dependent upon English ironmasters. As a result, the number of successful furnaces grew steadily over the first half of the century. This high level of development, largely kept secret from the British government, eventually enabled the colonies not only to challenge the most powerful nation in the world, but win.

For nearly a century iron production and its ancillary activities dominated Stafford County's financial, political, and social structure. Beginning in 1727 with the opening of Accokeek Furnace through the early 1820s with the closing of Rappahannock Forge, Stafford's two foundries mirrored the technological development of iron production in the rest of North America. Both furnaces touched the lives of every county resident who depended upon those works to supply them with both basic needs and luxury items. Accokeek and Rappahannock also employed many local residents, thus creating an economic base relatively separate from agriculture.

Over the years, little research has been done on Accokeek Furnace and what has been published on Rappahannock Forge has been based primarily on local oral tradition supplemented with a smattering of primary source references. As these two businesses operated during different time periods, they were long thought to have been independent of one another. Various contemporary records, along with the 1749-1760 business ledger included with this volume, prove that there were numerous ties between Accokeek Furnace, the Principio Iron Company of Maryland, and Rappahannock Forge. Perhaps the most significant of these ties were the many merchants and ships that dealt with both operations and are discussed here in some detail. Numerous individuals were involved in both businesses as well.

In compiling this work, the author has attempted to locate as many primary source documents as possible that pertain to the two works and the people associated with those facilities. Account holders listed in the ledger represent the counties of Fairfax, Fauquier, Prince William, Stafford, Spotsylvania, and King George and the towns of Alexandria, Dumfries, Falmouth, and Fredericksburg. To this has been added genealogical material, land and personal property tax information, newspaper notices, and explanations of the technologies common to the period. Considering the quantity of colonial records that have been lost, a surprising number of references to Accokeek Furnace and Rappahannock Forge survive. Sources are referred to within the text and a bibliography is included at the end of each article. Within the text will be found references to many places and points of historical interest in Stafford County. These sites have been marked on a map (Fig. 1-1).

The first chapter in this book deals with Accokeek Furnace and its parent company, the Principio Iron Company of Maryland. Included in this chapter are discussions of 18th century iron furnace technology and the social and economic impact of iron furnaces on the community. The second chapter is an in-depth study of Rappahannock Forge and its surprisingly extensive range of manufactured items. Finally, on a CD-ROM, included in a pocket inside the back cover, the reader will find a transcription of the only known surviving business ledger from Accokeek Furnace. Spanning the years 1749-1760, this ledger covers the furnace's closing years. Invaluable as a reflection of life in and around an 18th century iron furnace, the ledger also contains a wealth of genealogical information.

A key component missing from this study is a detailed archeological report on either site. The Virginia Department of Historic Resources conducted preliminary surveys at Accokeek (1983) and Rappahannock Forge (1973), but no comprehensive digs have as yet been carried out. It is the author's hope that this volume will spur an interest in such much-needed work. In the meantime the overgrown remains of these once-busy complexes lie all but forgotten in the forest, a mute reminder of Stafford's industrial past.

Acknowledgments

Special thanks to the Historical Society of Delaware for allowing me to transcribe and publish the Accokeek business ledger.

Thanks to Mrs. Connie Cooper of the Historical Society of Delaware for her many hours of patient assistance.

My appreciation to Alaric R. MacGregor for proof reading the text.

Thank you to the American Society of Arms Collectors who allowed me to use the photographs of the Rappahannock Forge musket and pistol originally printed in The Rappahannock Forge by Nathan Swayze.

Thank you to the dedicated and patient staff of the Central Rappahannock Regional. Special appreciation is expressed to Mercy Sais for her endless hours spent obtaining obscure interlibrary loan materials.

Special thanks to Walter V. "Pete" Roberts for making me reconsider Rappahannock Forge as a research project and for his many hundreds of hours tracking down seemingly endless leads.

I would like to extend my appreciation to the dedicated and cooperative staffs of the Library of Virginia and the Virginia Historical Society.

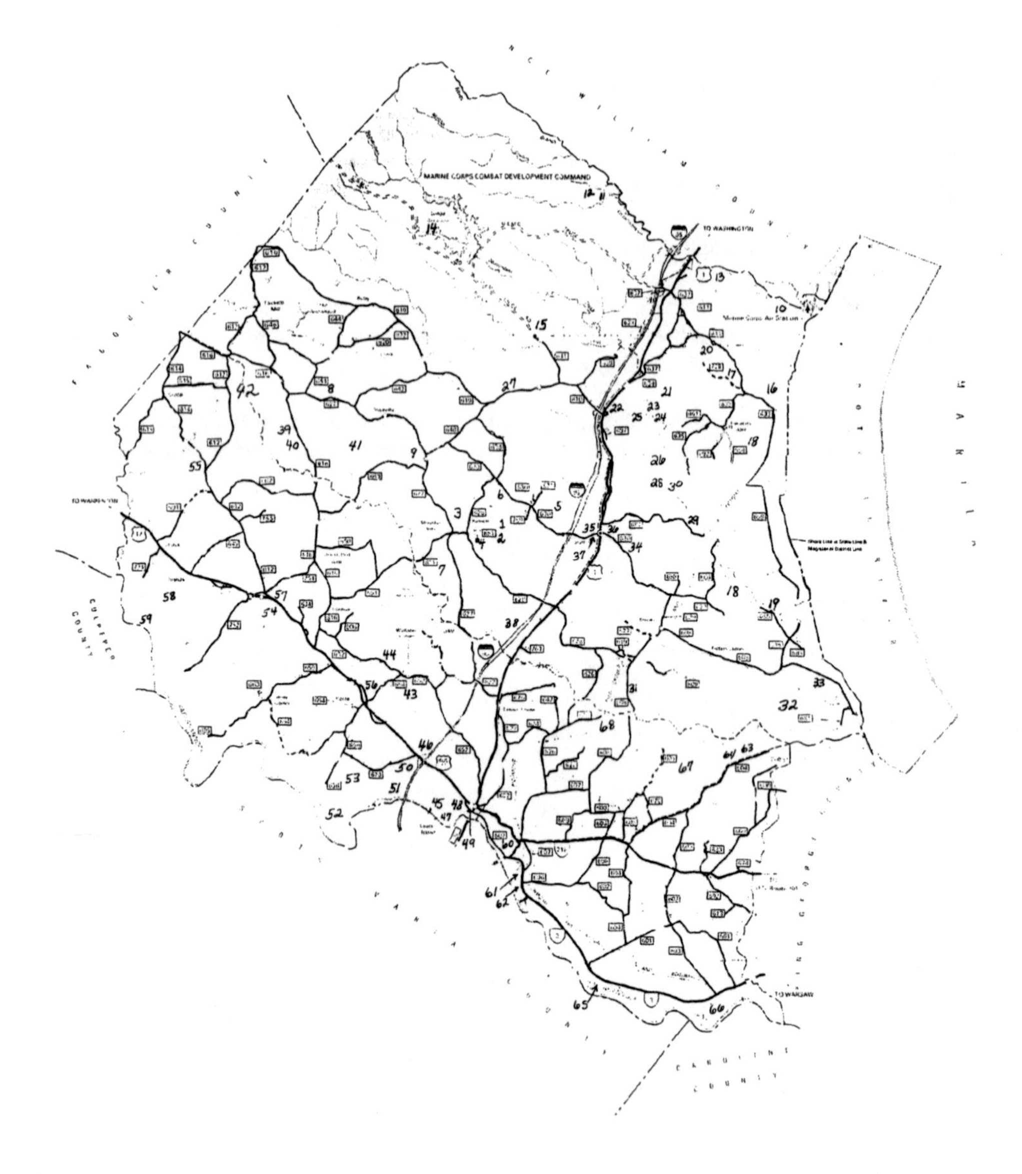

Fig. 1-1
Locator map for historic sites related to Stafford County ironworks.
Map courtesy of Virginia Department of Transportation

Map Key

1—Accokeek Furnace
2—Woodbourne
3—Mine Bank
4—Ramoth Baptist Church
5—Robinson's Quarry
6—Colonial Forge High School
7—Baxter Tract
8—Clover Hill
9—Reddish Hill
10—Dipple
11—Fristoe's Mill
12—Chopawamsic Baptist Church
13—Chopawamsic Farm
14—Mount Horeb
15—Windsor Forest
16—Richland
17—Brent's Mill
18—Brent's Quarter
19—Thorny Point
20—Bloomington
21—Palace Green
22—Aquia Church
23—Woodstock
24—Aquia Warehouse
25—Stony Hill
26—Coal Landing
27—Woodford
28—Concord
29—Hope Patent
30—Sandy Level
31—Potomac Church
32—Crow's Nest
33—Marlborough
34—Spring Hill
35—Stafford Courthouse
36—Garrard's Ordinary
37—Alexander's Ordinary
38—Carmora
39—Poplar Grove
40—Hampstead
41—Ludlow
42—Parke Quarter
43—Ellerslie
44—Liberty Hall
45—Rappahannock Forge
46—Stanstead
47—Mortimer's Island (Lauck's Island)
48—Ingleside (Richards' Hill)
49—Falmouth
50—Harwood Branch
51—Seine Pocket Farm
52—Greenbank
53—Woodend
54—Hartwood
55—Spotted Tavern
56—Berea Baptist Church
57—Pickett's Ordinary
58—Richlands
59—Richards' Ferry
60—Chatham
61—Pine Grove
62—Ferry Farm
63—Belle Plains
64—Cave's Warehouses
65—Traveler's Rest
66—Snowden
67—Salvington
68—Bellair

Chapter 1

Accokeek Iron Works

c.1694—Thomas Norman, by patent
c.1720—Augustine Washington
1743—Lawrence Washington
1752—Sarah Washington
c.1754—Col. Augustine Washington
____--William Augustine Washington
1777—James Hunter
1804—Francis Foushee
1804—William Foushee
1826—R. C. L. Moncure, Trustee
1826—Robert Dunbar
1826—James Alexander
c.1837—R. C. L. Moncure
1847—John Moncure
1886—George W. Black
1916—J. Clarence Fisher
1918—Julien V. Brooke
1920—L. T. Hitt
1923—J. W. Masters
1959—Hunter H. Simpson
1961—Louis Carusillo
1965—Edward J. P. Duffey
1969—Robert O. Ryan and Thomas L. Heflin
1974—Wayne E. Gordon
1981—Stafford Associates Limited Partnership
1989—Stafford County
1995—George Washington Boyhood Home Foundation

The birth of the Chesapeake iron industry was largely a result of overseas politics. Sweden, the source of much of England's iron during the early 18th century, had been routed in the Great Northern War. By 1716, the Royal Navy was paralyzed due to a shortage of iron, pitch, tar, turpentine, hemp, and masts. When George I embargoed Swedish trade in 1717, England was forced to look to the colonies to provide many vital products traditionally produced in the Baltic.

England never intended for the colonies to produce finished goods. The mercantile theory behind the entire colonial venture was powered by the premise that the colonies would produce raw materials for transport to England where they would be manufactured into finished products in English factories. These products would then be sold back to the colonies at a profit, thereby keeping English workers employed, and the English economy strong. Robbins wrote, "Altogether, the purpose was for England to realize the maximum economic advantage from its Colonial holdings" (Robbins, Maryland, 6).

First and foremost among the concerns of Parliament were the needs and interests of the Crown and English workers and investors. England encouraged the colonists to search for minerals and other raw materials that could be shipped home to keep English workers employed in their manufacture. The first and greatest colonial export was tobacco and the colonial economy quickly became dominated by the plant. Various other industries were attempted in the 17th century including silk, hemp, grain, and iron production. It took nearly 100 years, however, for iron making to become economically profitable.

Iron is abundant in many parts of Virginia, most deposits occurring near the surface where they are easily discovered and worked. The manufacturing of iron products, so essential to early settlers, was attempted from the earliest days of the colony. As early as 1607-08 workers were smelting iron in the fledgling village of Jamestown. One of the first settlers in Jamestown was a blacksmith named James Read and excavations conducted there in the 1950s revealed a small smelting oven and forge. Indian raids and a lack of funds and personnel prevented the Jamestown venture from being successful (Condit 2). As early as 1608 samples of iron ore were shipped home to England where it was smelted and tested at Bristol. In 1620 110 iron workers from Warwickshire and Staffordshire and 44 iron workers from Sussex were sent to Virginia to staff a newly-built furnace on Falling Creek in Chesterfield County just below the present site of Richmond. Unfortunately, on March 22, 1622 Indians fell on every English settlement along the James River massacring over 350 colonists including all 27 adult workers at Falling Creek. In addition to killing the workers, the Indians destroyed the buildings and reportedly heaved some of the machinery into the nearby river (Hudson, "Augustine Washington, 2611). Thereafter, colonial efforts in Virginia were concentrated in the tobacco industry and, for nearly 100 years, iron manufacture was largely ignored.[1] Not until Alexander Spotswood's[2] (1676-1740) arrival in Virginia in 1710 was the iron industry revitalized in that colony.

Almost immediately upon his arrival in the colony, Spotswood saw the potential in iron and sought to utilize public funds to encourage its manufacture. Parliament, under pressure from English iron producers who feared competition from colonial furnaces, refused to authorize any public funds for such purposes. Eventually, English iron manufacturers agreed to allow the production of pig iron in the colonies, but lobbied vigorously against any legislation that might allow bar iron or cast products to be made. With no hope of public encouragement, Spotswood decided to invest his own money in iron manufacture and, by about 1718, his Tubal Furnace was in blast. This facility was located near the confluence of the Rappahannock and Rapidan rivers about twenty miles southwest of Fredericksburg. Shortly thereafter, he built an air furnace near Massaponax, the first of its kind in the New World and the first that utilized mineral coal as opposed to charcoal. Massaponax was located about five miles below Fredericksburg on a neck of land at the outlet of the Massaponax River. This area later became known as New Post and is part of Spotsylvania County.

The first quarter of the 18th century was an age of great interest in minerals and the perceived fortunes that could be made by their exploitation. Several companies simultaneously began producing iron in the Virginia/Maryland region. In 1720 two iron companies formed in Bristol, England, the Principio Iron Company and the Bristol Iron Company. The latter consisted of investors Samuel Dyke, Jeremy Innys, John King, John Lewis, Walter King, Lyonel Lyde, and John Templeman of England and John Tayloe[3] (1687-1747) of Virginia (Heite, Pioneer Phase, 139). Shortly after its formation, the company authorized John Tayloe and John Lomax[4] (1675-1729) to act as their agents and purchase land in their names upon which they were to build an iron furnace (Brydon 98). Tayloe became a full partner in the business in 1729

---

[1] Other colonies were also experimenting with iron smelting. The earliest successful works in British America began production c.1645 at Saugus, Massachusetts.

[2] Alexander was the son of Dr. Robert Spotswood and his wife Catherine Elliot. He sailed for Virginia in 1710 and in 1724 married Ann Bryan. He served as Deputy Postmaster General of America in 1730 and died at Annapolis, Maryland.

[3] John Tayloe owned land in Charles County Maryland (Nanjemoy), and in the Virginia counties of Richmond, Essex, and Prince William. He married Elizabeth Gwyn, the daughter of David Gwyn and the widow of Stephen Lyde. John served as a member of the Council from 1732 until his death.

John's son, John Tayloe, Jr. (1721-1779) built Mt. Airy in Richmond County and signed the Treaty of Lancaster (1744) with the Indians of the Six Nations. He was also a member of the Ohio Company, trustee of the towns of Dumfries and Quantico, and a breeder of fine racehorses. John, Jr. was appointed to the Council under Lord Dunmore in 1757 and, following the reorganization of the state government, was elected to the first Council of State under Governor Patrick Henry. He resigned from this position on Oct. 9, 1776. John also owned the Neabsco and Occoquan iron furnaces in Prince William County. In 1747 he married Rebecca, the daughter of Col. George Plater of St. Mary's County, Maryland.

[4] There were ties between Bristol and Accokeek furnaces, though the loss of records prevents a thorough understanding of the connection. Contained in the land grant books is a patent dated Aug. 12, 1726 to Col. John Lomax of Essex County. He patented 200 acres in Stafford on Accokeek Run "adjoining land Michael Dermot sold Capt. Washington," and the lands of Harvey and Robert Carter.

(Richmond Court Records, Deed Book 8, pp. 488-491). At that time Tayloe hired Ralph Falkner[5] (died 1787), a former Accokeek employee, as manager at Bristol.

In 1721 the Bristol Iron Company constructed a furnace on what became known as Bristol Mine Run, now the dividing line between Westmoreland and King George counties. John Lomax was the ironmaster who oversaw activities there (Hudson, "Augustine Washington," 2613). These works were located on John Foxhall's old mill property that had been previously owned in the 1670s by Maj. William Underwood and were located near Leedstown. The furnace was built near deposits of bog ore,[6] a granular iron-bearing rock deposited in shallow veins just a few feet above the banks of the Rappahannock River. An unknown quantity of iron was produced here, much of it in the form of pigs which were transported to England. In all likelihood, Bristol also made castings such as pots, pans, firebacks, etc.

Exactly how long Bristol was in operation is uncertain. Dr. Peter King of Stourbridge, England has collected the records of English purchases of colonial iron and found that Bristol exported iron to England as late as 1753. However, court records indicate that a "riot" by employees, angry at the company's failure to pay them their wages, ended the furnace's production in 1729 (Eaton 11). Today, all that remains of Bristol is the furnace foundation concealed by a tangle of vegetation. Though much of Bristol's history is now lost, there are some tantalizing clues that point to a connection between that works and Accokeek. In 1726 Richard Brooks of Stafford leased to John Tayloe "all the land in County Stafford and on Potowmack runn containing 100 acres being part of 320 acres devised unto Richard Brooks by his Father Thom. Brooks deced and is to be bounded by Potowmack run aforesaid the land of Waugh and the land of Colo. Carter and so farr back takeing the whole breadth of the said 320 acres includeing the Iron mine stone and all the land faceing the runn" (Stafford Deed Book J, p. 284). This land was in the immediate vicinity of the Accokeek iron mines and was likely purchased to provide Bristol with ore.

[5] In 1725, while Accokeek was still under construction, John England hired Ralph Falkner as a clerk and he worked there as late as 1728. He seems to have been a capable individual for, after going to work for Bristol, Dr. Charles Carroll (died 1755) of Maryland inquired of him how best to build a furnace. By 1735 Ralph had removed to Prince William County where he managed Tayloe's Neabsco Furnace. In 1746 Ralph and his partners Edward Neale, Charles Ewell, and John Triplett contracted with Dr. Carroll to build a furnace on Carroll's land on Back River in Baltimore County (Robbins 155). This works came to be called Lancashire Furnace and was sold in 1751 to the Principio Company. In 1749 Ralph and his same three partners established the Occoquan Iron Works in Prince William County and, at about the same time, the four men patented land in Harford County, Maryland where they established Nottingham Forge.

[6] Iron deposits typically appear close to the surface and in wet areas. Also known as goethite, bog ore is a hydrous iron oxide. It is soft and porous with a dull or glassy luster, yellowish to dark brown in color. It forms by weathering and oxidation of other iron minerals and occurs as a precipitate deposited in bogs and springs. Colonial iron seekers used long rakes to pull lumps of ore from the bottoms of ponds or dug it out of shallow deposits in dry lake beds.

Fig. 1-2
Bog iron was easily dug from surface deposits. (Diderot, Vol. IV, Forges, 1st Section, Plate I)

Alexander Spotswood constructed his Tubal Furnace in Spotsylvania County just a few years prior to the erection of Bristol. When William Byrd (1674-1744) visited Spotswood in 1732 to discuss iron manufacturing, the former governor informed Byrd that at that time there were only four working blast furnaces in Virginia and not a single forge. These furnaces to which he referred were Tubal,[7] Massaponax,[8] Accokeek, and Fredericksville.[9] It is curious that he did not mention Bristol as the previously mentioned English records indicate that it was still in production at that time. Of the four furnaces he acknowledged, Spotswood owned the first two and had an interest in the fourth. Fredericksville was built in 1728 on the North Anna River. Spotswood was a partner in this venture as was Col. Henry Willis[10] of Fredericksburg. William Byrd wrote, "Colonel Willis had built a flue to try all sorts of ore in" (Wright 369).

During the 18th century, Virginia's iron manufacturing ranked third in economic importance behind tobacco and grain production. Over the years Parliament passed numerous acts to both encourage and limit the scope and production of American iron. In 1727 an act for the encouragement of adventureres in iron works prompted John Mercer of Marlborough to write a protest about the law's requirement that local

[7] Tubal Furnace was in operation from 1714 to c.1785. It was located in Spotsylvania County about 12 miles west of Fredericksburg on La Roche Run east of Mine Run and near modern State Route 620.

[8] Massaponax Furnace operated from c.1730-1785. It was located at the mouth of the Massaponax River on the south bank.

[9] Fredericksville Furnace was located between Douglas and Pigeon runs on the old Catharpin Road in the southwestern corner of Spotsylvania County. It was built by Charles Chiswell (1677-1737) who arrived in Virginia in 1704 and quickly became involved with Frederick Jones in frontier land speculation. He was a factor for the Royal African Company, the British slave monopoly. In 1706 Chiswell was appointed Clerk of the General Court of Virginia. Between 1726 and 1728 Richard Fitzwilliam, Surveyor General of the Customs for the Southern District, Charles Chiswell, and Larkin Chew, all patented land along Douglas Run in Spotsylvania County. William Gooch was also a partner in the venture. Other partners were Dr. George Nicholas and Capt. Vincent Pearse. The first founder at Fredericksville was James Rawlings who was followed by Robert Durham. The impounding of Lake Anna flooded the furnace site.

[10] Henry Willis (1691/2-1740) was born in Gloucester County, Virginia. He moved to Spotsylvania in 1730 and then to Fredericksburg where he lived at Willis Hill on Marye's Heights. He married Mildred Washington (c.1697-1747) as his third wife, the daughter of Lawrence (1659-1698) and Mildred Warner (died 1701). Mildred was the sister of Augustine Washington and the aunt of George Washington.

jurisdictions be responsible for building and maintaining public roads leading between the works, ore banks, and shipping points (Hening, vol. 4, pp. 296-297). Mercer and Peter Hedgman, one of the signers of the protest, were ordered to appear before the House of Burgesses who declared their petition to be a "scandalous and seditious libel, containing false and scandalous reflections upon the Legislature and the justice of the General Court and other courts of this Colony" (McIlwaine, Journals of the House of Burgesses, vol. 6, p. xxi). Mercer and Hedgman were found guilty of a misdemeanor, were reprimanded by the speaker, and fined. Their efforts were successful, however, as the amended version of the act passed addressed their complaints.

Throughout the 18th century, Parliament vacillated on its stand regarding colonial production of bar or wrought iron. While the need for bar iron had reached a critical level in England, iron producers, fearful of American competition, constantly lobbied Parliament demanding restrictions on imported bar iron. They sought to prevent the colonies from making anything beyond basic pigs, which were to be shipped directly home to England where English companies and workers would process them into the much-needed bars. A plethora of iron-related legislation was passed during the 18th century, most of which was simply ignored in the colonies. Legally, no forges with trip hammers, plating mills for making sheet or tin-plate iron, slitting mills for making nails, or steel furnaces could be built after passage of the Iron Act of 1750. By that time there were several such plants established and functioning profitably in the New World and the owners were certainly not going to close them as a result of Parliament's whim-of-the-moment. Most iron companies simply overlooked the restrictions, producing items that were necessary to local inhabitants and, therefore, generated income through their sale. The demand for iron was so great in Virginia that local blacksmiths consumed much of the metal produced here. Col. John Tayloe of Bristol Iron Works found that he could sell his entire output locally. Iron foundries provided the colonies with a wide range of implements, weapons, agricultural tools, and hardware necessary to support an agrarian lifestyle. The surviving Principio ledgers abound with all manner of manufactured iron products that were being sold to local consumers. These included but certainly were not limited to axes, mauls, maddocks, rakes, shovels, nails, cooking pots, wedges, hammers, plows, coulters, griddles, hoes, files, iron rings, chain, anchors, anvils, and knives.

Much of the iron produced in the colonies was shipped back to England in the form of pigs or bars. Pigs were long rectangular masses of iron which were made by draining molten iron into sand molds placed at the base of the furnace. A pig measured approximately 48 inches long by 6 inches wide by 3 inches thick and weighed from 55 to 60 pounds. Bars began as pigs but were processed at a forge, usually a separate operation from the furnace. The pigs were heated to a plastic stage and pounded out into bar shapes, either by hand or by water-driven trip hammers. This pounding aligned the atomic structure of the iron, forcing out impurities that would have made the iron brittle and difficult to work. The resulting bar or "wrought iron" could then be hammered into finished products. Wrought iron was the purest commercial form of iron and was similar to very low carbon steel except that it was never cast, but forged to the desired shape.

While England had been smelting iron since Roman times and, during the 17th century, had established a great number of furnaces, the ceaseless need for charcoal (as well as timber for construction of the Royal Navy's ships) had all but eliminated the once extensive English forests. Once it had been proven that the quality of colonial iron was equal to that produced in the Mother Country, production in America soared. By 1740 the colonies were annually exporting more than 2,500 tons of iron to England, most of that from Virginia and Maryland. Production continued to increase as demand grew in the colonies and in England.

The English government was unaware of the true extent of American iron production, the lack of awareness perhaps being a result of an orchestrated silence on the part of the colonial governors. Historian Arthur C. Bining wrote, "The development of the colonial iron industry from the second decade of the eighteenth century down to the Revolution was remarkably rapid. At the close of the colonial period there were more blast furnaces and a larger number of forges in the colonies than in England and Wales combined. The relatively high stage in industrialization reached by the colonists has not been entirely realized by historians. The colonists did not attempt to advertise their industrial activities nor did many of the governors inform the British government of the true state of affairs. The reports of almost all the governors to the Board of Trade, when such reports were sent at all, were inadequate and very incomplete, especially in regard to manufactures. While the Board suspected this and demanded accurate knowledge from time to time, complete information was rarely forthcoming" (Bining, British Regulation, 100). In the case of the Principio Company, the managers kept a separate set of ledgers to send home with their annual report. One of these, dated 1728 and on file in the Historical Society of Delaware,[11] is carefully labeled "This Ledger for Co. in England." This document is far more general than the carefully detailed ledger kept for "in house" use. There were few specifics listed as to what items the company furnaces and forge were actually producing. Instead, this special ledger outlined how many pigs and bars had been shipped to England and included a rough outline of expenditures and income at the furnace. Local demand for finished iron products was met with increased production at the furnaces and forges, despite the prohibition on the manufacturing of finished goods. As a result, iron quickly became a major industry in Virginia and by 1739 the colony was home to four blast furnaces, one air furnace and one forge. By 1781 Thomas Jefferson counted eight works in Virginia and reported that they annually produced about 4,400 tons of pigs and more than 900 tons of bar iron (Jefferson 28).

The early seeds of this important industry were in small production facilities such as Principio's[12] first bloomery in Maryland. This was a much simpler affair than a blast furnace as it required no water wheel and only a hand-operated bellows. Consequently, temperatures were not high enough to reduce the iron to a liquid state. A small quantity of iron ore was heated in a charcoal fire. The heat caused the iron components in the ore to separate into tiny crystals, which gradually began to unite into small metallic bodies, forming a porous, spongy mass shot through with microscopic particles of slag. Eventually, these metallic bodies formed a large mass called a bloom. While the iron particles were uniting, so were the impurities, most of which separated from the iron. Some impurities, such as silica, were reluctant to separate completely. As the bloom formed on the hearth, the ironworker prodded it with a long bar so as to allow the air blast to come into contact with all parts of the mass. When the bloom finally reached the desired consistency, somewhat like that of putty, it was removed from the furnace. A blacksmith then hammered the mass, forcing out those impurities that had not previously separated during the heating process. As the iron cooled, it was re-heated and the hammering continued. This process was continued until the resulting wrought iron was relatively free from impurities. Producing iron in a bloomery was not especially efficient as the slag formed during the process and removed by hammering contained a high concentration of iron. Further, a bloomery could produce only about 250 pounds of iron per day; hence, as the demand for iron increased, the early bloomeries were replaced by more efficient and productive blast furnaces.

Hand hammering of the bloom by a blacksmith was slow, heavy work that was soon mechanized. Heavy trip hammers, the hafts of which were depressed and released by cam-studded, water-powered axles, easily replicated the pounding of the smith. These workshops were known as refinery forges and were often a separate operation from the furnace. Pigs produced at a furnace were brought to the forge for processing into wrought iron bars. Prior to the use of rolling mills, these trip hammers were used to produce weapons and tools such as shovels, scythes, and mattocks.

Organization of the Principio Company

On Mar. 4, 1720 a partnership known as the Principio Iron Company was formed in Bristol, England. The original stockholders in Principio consisted of English industrialists and businessmen and included Sir Nicholas Hackell Carew, gentleman; William Chetwynd, gentleman; Joshua,[13] Osgood, and

[11] The Joseph S. Wilson Collection dates from 1724-1903 and includes 31 volumes of cashbooks, ledgers, waste books, day books, and sales books from the Principio Company.

[12] This bloomery was later converted to a furnace. It stood on the west side of Principio Creek between the General Pulaski Highway (U. S. Route 40) and State Route 7 in Havre De Grace, Maryland.

[13] Joshua Gee, an ironmaster and iron merchant of Shrewsbury, England, was largely responsible for the birth of the Chesapeake iron industry. In a 1717 pamphlet directed to Parliament, Gee suggested that the colonies, especially Pennsylvania, should receive Parliamentary encouragement to produce pig iron. He argued that colonial production of pig iron would protect British manufacturing interests and reduce the raw material required for a major British industry. He also said that the Virginia tobacco fleet could haul iron as ballast for a fraction of what it had previously cost to ship it as cargo from Sweden and Russia.

Samuel Gee, London merchants; John England (1680-1734), a Staffordshire ironmaster;[14] Thomas Russell, a Birmingham ironmaster and his sons William and Thomas; John Ruston, Stephen Onion, and Joseph Farmer. The Principio partners initially intended to open works in New Jersey and Virginia but established their first bloomery in Maryland. Their agent, John Copson of Philadelphia, purchased a 5,743-acre tract in Cecil County called Geoffarison. The warrant for this property was issued on Apr. 27, 1721 and the patent on May 24, 1722 (Robbins 190).

Principio was the first iron company to operate in Maryland, though their success and a ready supply of quality ore there set the stage for competition later in the 18th century. From a modest beginning, Principio became one of the largest iron producers in the colonies with blast furnaces and mills for making high-grade iron and steel products. At their peak in the 1760s, Principio was operating four furnaces and two forges along the Chesapeake Bay, i.e., Principio, Kingsbury, and Lancashire furnaces and Principio and North East forges. During this time, Principio furnaces and forges produced pig iron, bar iron, and castings such as forge hammers, forge anvils, plates for forge hearths, mantle pieces, firebacks, pots, skillets, firedogs, cart boxes, and a wide range of agricultural implements. Principio enjoyed some 60 years of profitability before being disrupted by the American Revolution.

Principio's initial representative in Maryland was Stephen Onion[15] (died 1754) who had been sent there to build a bloomery. Unfortunately, Stephen proved a poor manager and he spent most of his time establishing a store rather than building the iron works as he had been directed to do. In a letter dated Aug. 11, 1722 he reported to the company that the bloomery stack and buildings would be completed within a month, when, in fact, nothing had yet been built. Once the partners discovered his lack of progress, he was recalled to England, though he later returned to Maryland where he continued in their employ. John England, a brilliant ironmaster from Staffordshire, replaced Onion as administrator. Already advanced in years, England,[16] his wife, Alice (died 1723), and two of his brothers, Joseph[17] (1680-1748) and Lewis[18]

---

[14] The term "ironmaster" had several meanings. An ironmaster might be the owner of an iron company, individually, a shareholder in an iron company, or a member of a family that owned a company. It sometimes, but not always, referred to someone with a basic understanding of iron making.

[15] After leaving Principio's employ in the 1730s, Stephen remained in Maryland, going to work as clerk of the Baltimore Iron Works. In 1737 he set out on his own and, by 1744, had completed a furnace on the Little Gunpowder River in Maryland. The *Maryland Gazette* of Aug. 29, 1754 carried his obituary in which he was described as the "owner of Iron Works on Gun Powder River." Stephen married Deborah, the daughter of William Russell of Principio.

[16] John England was the son of John England (c.1650-1683) and his wife, Love (c.1650-1682) of Staffordshire England. In addition to John, Joseph, and Lewis, John and Love also had a son named William who remained in England. In 1690 John, Jr. married Alice Allen at Whittington near Litchfield, Stafford, England. They had issue:

--John—married 1719 Sarah Lloyd (born 1694) of Dolobran, England

--Allen of Alder Mills near Tamworth—married Hannah Bradford

--Joseph of Alder Mills—married 1710 Margaret Deal, the daughter of Samuel and Joanna Deale of Kent.

--Alice

--Ann—married ___ Forde

Allen and Joseph England inherited their father's share in the Principio Company and sold a half interest in it to John Price, Principio's London agent (Ruston, Thomas, Papers, deed).

Contained in the Thomas Ruston Collection in the Library of Congress is a deed relating to the disposition of John England's interest in the Principio Company and his lands in Stafford, neither of which survive among the Stafford court records. One deed, dated Dec. 31, 1741 states that John owned "one thenth part of Eleven twelfth parts of all Lands, furnaces, forges, mills works erections and buildings of the said partner at or near a place called Accokeep in the County of Stafford in the province of Virginia and of all iron, limestone, Ironmines stock negroes, horses, and Cattle, implem[ents], Utensills and appurtenances belonging to or used enjoyed with the said works at Accokeep aforesaid. And whereas the said John Price hath contracted and agreed with the said Allen England and Joseph England for the absolute purchase of one Moiety of half part of the right share and interest of said John England of the time of his death of and to the said lands as well freehold as leasehold and of and in all premises, forges and mills Erections, buildings (c.1681-c.1750), set sail from Bristol in April 1723. Seven weeks later, the Englands arrived in Philadelphia, then continued south to Maryland.

Upon arriving in Maryland, John found the situation far different from what had been represented to him in London by Joseph Farmer. A thorough assessment of the works so disgusted him that he resolved to go home. His letter to the partners in England carefully described the situation and his belief that the undertaking could only result in failure, chiefly for the want of a good source of iron ore. In a letter to Joseph Farmer John wrote, "instead of my coming alone thou ought to have come with me, or thyself first to have regulated matters about ye furnace and stove, for which I and ye Company will be great losers; for ye men which thou hast sent over, instead of going on with their work, are almost ready to mutiny, and for every 5s. worth of work which is done, I dare loudly affirm it lyeth ye Company in 20s. or more, which when I think of it, it maketh my heart ready to sink to think what I have brought myself to, and had no occasion to have done so as thou art very sensible of. I hope you will all confer together, and let me know what must be done in this undigested affair, for ye Lord permit me with life and health I intend to return in a little time. As for ye healthfulness of ye place ___ by what account I have we run ye hazard of our lives by staying here." The partners responded by asking him to stay on and bring some order and profit to the business.

Sickness was rampant in 17th century America and, shortly after their arrival, John and his wife were both taken ill. Alice died in September 1723. John lived but never fully recovered. According to a family genealogist, "The clerks who had been superseded by him were very jealous of his authority, thwarted him in every way; he was embarrassed for want of ready money to carry out operations, and his health was shattered; yet under all these depressing circumstances, which would try the stoutest heart, he never flattered in well doing but continued through it all in working for the company's good, and saved its property from waste and ruin" (England, The England Family, p. 7).

During this time, John also traveled to Delaware where he bought for himself a tract of ore-bearing land with the intention of starting his own iron works. The mine there proved less than adequate and he abandoned the idea, buying another plantation on White Clay Creek in 1726 where he built a mill known for years as England's Mill. While in Principio's employ, however, John resided at Richard Snowden's iron works on the east branch of the Patuxent River. Upon retiring from Principio, John moved to White Clay Creek. In 1729 he sailed to England to visit his family, returning to Delaware in 1730 where he died in 1734. His brother, Joseph, who established an extensive family in Delaware, inherited that property. None of John's children settled there.

John England proved to be a capable administrator and was a master of his craft. He served as Principio's administrator from 1723 until his death in 1734. The company stockholders gave him the responsibility of finding and evaluating potential furnace sites, locating suitable raw materials, designing and constructing new furnaces, forges, and the associated equipment and buildings, as well as overseeing the production, marketing, and shipping of the iron. Beyond matters of poor management and a drunken, less than cooperative workforce, John's primary concern was a shortage of quality ore. His search for iron carried him all along the Chesapeake Bay and, in 1723, to Stafford where he found Augustine Washington's ore-bearing land on Accokeek Run. Within four years Accokeek Furnace was in production and was beginning its 25-year reign as headquarters of the Principio Company. This sizable industrial complex was situated on the south side of Accokeek Run approximately 2.5 miles west of the present Stafford Courthouse at the east end of State Route 651 where it becomes Accokeek Furnace Road.

During the early 18th century, the area around Accokeek Creek was densely forested wilderness, though there were enough residents in the area to support two small grist mills. Discovering iron here was a relatively simple process. Locals frequently followed runs and branches looking for mineral deposits or

---

and works of and in the stock of Ironstone, ironmines, negroes, horses cattle ships, shipping, utensils and appurt[enances] to the said works or to either of them belonging to or used therewith, for the sum of £525."

[17] Joseph was born at Burton-upon-Trent, Staffordshire, England and died in Chester County, Pennsylvania. In 1710 he married Margaret Orbell (1685-1741), the daughter of Samuel Orbell of Deale, Kent, England. She died in Cecil County, Maryland.

[18] Lewis was born in Staffordshire, England and died in Cecil County, Maryland. He married Sarah England (c.1687-c.1750). She died in Cecil County, Maryland. In 1728 Lewis was listed in the Principio records as "Insolvent" and owed the company £8.8.2.

other interesting and useful materials. In the vicinity of the old iron works Accokeek Run remains littered with rusty rocks and chunks of gritty brown bog ore still may be picked up freely about the area.

Augustine Washington had ample opportunity to become aware of the iron deposits on Accokeek Run. Through his association with Bristol Iron Works Augustine probably became aware of Principio's plans for expansion. Augustine, who was then living on Pope's Creek in Westmoreland County, married Mary Ball (1708-1789) who had inherited some 600 acres between Accokeek and Potomac creeks. This land, located on the north side of Kellogg Mill Road (State Route 651) and on the east side of Potomac Run, provided at least some of the ore used in Accokeek's furnace. Augustine purchased several tracts that adjoined his wife's ore bank, thus creating the Accokeek Furnace tract.

As a result of disputes regarding overlapping boundaries and right of title, ownership of Mary Ball Washington's 600 acres was contested for years and warrants consideration here. The records of two 1805 court cases, Lewis vs. Commonwealth and Dejarnette vs. Lewis, provide some of the history of the tract. In his deposition, Robert Lewis (1769-1829) stated that "by Colour of the said Inquisition he is grieved and molested and that he from the possession of the said tract or parcel...containing six hundred acres of land, whereof a certain John Johnson...is supposed to have died seized...Robert Lewis saith that a patent issued to a certain James Harvey, from the Proprietors of the Northern Neck of Virginia for 600 acres of land being the land in the Inquisition aforesaid...which patent bears date the 1 day of Sept 1709" (Fredericksburg Court Records, Dejarnette vs Suddoth). James Harvey conveyed the land to John Hawkins on Oct. 10, 1711. Hawkins sold it on Sept. 10, 1712 to Richard Hughes (died c.1712) of Northumberland who devised it on July 13, 1714 to his stepson, John Johnson (died c.1720). John conveyed it on Oct. 1, 1720 to his sister Mary Ball, "then an unmarried woman."

When Mary died in 1789, she bequeathed the tract to her son, George. In 1794 George offered to give the property to his nephew, Robert Lewis (1769-1829). Several letters pertaining to this intended conveyance passed between the two men including one dated Apr. 29, 1783 in which George described the land as lying "near the Accocreek [sic] old Furnance [sic], and about eight miles from Falmouth...containing, as I have generally understood, about 400 acres of the most valuable pine in that part of the Country; but which, as I have been informed, has been much pillaged by trespassers." Washington informed Robert, "When you can ascertain the bounds thereof by a survey, for I have no papers to aid you in doing it, I shall be ready to convey to you my right" (Fitzpatrick, vol. 32, pp. 439-440).

Upon some investigation, Robert discovered that his uncle was not legally able to convey the property, as the law required that all of the descendants of his brothers and sister have a share in the land. Washington wrote to Robert, "I am not possessed of any papers belonging to it...but I should conceive that the tract is so well known that all the adjoining landholders are able to shew you the bounds of it." George had also heard that someone had talked of escheating the property, "but I never supposed injustice would prompt any one to such a measure." He suggested that Robert try to acquire it by that same process. Washington was unaware that the property had already been escheated, William Bell of Falmouth having acquired a patent on the property in 1790. In 1795 William sold part of the tract to Henry Suddoth who had placed a tenant on the land. After the Revolution, Richard Taylor lived on the tract rent-free for some twelve years before Suddoth forced him to begin paying rent.

In 1800, after George Washington's death and with the approval of the executors, Robert decided to re-claim the property. In 1803 he sold the 640 acres to Joseph Dejarnette who then attempted to collect rent from Henry Suddoth's tenant. How he did this is unclear as not until 1805 did Washington's executors actually deed the tract to Robert Lewis, that action being based upon Washington's earlier letter of gift. This resulted in a suit in 1806 that was heard in the Superior Court of Chancery for the Richmond District. Depositions for this case were taken at William Bell's tavern in Falmouth and are on file in the Fredericksburg Circuit Court office and they provide some interesting history on the tract. James Withers (c.1738-after 1808) "recollects often hearing his father John Withers who is now dead and many others call the land which is now in dispute Washington's land and the ore banks being part of the land purchased by John Jones from Robert Lewis was known by the name of Washingtons Ore Banks...The land whereon the said John Honey now lives was called Honey's old field." James also stated that John Honey had been permitted to live on the land in exchange for mining ore for the company (Fredericksburg Court Records, Dejarnett vs. Suddoth).

In another deposition in the same case John Seddon (c.1735-after 1808) claimed to have heard his father, Thomas Seddon (c.1686-1779), say that the tract "was conveyed to the mother of the late General Washington and the deponent has always understood that there was upon the said tract of land some race paths which were known by the name of Washington's race paths (Fredericksburg Court Records, Dejarnett vs. Suddoth).

A third deposition was given by John Murray. He stated that around 1749 he resided within one mile of the ore banks that "were under the direction of John Honey whose profession was a miner...These ore banks were called Washington's ore banks and this deponent knows of his own knowledge that the ore raised on these banks was in part if not wholly carried to the furnace of the Accokeek Company" (Fredericksburg Court Records, Dejarnett vs. Suddoth). It is not clear from the records how this case was finally settled.

Although Augustine didn't marry Mary Ball until 1731, he was unquestionably cognizant of the iron deposit on Mary's land. Around 1720 Augustine began purchasing land adjoining his future wife's 600 acres. Most of the land in the immediate vicinity of the furnace had been part of a 1696 grant of 1,150 acres to Thomas Norman (c.1625-c.1709) who styled himself in several deeds as a carpenter. During his lifetime, Thomas sold several parcels from the tract. On Jan. 9, 1701/2 he conveyed 130 acres to John Adams, a carpenter from Northumberland County who settled on his new purchase (Stafford Deed Book Z, p. 130). On Jan. 24 of that same year he conveyed 100 acres to John Russell (Stafford Deed Book Z, p. 129). In 1704 he sold another 143 acres to John Adams, by then living in Stafford, this parcel adjoining Adams' first tract and upon which he operated a grist mill on Accokeek Run (Stafford Deed Book Z, p. 235). Upon Thomas Norman's death c.1709, the Accokeek property passed to his son who divided the tract and sold it to several purchasers. Augustine Washington reassembled these tracts, several of which adjoined the land he acquired by his marriage to Mary Ball. At this point in time, the Washingtons were middle class planters but Augustine saw an opportunity to make his fortune in the iron business and he acted quickly to take advantage of that opportunity. In a relatively short time, Augustine acquired some 1,600 acres at Accokeek, giving him control of the ore body, roads, water supply, and the forest necessary for the production of charcoal for use as furnace fuel.[19]

According to a document included in Thomas Ruston's papers, the following tracts comprised the Accokeek land holdings:

Lands taken up by Capt. Washington:

Michael Dermott	240 acres
Clement Norman	125
John Spry	278
Henry Norman	100
John Adams	300
Richard Foulks	200
Took up by warrant	349
Took up by warrant	396
John Hopper	80
John Marr	100
	2176

Accokeek Lands bought and took up by Mr. John England:

Henry Norman	140
John Savage	430
Took up by warrant	197
Took up by warrant	82
Near Greenham's land	300
Near Norman's patent	300
	1449
Bought by Nathaniel Chapman	200
	3825

(Ruston, Thomas, "Principio Iron Works," p. 71)

[19] Northern Neck Grant Book 4, p. 97; Grant Book 3, p. 266; Northern Neck Survey Feb. 1, 1726, to John England for 430 acres; Northern Neck Survey Feb. 22, 1728, to John England for 156 acres.

Around 1724 Washington and England met to discuss a business venture at Accokeek. John England was impressed with the quality of ore on Washington's land and in early 1725 he contracted with Washington to take ore from the property. England's original intention was to ship the ore from Accokeek to Principio's furnace in Maryland. Washington's part of the deal was originally limited to being paid for the mining and transportation of the iron ore. England seems quickly to have decided that Accokeek had all the requisites for a furnace site—quality ore, a steady supply of water, plenty of timber, arable land for a company plantation, and nearby access to landings for shipping. He decided that it was economically advantageous to build a furnace at Accokeek rather than bear the expense of shipping the ore to Maryland.

On September 15, 1725 the partners wrote to England saying, "[We] give you hereby full power to build iron works on Capt. Washingtons land on ye terms you have agreed with him, and to use your own judgment and discretion in all things relating thereto" (Principio Partners to John England, letter) The partners continued by saying that they considered the Accokeek site "exceedingly well-chosen" and approved of the hiring of Ralph Falkner[20] as Accokeek's clerk. Funding for the Accokeek project came from the profits of the Principio Furnace in Maryland. Construction at Accokeek began nearly a year prior to the signing of a formal agreement with Washington and several months prior to England's obtaining legal title to the land. This was not unusual, however, as colonials rarely rushed to complete legal paperwork.

One of John England's first concerns was to select a group of workers to build and operate the new furnace at Accokeek. The Principio ledger reveals that in September 1723 he had Principio's tailor, William Martain, outfit the following workers. Each received a new shirt, coat, waistcoat, breeches, stockings, and shoes, as well as one quart of rum.

Whites	Negroes
John Foulks[21]	Jane
Samuel Tudman	Ann
Jasper Nockley	James
Michael Lewis	Jeme
Henry Pine	Nanny
William Evans	Patrick
Daniel Greenwood[22]	Prince
William Hinson	Tantarrow
Henry Price	Tom
James Warrington	Jenny
John Hartshorn[23]	Quash
Daniel Poole	Benjamin
William Sims	

Also sent from Principio at this same time was John Barker who finished his indenture at the new works. He later became a free workman for the company. Following soon after the initial group of workers was Peter Grubb, the stonemason who built the furnace. Some of the start-up costs for Accokeek included a

[20] At some point, Ralph Falkner returned to Maryland. The *Maryland Gazette* of Sept. 16, 1746 contains his notice offering to sell bar iron and a billiard table. At that time he was of "Port-Tobacco." In 1749 Capt. Charles Ewell (1713-after 1747), Falkner, Edward Neale, and John Triplett attempted unsuccessfully to open a furnace on Hooe's Run of the Occoquan River.

[21] In 1724 John Foulks was a salaried employee at Principio.

[22] Probably Daniel Greenwood, Sr. (died 1746) whose will was recorded in Westmoreland County. In his will he asked William Welsh to be guardian to his son, Daniel, and provide him with four years schooling. By the 1760s Daniel Greenwood, Jr. (born c.1732) was in Culpeper then Fauquier where he remained throughout the Revolutionary War. Daniel, Jr. served with George Rogers Clark and received land grants for three years of service.

[23] This may have been John Hartshorn (1712-c.1785), the son of Jonathan Hartshorn (born 1686) and Lucy Hempstead (died c.1749) of Norwich, Connecticut. John was born in Norwich, Connecticut and died in Cecil County. A John Hartshorn was listed in the 1742 Stafford quit rent rolls as owning 100 acres in that county. He continued on the Accokeek business ledger through the early 1750s. A later John Hartshorn was a merchant in Alexandria.

cheese provided by Joseph England and some earthenware. Joseph Growden sold the company a horse; Benjamin Turley's workers at North East made tools and consumed about 20 gallons of cider in the process. By the time Accokeek opened, the Principio Company had invested about £800 in the project (Principio Ledger 5, p. 60).

On September 19, 1725 William Chetwynd, principle shareholder in the company, wrote to John England regarding Accokeek Furnace. Chetwynd wrote, "I am of the opinion it would be proper to secure more land near Capn. Washingtons mine, because if ye mine is so considerable one as you mention it must answer very well to have two furnaces there" (Chetwynd, William to John England, letter). By the spring of 1726, England was patenting additional land in the area,[24] both for the company and for himself. Based upon the Stafford quit rent records, the company owned 5,854 acres,[25] not nearly enough to provide charcoal for one furnace, much less two. However, the Accokeek business ledger includes an entry in which it was recorded that in 1751 the company paid rents for Nathaniel Chapman (3,358 acres), Jane Williams (2,600 acres), and for its own 2,796 acres, making a total of 8,754 acres.

Augustine's first agreement with John England and the Principio Company involved only the ore to be mined from the tract. Washington agreed to supervise the mining of the ore and deliver it by wagon to the nearest landing on the Potomac River. The Principio Company agreed to transport the ore from the landing to its furnace at the head of the Chesapeake Bay for smelting. On July 24, 1726 Augustine and Principio completed a lease for the land. This document stated:

> doth sell all that said Washington's plantation and lands being at Accotink Run in Stafford County containing 1600 acres or thereabouts together with all the Soil thereof with all & all manner of Iron mines & Iron Oar & with 1 halfe of all other Mines Oars Mineralls...whether open or shall be discovered or not discovered within the limits of the said Plantation...said Washington doth grant unto England the free right & authority to sink cut & make any shafts pits in or upon the said lands in order to open or draw any of the mines mettellins oars & mineralls aforesaid & the earth stones and rubbish & oars out arising to leave in Heaps & the Ground in Holes & to refine and reduce the mines oars & minerals aforesaid at pleasure & to end & purpose to set up & erect any furnaces Buildings & Engines of other Erections whatsoever upon the said land...& to hold said Plantation lands & and all other...from the 10th day of June last past before the date hereof during full end & term of 1000 years from henceforth...Paying yearly to & each year unto Washington...the yearly rent of one Ear of Indian corn if the same be lawfully demanded...John England doth at own property charges cause to be erected & built in good workmanlike manner one good ffurnace capable for melting & testing Iron on some comodious piece of the Land hereby demised & thereon shall cause to be digg and made a good Trough for conveyance of water to this ffurnace which ffurnace shall be ffinished with Bellows Buildings & all other necessary appurtenances to such ffurnaces...and that within the time before limited for building of the ffurnace said England shall sustain the whole charges of repairs...during time of his continuance of the first blast to be made in the ffurnace...and England shall during the continuance of this present Lease set forth & appoint for said Washington...the full & equall Sixth part of all & singular such Iron that

[24] John England also purchased land for himself adjoining the furnace property. By 1730 he owned some 1,462 acres along Accokeek Creek plus 333 acres on Reedy Branch near Potomac Creek in the vicinity of the proposed airport site.

[25] There are no quit rent records that provide a total acreage for Accokeek during the 1720s or 1730s and the company had a long history of renting nearby or adjoining tracts in addition to its own acreage. According to the accompanying Accokeek ledger, in 1751 Accokeek paid quit rents on its own 2,796 acres as well as on 3,358 acres owned by Nathaniel Chapman and 2,600 acres owned by Jane Williams. This gave Accokeek a minimum of 8,754 acres that year. The rent rolls of 1773 and 1776 credit the company with 2,796 acres. The listing for 1768 is, perhaps, more accurate in that it lists Nathaniel Chapman as paying rents on 3,058 acres and Accokeek, itself, consisting of 2,796. Adding these two figures gives 5,854 acres that figure, of course, not including any property that the company had rented during the period in which it was in operation.

shall be melted or fforged at the place...Washington will defray the one full and equall sixth part of the charges of repairs to be necessarily made in or about the furnace...from time to time...after the first Blast forsd...England doth promise [he] shall not permitt any quantity of Iron oar exceeding 1 hundred Tons digg'd or hereafter to be digged won or gotten in or out of the plantation hereby granted...without the special Lycence of said Washington first had & obtained in writing...It is further covenanted that if either of said parties shall build any furnace or Ironworks for the melting & manufacturing of Iron within limits of the Colony of Virginia...he shall allow of the other in Partnership with him for all his Estate therein...if England shall at any time erect any ffurnace or fforge for the manufacture of Iron within the Kingdom of Great Britain he shall allow said Washington in partnership with him as to one sixth part of the proffits
(Lease, Augustine Washington)

In return for the use of his land, Washington received a 1/6 interest in the iron produced by the furnace and a tonnage fee for the ore that his workers dug from the mines. Principio had the right to build and use any necessary improvements and to take from the site all timber necessary to run the operation.

This lease was recorded in folios 524-529 of the now lost Book J of the court of Stafford County and was recorded during the session of June 10, 1728. The lease further stated that "since the making of the said lease, the said John England hath on the said premises erected an iron furnace, dam, and other works and buildings for the rosting of iron...the said John England, for and on account of the said works, hath purchased and taken in fee, or for years, or some other estate, several other parcells of land in Virginia aforesaid, amounting in the whole to one thousand five hundred acres or thereabouts." John England didn't present this contract to the Stafford Court until July 10, 1728. The reason for the delay is unknown, but by that time the signature of one of the witnesses, Peter Grubb[26] (1700-1754), stonemason, had to be certified in abstentia, at some expense to the company. John Toward[27] (died 1733) and Ralph Falkner also witnessed this indenture.

An unexplained dispute over money caused England and Washington to be excluded from all profits from Accokeek between May 31, 1726 and Mar. 31, 1728. An undated statement claimed that Washington was not responsible for "the Building the said Furnace." He was, however, to "pay his Proportion" of all other expenses. "But he having Neglected to advance any money towards buying Negroes or other Charges which he ought to have paid in proportion to his share and Interest on the said works," Augustine was excluded until such time as he paid his share. This statement says that John England, "having advanced nothing towards his share of the works heretofore," he, too, was to receive none of the profits (Heite, Accokeek, 10).

John England designed the furnace at Accokeek and oversaw its construction. Although Accokeek furnace was completed and capable of producing iron sometime in 1725, its first blast was delayed until 1727. On Nov. 9, 1726 William Chetwynd informed John England that the hearthstones wouldn't arrive in Virginia until that spring, there being no ship leaving Bristol until then. Chetwynd continued by recommending to England that "the pigs be weighed weekly and immediately sent to Bristol or anywhere near that place if can have shipping if not for London" (Chetwynd, letter). In 1727 as Accokeek began its

[26] Along with the contract, England presented the court with a certificate from Patrick Gordon, "Lieutenant Governor of the Province of Pensilvania and Counties of New Castle Kent & Sussex on Dellaware." Gordon had signed the document on Sept. 19, 1727 at Philadelphia and swore that Peter Grubb of Bradford Township in Chester County was a Quaker and had witnessed Augustine Washington and John England sign the contract. Peter was a first generation American, born in Chester County, Pennsylvania. By 1725 he was working for John England as a mason building the Accokeek Furnace. His name appeared on the Principio records for some years after Accokeek was completed, both as a customer and as a supplier. Grubb later moved to Lancaster County, Pennsylvania, now Lebanon County, to Durham Furnace. He is thought to have built the bloomery there c.1735 and the following year to have built Cornwall Furnace.

Peter married first Martha Bates Wall (died 1740) at which time he converted to the Quaker faith. He married secondly Hannah Mendenhall Marshall. His sons, Curttis and Peter Grubb, Jr. operated Cornwall Furnace in Pennsylvania (Dibert 6).

[27] John Toward was involved with Principio and Accokeek though his role is uncertain. At his death he left 800 pounds of tobacco for the building of a school at Potomac Church.

first campaign, forgemen at North East Forge made tools for the new furnace during which time they consumed 20 gallons of cider that was charged to the Accokeek account (Heite, "Pioneer Phase," 147).

Principio's capital was divided into twelve shares, John England owning a 1/6 interest in the company. Augustine Washington began his association with Principio as owner of 1/12 interest. In 1729 he sailed to England to negotiate face to face with the partners for a larger share. They reached an agreement, dated March 2, 1729, in which Washington held a 1/6 share in Accokeek and a ½ share in all mines other than iron. Evidently, the company was in hopes of discovering other types of ore and expanding their efforts beyond iron. This plan never materialized.

Accokeek Store

The establishment of a blast furnace resulted in the creation of a small village around the site and most 18th century furnace communities were remarkably similar and shared many common elements. In addition to the blast furnace, the Accokeek complex included at least twenty acres of mines, dams, canals, flumes, and related industrial buildings such as warehouses, a founder's house, stable, grist mill, store, storage sheds, and workmen's housing. John Wightwick's letter to John England of Oct. 2, 1730 also suggested establishing a company store.[28] Wightwick wrote, "And in such cases I believe it will be our interest to have a store there if you could find a proper person for that trust, because woodcutting, coaling, getting in provision, overlooking the furnace, keeping the accounts, and taking care of shipping home piggs will be a very full employment for [Nathaniel] Chapman, so that upon the whole I believe the Company would be glad to let that work to Captain Washington if we could be assured of the piggs being run Grey" (Wightwick to John England, letter). Shortly thereafter, a company store was opened and for many years catered to both employees and local residents.

The accompanying business ledger attests to the wide range of goods available from the store. Some of the items sold there were made on site, some were sent here from North East[29] or Kingsbury[30] in Maryland, and some were shipped directly from England. On Mar. 26, 1750 the *Pretty Sally*, Mathew

---

[28] The Accokeek ledger makes frequent references to the Fredericksburg, "Opecan," and Marsh stores. William McWilliams may have operated the store at Fredericksburg. The village of Opequon was established c.1735. The following year the Quaker Carter family of Bucks County, Pennsylvania settled on the east bank of Opequon Creek where it is crossed by the Berryville-Winchester highway (U. S. Route 7). On the west side of the creek they erected an ordinary and at the mouth of nearby Abram's Creek they built a flour mill and distillery. Numerous Quaker families settled themselves at Opequon and remained there until well into the 19th century.

Not only was Opequon a gateway to the west, it was also the site of one of Virginia's earliest Baptist churches. In 1743 a group of former members of the General Baptist Church at Chestnut Ridge, Maryland removed to Virginia and settled in this frontier village. The Baptist Church remained strong here for many years (Little 20).

A petition submitted by the inhabitants of "Opekkon" submitted an unspecified petition to the House of Burgesses. Dated Nov. 8, 1738, the petition was referred to a committee (McIlwaine, House of Burgesses, vol. 6, p. 330). For many years Opequon was a thriving community on the edge of the American frontier and Principio's management capitalized on the business opportunities there. Nothing is known of the Marsh store and there are too many areas in Virginia known as "Marsh" to determine the location of this store.

[29] The North East Forge site is on the west side of North East Creek on land once called Vulcan's Delight. During the 19th century, the McCullough Iron Company purchased North East and processed iron in a rolling mill.

[30] The Kingsbury Furnace site is in Herring Run Park within the bounds of the city of Baltimore. The furnace stood on the east side of Herring Run near its junction with Back River, just southeast of Micheo Kadison Cemetery. This property was purchased in 1734 and by 1744 the furnace had been erected. In 1781 when the State of Maryland confiscated Kingsbury, a complete inventory of the personal property was completed. This included a very large mercantile stock, a clerk's house, iron house, salt house, smith's shop, mill, wheelwright shop, nail store, kitchens, a landing house, store house, furnace with casting house and 2 dwellings (Principio, Kingsburg Furnace inventory).

Johnson, master, delivered a shipment of goods from England. These items included a range of fabrics, buttons, mohair, yarn hose for men and women, jackets, waistcoats, breeches, caps, men's' hats, Negro shoes, falls (shoes) for men and women, thread, "Hampen brown sprigs," dowlas, and "pack cloth and cord." After being inventoried, these goods were added to the store's stock.

The company ledger also reveals the names of a number of individuals who came south from Fairfax and Prince William counties to do business with Accokeek furnace and its store. One is left to wonder why these people bypassed Occoquan and Neabsco in favor of Accokeek. The most obvious possibility is that the three operations offered different amenities to their customers. If the fragments of business ledgers from Neabsco and Occoquan are representative, then Occoquan's store was far more limited than that at Accokeek (Fredericksburg Circuit Court, Lawson). These same Occoquan ledgers also include entries for some shoe making and repairs as well as a limited amount of blacksmithing done for people outside of the company's employ. Apparently these services were offered at Accokeek on a far greater scale.

Another important point learned from comparing business ledgers is that the various Virginia and Maryland iron works usually had more of a cooperative than competitive relationship. This was due to a combination of factors. During the first half of the 18th century, there were few men in the colonies who had any expertise in ironmaking. Only a handful could be coerced to leave the comfort of Britain and settle in the dangerous wilderness of the New World. High salaries and attractive incentives worked to convince a few to make the arduous voyage but there were still not enough to meet the demands of a growing iron industry. Consequently, these ironmasters moved from one facility to the next, overseeing the building of furnaces and operations and training men to perform the many jobs associated with iron production. Secondly, the bloomeries and furnaces of the Chesapeake region couldn't keep up with the nearly insatiable demand for iron, both here and at home. They often pooled their resources and made a little extra money by selling one another ore, pigs, and bar iron as needed. As a result, company ledgers include numerous entries recording accounts between different iron companies and furnaces. From papers included in the Virginia Colonial Records Project one learns that Occoquan and Neabsco maintained a connection with Principio's Maryland operations. Rappahannock Forge bought pig iron from both Prince William County furnaces and John Strode of Rappahannock wrote letters to Thomas Lawson advising him of the best places to procure oyster shells for flux. Occoquan's Charles Ewell, John Tayloe, and Peter Presley Thornton (1750-1781) maintained accounts at Accokeek. Many of the ships that hauled iron for Accokeek did the same for Neabsco, Occoquan, Rappahannock Forge, and Principio's four Maryland works. Thus, it appears that these seemingly autonomous businesses actually formed a well-coordinated network. The merchants who took castings and pig and bar iron on consignment, many of whom owned ocean-going schooners and had parent stores in England and Scotland, completed the network.

Another integral part of the furnace community was the plantation that produced the meat, vegetables, and fibers for all those associated with the works. It is the author's belief that the furnace plantation is the present day Woodbourne farm, once owned by John Moncure (1793-1876) and presently owned by Mr. and Mrs. Turner Ashby Blackburn. The land around Accokeek furnace is quite hilly and unsuited to cultivation. The fields of Woodbourne are flat or gently rolling and well suited to farming. When John Moncure acquired the property in 1847, there was already an old "office" attached to the dwelling house. Given that there were few trees standing in the area during the time in which the furnace was operating, the location of the office on this hill provided an unobstructed view of both the furnace and plantation. This house and office sat high upon the hill overlooking the fields and furnace site and it is quite possible that this was the company office in which John England worked. An ancient county road, still in daily use, leads to the farm. During Turner Blackburn's early childhood, there were scattered remains of old cottages along a low ridge near the furnace site. His grandmother referred to them as the slave cottages belonging to the furnace and said that they had pre-dated John Moncure's occupation of the farm.

Based upon company ledgers, this one plantation was unable to support the large workforce at Accokeek. In 1749 the furnace also rented a plantation from Christopher Knight and a plantation and orchard from William Jackson, a wheelwright employed at Accokeek. People with accounts at the company store often paid their bills with produce and other food items rather than with cash and Accokeek purchased food from the company's other furnaces and forges. The amount of food required to sustain the workforce at Accokeek was tremendous. Beyond what was produced on site, company ledgers record the following food purchases:

April 1731—paid Principio for 564 lbs. pork
April 1731—paid Principio for ½ bushel of peas and 18 bushels of salt
Nov. 28, 1749—bought a total of 6,358 lbs. pork from local farmers
Mar. 10, 1745—from North East Forge 12 bushels oats
Aug. 12, 1748—from North East Forge 61 ½ bushels corn
Sept. 3, 1748—from North East Forge 208 ½ bushels corn
Sept. 24, 1748—from North East Forge 300 bushels corn
Oct. 22, 1748—from North East Forge 1 hogshead of rum (104 gallons)
1 hogshead of molasses (104 gallons)
40 lbs. loaf sugar
Nov. 28, 1748—from North East Forge 300 bushels corn
June 23, 1749—from North East Forge 400 bushels corn
June 26, 1749—from North East Forge 500 bushels corn
Apr. 25, 1750—from North East Forge 1 cask molasses (91 gallons)
Apr. 30, 1750—from North East Forge 65 lbs. cheese
June 12, 1750—from North East Forge 1 gallon rum
Dec. 4, 1750—paid John Allen for 4 quarters beef (342 lbs.) and 23 hogs (2,591 lbs.)
Dec. 20, 1750—bought from Col. Thomas Harrison 19 hogs (1,990 lbs.)
Jan. 1, 1751—paid Peter Hedgman for 32 hogs (3,782 lbs.), 116 lbs. fat, and 13 hogs (1,492 ½ lbs.) "from Wm. Brent's Quarter"
Jan. 7, 1751—paid Peter Hedgman for 1,559 lbs. pork
Jan. 18, 1751—paid John Allen for 4,802 lbs. pork
Jan. 21, 1751—paid Richard Hodges for 544 lbs. pork
___, 1752—paid North East Forge for 517 lbs. bacon
___, 1752—paid Kingsbury Furnace for 1,791 lbs. salt pork

Charcoal Production

Charcoal was critical to iron production. Not only was this essential material used in the furnaces, but also by the finers and chafers who forged bar and pig iron into finished products and by blacksmiths who carried out the extensive repair work normally performed at an iron works. Because it was relatively cheap and produced about twice as much heat as wood, charcoal was the fuel of choice in American blast furnaces until the 1840s when it was replaced by coke. During the 18th century, however, mineral coal was being used in several Virginia furnaces. Alexander Spotswood was using mineral coal in his Massaponax Furnace in the 1720s (MacMasters 64). Cinders, a byproduct of mineral coal, are also scattered around both furnace sites at Rappahannock Forge. The Westham Foundry near Richmond, established at the beginning of the Revolution, also utilized mineral coal in the production of cannons. The use of coal for the production of iron castings such as those made at Massaponax resulted in superior quality products and was absolutely necessary for the casting of cannons. However, the use of mineral coal in blast furnaces was not common during this period.

An iron furnace required an ample supply of timber for making the charcoal used in the smelting process. The preferred varieties of wood were hardwoods such as hickory, oak, cedar, and chestnut and these deciduous woods yielded about one-sixth of their weight in charcoal. Thus, 6 to 8 tons of wood produced about 1 ton of charcoal. There are several methods of estimating and conceptualizing charcoal consumption in typical 18th century furnaces. It took approximately two acres of hardwood (about 12 cords or 1,536 cubic feet) to make one ton of smelted iron. Another way of expressing this is to consider that the average blast furnace consumed three tons of ore and approximately 300 bushels of charcoal every 24 hours. William Byrd noted, "The properest wood for [iron making] was that of oily kind, such as pine, walnut, hickory, oak, and in short all that yields cones, nuts, or acorns. That two Miles square of Wood would supply a Moderate furnace, so that what you fell first may have time to grow up again to proper bigness (which must be four inches over) by that time the rest is cut down" (Wright 348). All of this is equivalent to one acre of 20-25 year old hardwoods consumed by the furnace each day while in blast. Robert B. Gordon noted, "A typical mid-19th century blast furnace, the Buena Vista in Connecticut, each year burned 356,000 bushels of charcoal, made from 11,900 cords of wood, in 475 meilers [pits]. Since the average forest yield here was twenty cords per acre, the proprietors cut about 600 acres a year. They maintained a continual wood supply by cutting 12,000 acres in a twenty-year rotation" (Gordon, American Iron, 40).

By the early 18th century, most of the great English forests were gone, having been used not only by the iron furnaces, but in the construction of ships for the British navy. It may be argued that this latter use had had a more profound impact on the forests than did iron working. Charcoal makers were among the earliest conservationists. English colliers had worked under mandatory conservation rules since the 13th century and had developed a system of harvesting timber known as coppicing. After the original forest had been cut, livestock was kept out of the area in order to prevent their eating the tender new shoots sent up from the stumps. Carefully managed, a forest thus cut renewed itself and was ready to cut every 20 to 30 years. In addition to making the forests self-perpetuating, wood cut from coppiced forests was smaller in size and seldom required splitting. Because of its smaller and more uniform size, coppiced wood also assured the collier of even burning in the charcoal pits.

Charcoal making was the most labor-intensive part of iron production and second in cost only to building and outfitting the furnace. At many furnaces, half of the workforce was in some way involved in charcoal production. When the furnace was out of blast for relining or other repairs, all the men who could be spared were sent to the forest to cut wood. Overseeing this was the collier, a highly skilled worker who was largely responsible for the economic life or death of the furnace. In 18th century America there were very few professional colliers[31] and, with the exception of the furnace manager and founder, they received the highest wages of all the workers at the furnace. One collier, with the help of five or six assistants, could simultaneously tend 15 to 20 charcoal piles also called pits or meilers.

Trees were normally cut in winter, allowed to dry and charcoaled in spring and summer. Once cut, the limbs were removed and the trunks cut into four-foot lengths. These were hauled via horse, mule, or ox-drawn sled to the charcoaling area where they were stacked in piles to dry. After drying, the wood was split into 1 to 4 inch widths called lapwood and 4 to 7 inch widths known as billets.

Charcoaling was done on an area of level ground called a hearth. It measured about 30-50 feet in diameter and had a slightly raised center (similar to a pitcher's mound) to encourage drainage. A ring of soil, cleaned of stones and organic matter, was placed around the perimeter of the hearth. The first step in constructing the pit was the placement of the fagan, a center pole about 18 feet long and about four inches in diameter. A triangular chimney made of 18-inch long split wood surrounded the fagan. The hearth was then covered with split wood, a layer of bark or moss, and a layer of sandy clay. Concentric rings of billets were placed on end around the chimney. The billets were flared outward at the base to provide the proper pitch, the entire pile being shaped like an igloo. The layers of billets sometimes rose to three or four tiers in large burns, the average pit being made up of 25-30 cords of wood. The object of piling the wood in this manner was to promote air circulation from within the pile. Lap wood and a chinking of charred wood were used to fill in the spaces around the billets. The entire pile was then covered with lap wood and small branches and sealed with a layer of wet leaves, earth, charcoal dust, or turf with the grass side turned inward. The seal prevented air from entering the pit except through holes made by the collier. The thickness of this outer layer was critical to the success of the burn. Too thin and the fire would break through and consume the entire pit. Too thick and the explosion of trapped gases might destroy the pile and injure workers. The collier and his assistants kept a close eye on the steaming, smoking pile to guard against blowouts that might lead to a flaming of the pile. When blowouts occurred, dirt from the peripheral ring was used to seal the holes and stifle the fire.

After completion of the pile, the fagan was removed, thus creating a chimney that was filled with kindling and burning wood chips to ignite the stack. Once the fire was established, the top vent hole was covered with a piece of turf. The first stage of the burn was called "sweating." The collier opened a narrow ventilation band around the base of the pit, allowing the escape of water that condensed inside the pit. He also opened the top of the seal to prevent steam explosions. When the sweating stage was over, the open band at the bottom was filled in using dirt from the peripheral ring and the pit was allowed to burn slowly for three or four days while the heat carbonized the wood. The collier then opened a ring of vent holes

---

[31] Principio ledgers from 1730 provide wages for some of the company's employees. Woodcutters were paid 2 shillings, 9 pence per cord of hardwood; colliers received 6 shillings, 7 pence per load of charcoal. Forgemen earned £1 per ton for blooms and a white furnace foreman was paid 20 shillings per ton of iron smelted (Hudson 2615).

about halfway up the pit. The smoke that escaped from these holes was initially a yellow-gray color. A carbonization progressed, the color of the smoke changed to a blue-gray. Gradually, as moisture an volatiles left the wood, the pile shrank until it was about one-third of its original size. If the burn proceede properly, the fire burned from the top down.

Fig. 1-3
Construction of charcoal pits (Diderot, vol. 1, Oeconomie rustique, Charbon de Bois, Plate I).

Fig. 1-4
Burning charcoal. The workman in Fig. 4 lights the fire from the top. The pit in Fig. 5 is burning well. As the burn proceeds, workmen begin raking charcoal out of the pit (Fig. 8) (Diderot, vol. 1, Oeconomie rustique, Charbon de Bois, Plate II).

Depending upon the size of the pile and the nature and dryness of the wood, it normally took ten days to two weeks for the pit to "come to foot" or for all the billets to char. The best results were obtained from wood that had been seasoned to a moisture content of 20% or less. As the burn continued and the smoke and steam subsided, raking out began. Long-tined iron charcoal rakes were used to pull finished charcoal from the top and perimeter of the pit. The collier began raking coals from the top of the pit before the entire pile had been carbonized. Perfect charcoal made a distinctive tinkling sound as it was raked out into small piles to cool. Raking out was a tedious process because charcoal held fire for a long time and only a small amount could be raked at once, the rest of the pile being left to continue charring. After raking coals from the periphery of the pit, dirt from the ring was thrown over the remaining pile and the burn continued. After the colliers determined that no fire remained in the coals already raked out, the fragile pieces were gently gathered into shallow oblong baskets made from thin strips of green wood. These were loaded onto coal carts and transported to a coal house, also called a stockhouse, near the furnace. Because charcoal was extremely susceptible to moisture, it was essential to keep it dry.

Charcoal was too bulky to store in quantity and would crush under its own weight when bags or baskets of the material were stacked. It also lost much of its quality if stored too long, though charcoal houses were built near the furnace to house it and keep it dry. Thus, it was essential that cutting, coaling, and hauling be carefully timed so that the furnace was provided with a constant supply of fresh charcoal during the blasting season.

It took thousands of acres to produce enough timber to sustain a blast furnace indefinitely. According to the 1768 quit rent records, Accokeek had only 2,796 acres, though they rented a substantial quantity in addition to what they owned. Furnace managers also purchased cordwood from outside company-owned lands. A ledger entry dated July 31, 1750 records the company's purchase of 433 ½ cords from various local residents. A North East Forge ledger entry of July 1, 1754 records the sale of 1,399 ½ cords of wood to Accokeek, this being shipped from Maryland.

The life of a collier was lonely and harsh. Near the pits were built temporary shelters or huts. These were about ten feet in diameter and were made of 16 poles 12 to 14 feet in length. The poles were assembled into a cone shape, fastened together at the top. This was covered with sacks to make a coarse lining for the interior of the hut. Starting at ground level, leafy twigs were arranged around the hut to act as "keys" to secure a turf wall. Blocks of turf were stacked like bricks in horizontal courses, grass side uppermost. The whole of the exterior of the structure was covered with this turf, including the top, the only opening being a small door. During the building process, a small "porch" or overhang was extended above the opening to deflect precipitation. Sacks were hung across the opening and acted as a door. Inside the hut two rough bedframes were placed, one on either side of the door. These were covered with mattresses made of straw-filled sacks. A brazier of burning charcoal, used for cooking and heat, was placed under the porch overhang but was never left there overnight because of the danger of fumes. It was in such dwellings that the colliers spent nearly nine months of the year.

Iron smelting and its attendant charcoaling have long had the undeserved reputation of fouling the environment. In reality, 18th century slash-and-burn farming had a greater impact upon the soil and waterways than did iron manufacturing. Robert B. Gordon argues that there were only limited long-term environmental effects from mining, timbering, charcoaling, and smelting. As previously stated, ironmakers were the first to concern themselves with forest management. During the 18th century, most iron was taken from shallow pits, trenches, or simply raked from the bottoms of ponds. Rarely were shafts driven into the earth and, following mining operations, the pits filled with water to become sanctuaries for wildlife. While it is true that charcoaling produced pyroligneous acid that was potentially a source of local pollution, the coaling pits were scattered over such a wide geographical area, that the chemical produced little long-term environmental impact. Within a generation of the cessation of mining and coaling, coppiced trees had become mature forests and there was little evidence of any prior industrial activity (Gordon, Landscape, 117).

By contrast, farming methods of the period had a far greater impact upon the environment. Typically, new fields were opened by strip cutting the trees. Smaller vegetation was removed by burning and tobacco and corn were immediately planted between the remaining stumps. There was no thought to crop rotation or to contour plowing. Corn and tobacco were planted season after season until, within five or six years, the soil was so exhausted that it could no longer produce. When a planter depleted his fields, he simply opened another patch of woodland and began anew. Erosion was such a problem that within 40 to

50 years of their creation, ports such as Dumfries and Falmouth were so silted that ships could no longer dock. Telltale evidence of early farming remains visible today in the form of silted tributaries while old iron working sites are all but invisible.

Furnace Technology

Man's earliest efforts at iron making were carried out in a bloomery, a type of forge in which small quantities of the metal were produced by heating ore and pounding out impurities. The technology for making iron changed little from 500 BC until the development of the blast furnace in the 15th century. This revelation occurred in Liege, now Belgium, and it not only greatly increased the quantity of iron that could be produced, but introduced a new form of the metal called cast iron. The first blast furnace in England was erected at Newbridge, Sussex in 1496. The technology developed during this period remained largely unchanged throughout the 18th century, the next major improvements not occurring until 1840 with the utilization of coke as fuel.

The physical layout of most 18th century furnace sites followed a predictable pattern. These complexes included a furnace, casting and bellows houses, a dam, head and tail races, a grist mill, barns, smoke house, slag dumps, flux piles, charcoal houses, ore warehouses, a company store, roads between the furnace and waterfront landings, warehouses, mines, ore roasting facility, ironmaster's house, workmen's housing, a plantation, and shops for wheelwrights, cartwrights, and blacksmiths. Very often, ironworks also included bakehouses, spinning houses, and a storage facility for gunpowder. Industrial buildings were clustered around the furnace which was built below an impounded creek or run. Charcoal, ore, and flux houses were built on higher ground above the furnace to allow for easy charging. Slagheaps were normally downstream or across the run, out of the way but not far from the furnace. Residences and ancillary shops were within walking distance from the furnace and near the main road (Heite, "Pioneer Phase" 162).

Mine pits were normally located within a mile or two of the furnace. Accokeek's mines were located on the north side of Potomac Run about a mile northwest of the furnace. Archeological examination of these mines should reveal the presence of roads, spoil heaps, and miners' lodgings. While much of the bog ore typical of the area could be broken loose with picks and shovels, blasting was also a common means of breaking free the ore and entries in the Accokeek business ledger indicate that some blasting was done there. Generally speaking, most iron ore trenches and pits rarely exceeded 40 feet in depth, the iron being laid down in surface deposits. Some furnace sites also had ore roasting areas; this process helped remove some troublesome impurities. Ore was roasted at Fredericksville Furnace, but it is unknown if Accokeek ore received the same processing.[32]

Also required for successful furnace operation was a steady supply of running water to power the bellows and gristmill. Accokeek Run provides a fine, steady flow of water, making it ideal for powering a water wheel. The typical bellows wheel was usually a large affair, some 25 feet in diameter. Few technological advances occurred in iron manufacture during the 18th century and overshot bellows wheels, though not unheard of, were rare. Overshot wheels turn by water flowing over the top of the wheel, the weight of the water substantially increasing the torque of the wheel. Powering the bellows that pumped air into the furnace required 15-20 horsepower. A typical 1,300-pound trip hammer used in a bloomery or finery forge required 21 horsepower to make 110 stroke per minute with a 2.3-foot lift. A smith shaping axes or similar tools using a tilt hammer used 20-35 horsepower (Gordon, American Iron, 52). Based upon the layout at Accokeek, it appears that either a breast-shot or undershot wheel powered the bellows.

The furnace bellows were an interesting bit of technology. Massive affairs, they were made of two-inch thick, wedge-shaped boards lined with iron. They were normally 18 to 20 feet long by 4 to 6 feet wide at the rear, narrowing to half that width at the front. The sides of the bellows were made from 18 to 20 bulls' hides. The leather sides required frequent lubrication to prevent cracking, butter, or tallow fulfilling that purpose. Most furnaces had two bellows placed side by side. A water-powered camshaft was designed

---

[32] The purpose of roasting the ore was "to expel the volatile, useless, or noxious substances, as water, vitriolic acid, sulphur, and arsenic; to render the ore more friable, and fitter for the subsequent contusion and fusion" (Encyclopedia, vol. XI, p. 460). Roasting was "performed by kindling piles, consisting of strata of fuel and of ore placed alternately upon one another, or in furnaces similar to those commonly employed for the calcination of limestone" (Encyclopedia, vol. XI, p. 468).

so that as one bellows was compressed, the other was opened, thereby maintaining a continuous blast of air. These leather and wood bellows had long been used in English blast furnaces but were difficult to manufacture and didn't last long under the extremely heavy use to which they were subjected. By mid-century, English inventors were experimenting with variations on the traditional furnace bellows. Leaders in the field included Isaac Wilkinson, James Knight, and John Smeaton and by 1762 blowing tubs were available to furnace operators (Heite, "Pioneer Phase," 175). Essentially, these were cylinder and piston air pumps and, like the old-style bellows, were used in pairs. The pistons worked alternately and produced a stronger and steadier blast than traditional wood and leather bellows. Blowing engines were not installed at any Principio works until the early 1770s, long after Accakeek had ceased operations.

Accokeek Creek was dammed to control water flow to the wheel powering the bellows; remains of a silted-in pond are clearly visible on the northeast side of the creek. Water was collected in the pond behind the dam. When it became necessary to operate the bellows, the water was released from the pond. Flowing down a sluiceway cut into the rock, remnants of which are still visible, the water turned the bellows wheel. A raceway approximately 90 feet long was excavated directly into the bedrock. The Accokeek site has suffered from repeated floods and there are no known contemporary drawings of the facility. Further, no intensive archeological study of the site has yet been conducted. Thus, any attempt to interpret the surviving features is purely speculative. The existing raceway is intriguing. It appears to have carried water from the dam toward the furnace, but made a right-angle turn some distance before reaching the water wheel (see Fig. 1-9). At present, this raceway opens on the southern edge of Accokeek Run. Not noted on the figure is a man-made flat surface cut into the rock at this end of the raceway. Perhaps the water flowed through the stone raceway, then made a second 90-degree turn into a wooden flume that carried the water to the bellows wheel.

The water from the raceway discharged into a sand and gravel flood deposit in the creek. Although there are substantial sand deposits along Accokeek Run, this sand was not suitable for iron casting. Very special, fine-grained sand was required for this process. The source of the casting sand used at Accokeek is unknown. On the east side of Bethel Church Road (State Route 600) is a very old sand pit that has supplied sand for iron and steel casting for many years. This sand is actually the remains of a prehistoric sand dune and is of a consistent fine and even texture. To the author's knowledge, this is the most convenient source for fine quality casting sand, but no records remain to indicate where Accokeek management obtained this most vital part of the iron making process and there is no mention of it in the ledger.

At the heart of this industrial complex was the furnace itself. Whenever possible, furnace sites were chosen for their proximity to a hill. The typical 18th century English blast furnace was a truncated, four-sided, pyramidal structure built of stone or brick and was hollow down the center. Its 24-foot square base stood on footers about 26 feet square. The stack averaged 25 to 28 feet in height and was 15 to 15 ½ feet square at the top. Any material except limestone could be used to build the stack and some furnaces were constructed with bricks. Scattered cut sandstone blocks around Accokeek suggest that that furnace may have been built with this locally quarried material. There are numerous sandstone quarries along Rocky Run and Aquia Creek. The quarry on Rocky Run (later known as Robinson's Quarry) was closer to the furnace than those on Aquia Creek, being located only about 2 ½ miles away. One of the roads leading to the furnace passed directly by Robinson's and this quarry was long noted for its fine quality stone. At several points the outside of the stack was girt with iron bands to prevent cracks caused by expansion and contraction as the furnace heated and cooled. The furnace typically had two arched openings built into the lower sides of the structure. Through one of these the tuyere carried air to fan the burning charge. The second was a work arch that permitted access to the bosh for the purpose of manipulating the iron during smelting and tapping of the molten iron (Fig. 1-5).

Proper construction of the foundation was critical to the furnace's success. Workers usually dug down to firm footing so as to support the weight of the stack. Some builders laid wooden sleepers in the bottom of this excavated pit. The next step was to dry-lay a stone foundation a foot or two wider than the stack in each direction. This stone base usually contained interstices or spaces and, sometimes, a drainpipe to ensure drainage of ground water. Water trapped in the masonry could result in devastating steam explosions, so builders were meticulous about drainage.

The inner walls of the furnace, which were subject to extreme heat, were constructed of either heat-resistant brick or rough-cut stone. The bulk of the stack was kept as great as possible to provide insulation and heat storage. The inside of the furnace, into which were dumped carefully measured layers of ore, fuel, and flux, was shaped like an inverted pear. The upper part of the chimney, called the throat, widened as it neared the bottom. About two-thirds of the way down, the sides angled in at a point called the bosh, a combustion chamber built of firebrick or slate and open at both the top and bottom. This oval-shaped compartment was insulated from the walls of the stack by a layer of sand or crushed stone. Below the bosh, the chamber surface dropped vertically, forming a crucible where the molten iron settled and was periodically drawn off through a hole low in the work arch.

The top of the stack was floored, sometimes with stone, sometimes with iron plates. On top of this floor was a chimney with a hole in the side next to the charging bridge. The chimney directed flue gases up and away from workers charging the furnace. Early furnaces often lacked this chimney, the charge being dumped directly into the gaping furnace mouth. Periodic belching of the furnace often ejected quantities of scorching gasses, rocks, and particulate matter into the faces of the workmen, resulting in innumerable deaths and injuries.

The hearth was the most important part of the blast furnace. It consisted of a stone floor about 7 feet by 4 ½ feet by 1 foot thick. The stones making up this base were laid on a bed of sand. On top of the base was built a low stone wall, usually about three stones high. All cracks and spaces in the stonework were mortared with fireclay (Heite, "Pioneer Phase," 167-170).

Many early American ironmakers imported their hearthstones from England, sometimes delaying the first blast for a year or more while they awaited shipment. Although there were sandstone quarries very near Accokeek, John England also ordered at least some of Accokeek's hearthstones from England. In November 1726 William Chetwynd advised England that delivery of the Accokeek hearthstones would be postponed until the following spring, there being no ship leaving Bristol until that time (Chetwynd, Letter). While this seems an unnecessary waste of time and money, poor quality stones could ruin a blast or a furnace. The replacement of hearthstones was a necessary part of furnace maintenance. Sometimes this was done each season or every couple of seasons, depending upon the quality of the stone and the condition of the furnace. A large, successful furnace was capable of producing about 800 tons of iron per year and about 1,600 tons of slag. Smaller or less productive furnaces might yield half that amount.

In addition to the two large arches there were several small openings built into the furnace. These were kept plugged with clay dams that could be removed for periodic tapping of the molten iron and waste slag. As the ore melted, the iron settled in the crucible in the bottom of the furnace. The rock matrix, also liquefied but lighter in weight than the iron, floated on top of the molten metal. This dark green glassy "slag" was periodically tapped from holes located a bit higher on the furnace wall and was dumped nearby as waste material. Slag dumps are common indicators of old furnace sites. At Accokeek, the slag was hauled across the top of the dam to the opposite side of the creek where it was dumped and eventually formed a small hill. Today this hill is covered with a heavy growth of timber and resembles every other hill in the area. However, scraping away the thin layer of leaves reveals the slag beneath. Pieces of slag still litter the furnace site and creek bed and children playing in the creek have long believed that it proved the existence of a local volcano!

If the terrain didn't permit building the furnace into the side of a hill, then a hill was built next to the furnace. A bridge was then constructed from the top of the hill to the top of the furnace to allow easy charging. This bridge was very often covered to protect the workers and supplies from the weather. Supplies of ore, fuel, and flux were kept in baskets on the top of the hill, usually in simple wooden buildings and workers, called "fillers," rolled barrows or carried the baskets across the bridge to load the furnace from the top. During the smelting season, charging continued 24 hours a day. The fluxes employed in iron smelting varied with the nature and amount of impurities in the ore. These impurities included sulfur, arsenic, and phosphorus. Sulfur caused the metal under the hammer to crack and crumble when worked hot, while phosphorus caused it to crack or crumble when worked cold. Iron ores containing silicate of alumina required a flux of lime. Lime added caused the impurities to separate from the iron and to dissolve into the liquid slag which was drained from the furnace during the smelting process and disposed of. In the Chesapeake region oyster shells were often used as flux, primarily because of their availability. The limestone or oyster shells were sometimes calcined or heated prior to use, a process carried out in a kiln located near the furnace. Whether or not shells used at Accokeek were calcined is unknown. Shells are still easily collected on site.

Before use the furnace had to be seasoned, a lengthy process by which every drop of moisture was gradually removed from the masonry walls. Once the furnace was actually put into blast, the slightest bit of moisture remaining in the walls might cause the furnace to explode. Robert Plot, a 17th century observer of iron making, wrote, "should one drop of water come into the Mettall, it would blow up the furnace, and the Mettall would fly about the workmens ears" (MacMasters 39). This quaint description belies the deadly nature of such an explosion. The seasoning process began by filling the entire furnace chamber with charcoal, which was lighted from the top and allowed to burn down to the level of the tuyere. It was then re-filled and allowed to burn back up to the throat. This fire was kept burning for about three days. A cold air blast was gradually introduced through the tuyere thereby slowly increasing the temperature inside the furnace. It took a week to ten days of constant burning before the furnace was properly seasoned and the heat was high enough to make iron (about 1,400 degrees Centigrade). Once the desired temperature was achieved, fillers began charging the furnace with alternating layers of charcoal, ore, and flux. The founder increased the charge very gradually until the furnace was up to full production. Assuming the fire could be maintained, smelting continued nonstop six to eight months, providing there were no technical difficulties and the operators were able to maintain a constant supply of ore, fuel, and flux, and a steady flow of water. This period of smelting was known as a campaign. Principio's Maryland furnaces generally blew from late December until June or early July. The average lifespan of the interior masonry was bout 30 weeks, after which time the furnace required emptying and relining.

Once the furnace reached the desired temperature, more ore, charcoal or pit coal, and flux were dumped in from the top in alternating layers. A typical charge consisted of 30 bushels of charcoal (about 600 pounds), 1,000 pounds of ore, and 50 pounds of flux. Every 15 minutes or so more ore, fuel, and flux were dumped into the top of the furnace. This was a dangerous task as the workers had no way of knowing when flames, smoke, and chunks of hot charge might erupt through the charging hole and into their faces (Gordon, American Iron, 120).

The casting house was a long, low building with a roof and sides that abutted the furnace. The floor of the casting house was covered with a thick layer of sand, scorched and blackened from use. Dug into the sand were several long shallow trenches (called sows) running the length of the building. At right angles to these trenches were numerous short trenches called pigs. Into the bottom of each pig mold was usually pressed the name of the manufacturing furnace and, perhaps, the year. When the founder determined that the iron was ready to be tapped, he opened a clay plug in the side of the furnace wall allowing the glowing molten iron to flow down the sow trenches and into the pig molds. This tapping operation was carried out by the guttermen under the supervision of the founder. Once cooled and hardened, workers used sledge hammers to break the pigs free from the sows. The pig iron was then stacked to await shipment either to England or to a forge where it was processed into bars of wrought iron. Sand molds were also used to make a variety of cast items such as pots, pans, Dutch ovens, fire backs, shot, tools, farm implements, stoves, etc. For these items, the molten iron was collected by potters using ladles and was poured into the molds.

Early Blast Furnace

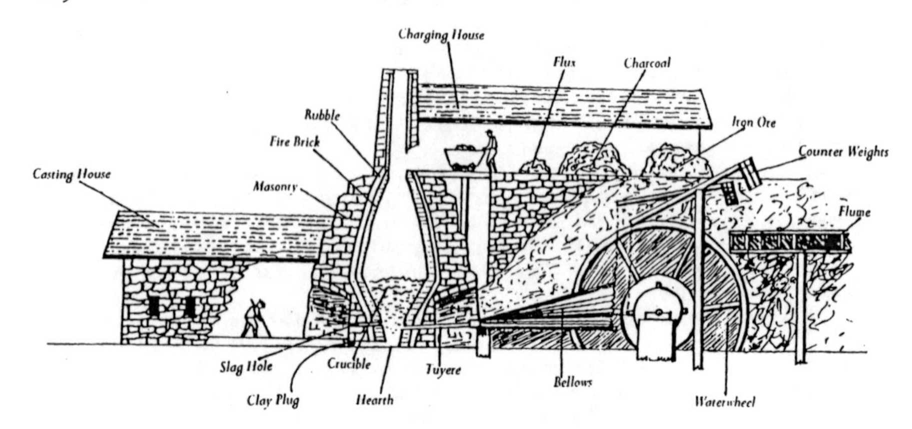

Fig. 1-5
Diagram of a Charcoal-Fired Blast Furnace

Fig. 1-6
Casting molten iron from the furnace into sand molds. The young boys pictured are skimming slag from the liquid iron. The workman to the right is chipping off scale from a section of cast pipe (Diderot, Vol. IV, Forges, 3rd Section, Plate IX).

The process of smelting iron in a charcoal-fired blast furnace was relatively simple, yet workers were always in eminent danger. Fillers were always at risk of flames and hot particulate matter being belched into their faces as they dumped the charge into the top of the stack. The most feared threat, however, was from explosions. A constant concern, explosions might be caused by blockages in the venting or by the charge settling unevenly and creating pockets of volatile gases. If the furnace exploded, it violently vomited all its contents, molten and solid, from every opening. Not only were nearby workers killed, but the explosion spread fire all about the area. Explosions often occurred with little warning, though workmen knew to watch for flames that alternately shot upwards, and then ominously disappeared. If the furnace commenced behaving in such a manner, all in the vicinity knew to run for cover.

Joshua Hempstead[33] visited a Cecil County, Maryland iron furnace in 1749 and described the operation of a blast furnace as he witnessed it:

> ...the Large Bellows 2 pr go by water & the fire goes not out after it is once blown up until the Season of ye year comes about. the furnace I suppose is 20 foot high or more & is fed with oar [ore] & coal &c at the Top as if it were the Top of a Chimney all put in there. there they bring in Horse Carts the oar and the Coal & oyster Shels & there Stayd two men Day & night. the Top of ye furnace is about breast high from the floor where they Stand to Tend it & ye flame Jets out Continually and is Extinguished by the oar Coal & Shels as they feed it. Each Couple Tend 24 hours in which time they Run or Cast twice. they have Small Baskets that hold about a peck & half & they put in a Cart in number of Baskets full of oar & a Certain Number of Baskets of Coal and a Certain Number of Baskets of oyster Shels. all in Exact Proportion and as the materials Consume below in the furnace they filled up at the Top & out of the Bottom beside the Iron yt is drawn off near a Day there is vast Quantities of Glass that Runs out Every now & then & is Tough & hangs together like an ox Hide & they dray it away with such a hook as the Tanner pull up hides with & when it is Cool is as Brittle as any other Glass & they Cart it away & bestow it in waste places to mend the Cartways & Dams Even as Small Stones. (there is one man besides the 4 that Tend by Course that is Constantly breaking the Rock oar Small with a Large Hamer or Sledge) which lyes like a little hill near the Coave where it is Landed out of the Large Boats Something biger than our ferry Boat (Hempstead 348).

On Oct. 2, 1730 John Wightwick, one of the English partners, wrote to John England to discuss several matters including a commentary on the quality of Maryland and Virginia pigs:

> The price of piggs is very low at present, we hope in a short time it will be better. The potomack [Accokeek] piggs will not reach 5l. 15s. they being cheap piggs are fittest for founderys, and I dont doubt but we might sell a large quantity yearly in London for that use. We have sold some of these last arrived, but the dealers complain they are too white, if they were grey they would go off much better.
>
> The way to distinguish grey piggs from white is this: taber upon them with a hammer and if they have a deadish leaden sound they are grey, they will also be pretty level in the belly of the pigg, & the edges of them will be blunt. If they are white, by hammering on them they will ring much more, the undersides of them will be hollow & the edges much more sharp, and if you strike the white ones at the end they will break short.
>
> This information I had from a Founder whom we sold a parcel to; and this difference is not so much owing to the ore as the fluxing, for we have pretty near an equal quantity of grey piggs mixed with the white which inclines me to think this must be some mismanagement of the Founder that they are not all grey. I therefore hope you will consider and give directions to the Founder accordingly, and let the Hearths at both works

[33] Joshua Hempstead (1678-1758) was a resident of New London, Connecticut. There he was a judge, farmer, surveyor, shipwright, carpenter, and stonecutter. For over 40 years he kept a diary of his travels, observations, and important events.

be always set Burrow and not Transheer, for white piggs will do us great prejudice in the sales.

In that same letter Wightwick said of the operation at Accokeek:

> We have received since you left [Capt. Washington's] Ltres. of ye 27th, Apr. 4th June 10th & 10th July and are well pleased with the clear account he gives us of ye state of that work. There is a pretty good blast made and it was not his fault that it was not a better, and that he hoped to be able to ship it all home this season of which we have already recd. 276 tons. We are pleased with his agreement, with the Colliers, with the reduction of the price of woodcutting and his new manner of cording the wood which will be greatly to our advantage, the same being you know always done in England.
>
> We are also glad to hear he is in such forwardness for another blast. He says he thinks it will be cheaper to send Hearths from England than to bring the stones 20 miles land carriage from a quarry (where he says there are good ones) at that distance from the works but I hope you will consider that part thoroughly because sending hearths from hence is very chargeable in freight, as well as trouble some to get them and put them on board (Swank 452).[34]

William Byrd (1674-1744) spent much of 1732 traveling to various iron works and writing a journal that he called "A Progress to the Mines." Byrd seemed intent upon starting his own iron works, but never quite accomplished that goal. Byrd spent considerable time with Alexander Spotswood (1676-1740) who, by 1732, owned two working furnaces. Spotswood outlined to Byrd the labor needs of a successful furnace:

> [Spotswood] said there ought to be at least an hundred Negroes employed in it, and those upon good land would make corn and raise provisions enough to support themselves and the cattle and do every other part of the business. That the furnace might be built for £700 and made ready to go to work, if I went the nearest way to do it...That if I had ore and wood enough and a convenient stream of water to set the furnace upon, having neither too much nor too little water, I might undertake the affair with a full assurance of success, provided the distance of carting be not too great, which is exceedingly burdensome. That there must be an abundance of wheel carriages shod with iron and several teams of oxen provided to transport the wood that is to be coaled, and afterwards the coal and ore to the furnace, and last of all the sow iron to the nearest water carriage, and carry back, limestone and other necessaries from thence to the works; and a sloop also would be useful to carry the iron on board the ships, the masters not always being in the humor to fetch it.

Byrd then enumerated the people employed at the furnace, "viz.: a founder, a mine-raiser, a collier, a stocktaker, a clerk, a smith, a carpenter, a wheelwright, and several carters. That these altogether will be a standing charge of about £500 a year. That the amount of freight, custom, commission, and other charges in England comes to 27s. a ton. But that the merchants yearly find out means to inflame the account with new articles, as they do in those of tobacco. That, upon the whole matter, the expenses here and in England may be computed modestly at £3 a ton. And the rest that the iron sells for will be clear gain, to pay for the land and Negroes, which 'tis to be hoped will be £3 more for every ton that is sent over" (Wright 360).

Although William Byrd never visited Accokeek, he included a description of the works in his journal. He stated:

> England's Iron Mines, called so from the chief manager of them, belongs to Mr. Washington. He raises the ore, and carts it thither for 20 s[hillings] the ton that it yields.

[34] The Principio ledger of Sept. 29, 1734 also records that the company paid £12 "For Freight of Hearth Stone from Rappahannock to Accokick from thence here and charges." In that same ledger is another entry dated Nov. 23, 1734 in which James Markham brought to Principio "a set of Hearth Stones from Virginia."

The furnace is on a run that discharges into the Potomac. And when the iron is cast, they cart it about 6 miles to a landing on that river. Besides Mr. Washington and Mr. England there are several persons in England concerned in these works. Matters are very well handled there, and no expense is spared to make them profitable, which is not the case in the other works I have mentioned (Wright 368-369).

Labor

Life at an iron works was difficult for both white and black workers. Michael Robbins aptly described working conditions at an iron works of this period saying, "There is hardly a way that one could now appreciate the combined impact of working a night shift without electricity at a furnace located adjacent to a malarial swamp, perhaps working elbow to elbow with African slaves and perhaps with a stomach full of salt pork, corn meal, and 'strong beer'" (Robbins, Principio, 239). Iron smelting was a labor-intensive endeavor and labor issues were a constant concern for furnace managers. Smelting was also a highly technical process that required a wide variety of both skilled and unskilled workers. During the first 20 or 30 years of serious ironmaking in the colonies, skilled ironworkers were extremely difficult to find and keep and those without the requisite skills were notoriously poor workers. Initially, Principio balked at utilizing slave labor in their iron making operations though their reasons are unclear. Principio's earliest unskilled labor force was composed of free and indentured employees, including convicts. The furnace also employed a number of unskilled workers who dug and broke the ore, hauled the ore, flux, and fuel to the furnace, drove the teams, repaired fences, dug ditches, worked the plantation and performed the endless labor required to keep the furnace operating. This workforce was comprised essentially of whoever was available. Upon his arrival in Maryland, John England began writing a series of letters to the partners describing the horrendous difficulties that he faced with his laborers. No doubt, many of these men were convicted felons who had been sentenced to transportation to the colonies and their honesty and work habits were already doubtful. Between 1614 and 1775 some 50,000 Englishmen were sentenced by legal process to be transported to the American colonies, a sizable number of whom found work at iron furnaces. Drunkenness amongst the workers was a perennial problem resulting in constant fights, careless work, and absenteeism. Principio's hesitation over utilizing slaves was short-lived. The severe shortage of dependable white labor seems to have forced Principio to reconsider the makeup of their labor force for in 1721 they bought 18 slaves. By 1723 company ledgers listed 36 white servants and 13 blacks. On Sept. 19, 1725 William Chetwynd, principle shareholder in the company, wrote to John England saying, "You have full power to buy [for Accokeek] at any time 40 or 50 blacks, or what number you think proper and draw for ye money" (Heite, Historic, 8). Skilled labor for the Accokeek project came from the Maryland operation including Peter Grubb who was the stonemason at Accokeek.

Successful operation of a blast furnace also required a variety of skilled workers including founders,[35] guttermen, potters, finers, chafers, hammermen, millers, colliers, blacksmiths, keepers, and various skilled assistants. Founders were responsible for the actual operation of the furnace and oversaw the other skilled and unskilled laborers involved in the smelting process. Guttermen set up and tended the sand trenches and molds used to form pigs and other cast objects. Potters made and filled closed molds for hollowware such as kettles. Finers worked in a forge,[36] reheated pigs, and hammered them into blooms or

---

[35] In his will George Williams (died c.1750) described himself as "George Williams of County Stafford Founder." He made bequests to his various heirs and then stated, "I have a considerable sum of cash due me from the Accotink [sic] Iron Mines Compa and others" (Stafford Will Book O, p. 89). George named Nathaniel Chapman, Accokeek's manager, as one of his executors. His estate inventory listed a considerable quantity of iron and charcoal-making tools (Stafford Will Book O, p. 114). During the 1720s, George Williams was listed in company ledgers as a founder at Principio's main furnace in Maryland. This branch of the Williams family owned land in the north end of Stafford County (now part of the Marine Corps reservation) and married into the Hore family from the same area.

[36] Eighteenth century blacksmiths' forges were fairly standard in design. Normally a massive brick structure, the forge consisted on a square chimney that extended up through the roof to a distance of about four feet above the ridgepole. Built off one side of the chimney was a brick firebox, about waist high and measuring about five feet from front to back and up to eight feet wide. Set into the top of this brickwork

anconies (dumb bell-shaped bars with a knob on each end. Chafers also worked at the forge and assisted the finer by heating the pigs and bars to the necessary temperature for hammering. Hammermen were skilled forge workers who were responsible for hammering the pigs into bars and keeping in operation the ponderous trip hammers that did the actual pounding. Millers were responsible for maintaining the water-powered machinery of the furnace and forge. The colliers were charcoal makers, and the blacksmiths made a wide variety of items from the processed bars and did repairs on the many metal parts of the furnace equipment. Keepers maintained the store and company inventory. Also constituting part of the skilled labor force were wheelwrights and cartwrights. William Byrd learned from Alexander Spotswood that it required from 100-120 slaves, including women who cooked and sewed, to run a charcoal blast furnace. If all workers, skilled and unskilled, worked together towards the common goal, the chances of a successful operation were substantially increased.

Most 18th century iron furnaces had similar site plans and labor forces. Papers on file in the Fredericksburg Circuit Court provide a listing of the numbers of workers employed at Neabsco Furnace in Prince William County and these make an interesting comparison to what is know of such matters at Accokeek. In 1755 Neabsco had

> a furnace a water grist mill on the furnace dam, One Blacksmith shop with two fires and with four blacksmiths to work therein, One Wheelrights shop with a man on wages and a slave to work under him two Ore Vessels with white skippers on wages, navigated by servants and slaves belonging chiefly to the Works, two horse teams composed of six horses each, drove by a white man on wages and a slave, one Ox team drove by a slave, Slaves to cut wood and to burn coal; stock takers, founders, fillers, Ore burners and other labourers to the number of about forty or fifty. Exclusive of these there were two quarters with about eight or nine hands at each with Overseers, a mine bank in Maryland with, (I believe) about twelve or fifteen hands and an overseer, Ore carts and teams to hawl the ore to the landing, with flatts &c to take it down to the sloop which lay at a distance from the landing, but in the course of a few years, probably about the year 1760 and 1761, the business increased partly by purchases and partly by drafting slaves from other parts of Colo. Tayloes estate, Settling and Stocking the plantations purchased, erecting merchant mills, taking the collection of rents and granting leases to the tenants on the Kettocton lands. Purchasing some, and renting other ore banks in Maryland, building two and purchasing one other Ore Vessel, superintending a little retail store of Wet and dry goods, principally intended to assist in paying off wagonmen and procuring provisions (Fredericksburg Court Records, Lawson vs. Tayloe).

From that same suit is another deposition describing the work force at Neabsco in 1778. According to the deponent, the works at that time consisted of

> fourteen Wood cutters besides white men hired at different times, Six Colliers and a hired Overseer, five blacksmiths, two ship Carpenters, two wheelwrights, two Coopers, two shoemakers, one tanner, three house carpenters, a grist Mill kept by a Negro, a Merchant Mill kept by a hired Miller and a Negro, two horse teams and one Ox team drove by negroes, six or eight hands employed in Manufacturing Cloth and linnen, several under inoculation for the small pox, a Schooner navigated by five hands, a Sloop navigated by a hired Skipper and four hands, a schooner navigated by four hands, two hands employed in a flat bringing Ore to the landing, three quarters with overseers and about twelve hands to

---

was the hearth, a square bin that filled the inside dimensions of the brickwork and was about twelve inches deep. A slab of iron formed the bottom of the hearth, the iron slab having a hole in the center that accommodated the tuyere, or air nozzle attached to a bellows. Near the corner of the forge was a heavy wooden crane, often equipped with a traveling carriage and pulley, by which heavy pieces of iron were lifted into the hearth. The bellows was normally mounted behind the chimney. Based upon the large quantity of blacksmithing and repair work carried out at Accokeek, there were probably several such blacksmith's forges on the site. The buildings that housed the forges often contained other water-powered equipment such as grindstones and polishers.

each, one quarter with an Overseer and three hands. I cannot exactly say how many hands were employed at the Mine bank in diging ore, hauling it to the landing and working the plantation there, tho' I believe there could not be fewer than twenty (Fredericksburg Court Records, Lawson vs. Tayloe).

Surviving early records provide the names and occupations of some of those employed at Accokeek but certainly not all. On Feb. 1, 1808 John Murray stated that he had worked the Accokeek iron mines from the time he was a young boy until about the age of 21 (Fredericksburg Court Records, Degarnett vs. Suddoth). In 1783 and 1784 John was listed on the Stafford personal property records. A notation made on the 1787 tax rolls states that he was then living in Prince William though he continued on the Stafford rolls through 1790. In his deposition Murray stated that John Honey had been overseer of the Accokeek mines during the time that he had worked there. John Honey was not listed in the Stafford land or personal property tax records, suggesting that he probably lived in company housing. However, John's name appears repeatedly in the accompanying ledger. The first member of the Honey family listed on the Stafford tax rolls was William Honey who appeared in the 1785 list of inhabitants residing south of Potomac Run. Descendants of the Honey family remain in Stafford today.

The names of other workers are revealed in the Accokeek ledger. Between 1749 and 1750 the following employees were named:

Hugh Adie[37] (carpenter)
William Adie[38] (1729-1797) (carpenter)
John Allen (died 1750) (assistant manager & store manager)
William Bannister, Jr. (woodcutter)
Alexander Beach[39] (before 1655-c.1762)
Thomas Betson[40] (skipper)
Reubin Boyce (sailor)
George Browne (mason)
Cuthbert Byram[41] (woodcutter)
Peter Byram, Jr.[42] (woodcutter)
William Byram[43]
Jonathan Chapman (died c.1749)
Mason Combs[44] (1714-1784) (woodcutter)
John Conner[45] (collier and plantation overseer)

---

37 Hugh was included on the 1742 Stafford quit rent rolls, though no acreage was listed.

38 On the early Principio ledgers of 1723-1724 is listed William Adie, carpenter, who was paid £10 per annum. A William Adie was included in the Stafford quit rents of 1768 at which time he owned 750 acres (Bloomington).

39 Alexander was the son of Peter Beach. In 1722 when he executed a deed to Thomas Hopper, Alexander was residing in Hanover Parish in King George County. He was included on the 1723 quit rent rolls as owning 400 acres but a margin notation reads, "Can't hear of him."

40 Thomas was employed as a skipper and was included on the ledgers of both Kingsbury Furnace and North East Forge. In 1758 he was listed as skipper of the *Betsy*. According to The Register of Overwharton Parish, in 1748 Thomas married Jane Merringham in Overwharton Parish.

41 Cuthbert owned and resided on property that is now the site of Colonial Forge High School. Cuthbert, Benjamin, and Peter Byram were the sons of Thomas Byram and the grandsons of Abraham Byram and Dorothy Clay.

42 A Peter Byram was included on the 1742 quit rent rolls with 450 acres. He was also on the 1768 rolls with 250 acres. He married in 1752 Martha Horton.

43 William married in 1747 Sarah Gough.

44 Mason was the son of John Combs (c.1662-1717) of Stafford and Ann Mason (1670-1708). Mason was born in Overwharton Parish and died in Surrey County, North Carolina. He married in 1732 Sarah Harding Richardson (born 1714). Sometime in 1751 Mason and his family removed to Augusta County, Virginia, now Frederick County.

45 Listed on the Principio ledger in 1754.

John Corbin[46]
William Corbin (woodcutter)
William Courtney[47]
James Cumberland (sailor)
John Dagg[48] (died before 1755) (carpenter)
Thomas Ferguson (skipper)
Benjamin Fletcher (collier)
Isaac Fowler (woodcutter)
John Fritter[49] (woodcutter)
Robert Garrett (tailor)
William Holdbrook[50] (collier)
John Honey[51] (miner)
William Humes (blacksmith)
Grace Jackson[52] (died 1754)
Thomas Jackson[53] (wheelwright & carpenter)
William Jackson[54] (died 1749) (wheelwright)
John Kite (carpenter)
Christopher Knight[55]
John Lemon (miner)
John Minor[56] (died 1751)
Henry Nelson[57] (c.1700-1749)
John Ogleby (woodcutter)
John Ralls, Jr. (1725-1763) (blacksmith)
Anthony Rhodes[58]
George Salmon (c.1715-after 1785)
William Scott (blacksmith)
John Smith (skipper)
Henry Sturder (sawyer)
Henry Taylor (waggoner)
Charles Thornton (woodcutter)
Henry Turner (collier)

---

[46] John Corbin married in 1749 Frances Fant.

[47] Probably the William Cotney listed in The Register of Overwharton Parish who married in 1745 Mary Barbee. Several generations of the Stafford Courtneys were blacksmiths.

[48] Died in Stafford before 1755.

[49] Probably the John Fritter (sometimes spelled "Flitter") who married in 1755 Bridget Riggins. A John Fritter was mentioned in the will of John Peyton (1691-1760) of Stony Hill who left land in Fauquier County to his son, Yelverton (1735-1795). By the time of John's death, Yelverton was residing on the tract, but John noted that it had formerly belonged to John Flitter.

[50] William married in 1744 Elizabeth King.

[51] According to the Accokeek ledger, on Jan. 12, 1749 John was provided with 16 ½ pounds of powder "to blow up ore." John married in 1748 Hannah Bussel. She also worked at Accokeek in some capacity. The Bussell family seems to have been involved later at Rappahannock Forge.

[52] Grace Maulpis was the wife of William Jackson (died 1749) and the mother of Thomas Jackson.

[53] After Accokeek closed, Thomas returned to Maryland and in 1754 was listed on the Principio Furnace ledger.

[54] William Jackson was listed in the 1742 quit rent rolls as owning 150 acres in Stafford.

[55] A Christopher was named on the 1724 List of Tobacco Tenders along with Isaac, Leonard, and Epraim Knight. Together they worked 8,270 plants. In 1785 a Christopher Knight was included on the List of Inhabitants and Buildings and lived on land adjoining Peter Knight.

[56] John was included on the 1742 quit rent rolls with 200 acres.

[57] In 1742 Henry married Jean Gwodkin.

[58] Listed on the Principio ledger in 1754. He had an account in 1757 with North East Forge and probably worked there as well.

George Williams[59] (died 1750) (founder)

The business ledger lists the names of 70 slaves and several indentured servants at Accokeek. Some of these belonged to the Principio Company, but most were owned by Nathaniel Chapman or George Williams. Some of these were skilled workers who earned much higher wages for their owner than did basic laborers. From 1749-1750 Nathaniel's slaves and their occupations were:

George (collier)
Juba (miner)
Sillah (house servant)
Toney (carpenter)
James (carpenter)
London (founder)
Will (miner)
Tom (miner & sailor)
Jamey
Ceasar (collier)
Tom Wood
Jude (miner)
Port (founder)
Sambo (miner)
Sam
John Garner
Jack

George Williams was listed in the company ledger as owning Dublin and Tom. After George's death, his wife, Jane (Pope) Williams, received compensation for work done by her slaves/servants.

One of the most interesting features of the accompanying ledger is that some of the slaves at Accokeek maintained accounts with the company store and, at times, paid cash for their purchases. Early on, ironmasters discovered that harsh treatment of their slaves and forced work beyond a given number of hours per day was counterproductive and only resulted in work slowdowns, sabotage, and runaways. In studying the records of the Principio Company, Etna Furnace, Elk Ridge and the Patuxent Iron Works, Ronald L. Lewis found numerous references to incentive systems successfully employed by iron masters to encourage productivity amongst slave workers. One method was a type of "wage" or "production allowance" offered to a few highly skilled slaves for performance beyond the required task or quota. Often, slaves who worked in positions of supervision over other slaves, perhaps founders or forgemen, might receive a shilling per ton of iron processed. This particular reward system affected only a few black hands. Of greater impact to a larger number of slaves was the payment of cash or credits at the company store. This was done for those slaves, both skilled and unskilled, who worked extra hours beyond what was required of them. Known as the "overwork system," this was a common motivational device used at many iron works throughout the slave era. According to Lewis, "The overwork system attempted to make the industrial slave a disciplined and productive worker by merging his physical and economic interests with those of the ironmaster. In turn, this reduced the need for physical coercion, which might do more harm than good to the ironmaster's production goals." In addition to the slaves' ability to earn cash or credits by means of extra work, they were provided with space for a garden and could earn additional money by selling produce they grew (Lewis, Slavery, 142). Slave accounts at the Accokeek store show that black workers purchased a wide variety of goods from staple to luxury items. Slaves also bartered for goods from the store by trading produce, eggs, fowls, etc., for desired merchandise.

After Accokeek closed, Nathaniel Chapman returned to Maryland and many of his slaves moved to Kingsbury where they appear in the account books of 1768-1770.

---

[59] The founder was one of the highest paid positions at the furnace because the success of the operation largely rested upon his expertise. On Aug. 26,1749 George was paid £121.4.0 for making 404 tons of pigs. He worked at the Principio Furnace in Maryland through September 1728 when the company settled his account and he was sent to Accokeek as replacement for the original founder, John Barker. George Williams married Jane Pope and was included on the 1742 Stafford quit rent rolls with 1,950 acres. This land was in the upper end of the county.

George may have been the son of George Williams, blacksmith, and his wife Sarah from Apoquinimunk Hundred, New Castle, Delaware. In 1730 this George Williams sold land in New Castle County and moved to Frederick County, Virginia. By 1741 he was a resident of Prince George's County, Maryland (O'Dell 80).

Use of Mineral Coal

The high cost of iron and steel in the 18th century was due to the limited quantity of iron that could be made in charcoal-fired blast furnaces. Charcoal was the cheapest and most readily available blast furnace fuel and could be used to produce nearly any desired iron implement with the exception of cannon. Charcoal was fragile, however, and if the furnace stack was too high, the weight of a full charge of ore, fuel, and flux crushed the charcoal into powder, blocking air circulation and making smelting impossible. Even assuming, however, that the charcoal didn't crush into dust, it burned quickly, making it difficult to maintain a consistent temperature in the furnace.

Though not widely adopted until about 1840 as a blast furnace fuel, pit or mineral coal was occasionally used in colonial iron furnaces. Though more expensive and not as readily available as charcoal, mineral coal didn't burn out as quickly, produced a hotter flame, and was tolerant of heavy charges without being smothered by its own weight.

Coal was discovered in various places throughout the colonies but Virginia mines were probably the first to be worked. In 1704 coal deposits were discovered along the James River near the later site of the city of Richmond. It is said that a duck hunter who shot his dinner from a high bank overlooking the river made the discovery. The man and his dog scrambled down the hill to retrieve the duck from the water and in so doing discovered an outcrop of coal in the bank. When word reached William Byrd of Westover, he quickly patented the property; thus, the Virginia coal mining industry was born. On June 13, 1766 Samuel Duval advertised mineral coal for sale at Rockett's a lower landing at Richmond, for 12 pence per bushel, "equal to Newcastle coal" (*Virginia Gazette*, June 13, 1766).

One of the problems inherent with the use of mineral coal in a blast furnace was that sulfur in the coal contaminated the iron, making it brittle. All coal contains sulfur to some extent and iron smelted with coal will have sulfur in it. This wasn't critical if the iron was used in most castings, but the finery process for converting the cast pigs into bars of wrought iron didn't remove the sulfur. Such iron was useless because, at red heat, it became "hot short" or "red short" and crumbled when the smith attempted to work it. In 1709 English ironmaster Abraham Darby developed a process for coking the coal. This was a process similar to charcoaling by which the sulfur was removed from the coal prior to its use in the furnace.

Alexander Spotswood quickly recognized the value of mineral coal. By 1732 Spotswood had constructed an air or reverbatory furnace on Massaponax Run in an area now called Pratt's Rock. This unique structure was the first of its kind in the New World. Unlike the typical blast furnace, the air furnace used mineral coal as a fuel. While Henry Cort is generally credited with perfecting the use of mineral coal in furnaces in 1784, Spotswood was using it successfully more than 50 years earlier and nearly a century before it came into general use in the United States (MacMasters 64-65). This state-or-the-art facility was used primarily for re-melting pigs made at Tubal Furnace. From the re-processed iron, Spotswood made a variety of cast items including fire backs, firedogs, pots, and pans. Because the iron thus produced had fewer impurities to cause brittleness, his products were considered to be of superior quality (MacMasters 68). Massaponax Furnace operated until 1758 when Alexander's eldest son, Col. John Spotswood (1725-1758) died and iron production began at Rappahannock Forge.

The surviving Principio records don't reveal any information about the use of mineral coal in their furnaces and we are left to speculate about Accokeek. Considering the fact that Principio was the largest iron company in the English colonies and that coal was being used successfully at nearby Massaponax Furnace, coal may also have been used on a very limited basis at Accokeek. The two furnaces were within 25 miles of each other and it is unlikely that Principio would have been outdone by Spotswood's operation.

On Sept. 11, 1758 the Principio partners informed Nathaniel Chapman, "...the circumstances of the iron trade are very variable here, which we can observe in one instance; they have in this Kingdom got in a way of Running Oar into Pig Iron with Pit Coal Coaked, and at a charge under what we can make it in America. If this method should succeed in making good Iron, Pigs must be at a very low price, and will urge us upon other methods of Converting more Pigs into Barrs" (Principio Partners to Nathaniel Chapman, letter). Nathaniel's answer is unknown, but until the 1840s charcoal continued as the primary fuel in American blast furnaces. There were various reasons for this, not the least of which was the abundance of inexpensive charcoal to be had on the extensive tracts the companies had already invested in. Louis Hunter believed that American ironmasters were reluctant to shift to pit coal because agricultural implements fashioned from charcoal-produced iron were better adapted to local needs. He also claimed that smiths found that iron made with charcoal was tougher, more malleable, and easier to weld than that made from mineral coal (Hunter 215).

The naming of Coal Landing on Aquia Creek is a curiosity that seems to be in some way connected to Accokeek Furnace. The earliest surviving deed that makes reference to Coal Landing is dated May 8, 1786. This instrument records the sale of one acre from George Brent, Esquire (1760-1804) to Robert Stuart of Baltimore "near the Cool Coal Landing" (Stafford Deed Book S, p. 384). The reference to the landing is used as a surveyor's landmark indicating that the landing was well known by this name prior to the time of the deed. To date, the author has been unable to locate any documentary material to support or refute the use of mineral coal at Accokeek, even on an experimental basis. Another fascinating reference to coal involves Coal Trips, the Aquia Creek plantation belonging to Col. Elijah Threlkeld (1744-1798). Coal Trips is a very old name and the road leading from Accokeek to Thorny Point (now Rt. 630—Courthouse Road) passed directly by it. As yet, the author has been unable to determine the meaning of "coal trips."

Shipping

William Byrd's 1723 description of Accokeek Furnace refers to the transport of Accokeek pigs to Maryland by way of the Potomac River. The old road, traces of which are still used and remain on Stafford County maps, ran west from Coal Landing, followed Rocky Run past Robinson's stone quarry, turned south, crossed Courthouse Road (State Route 630) just a few hundred yards east of Eddie's Repair Shop, and led to the furnace.[60] Along this road supplies were carted to the furnace and the pigs hauled back to Coal Landing. There they were loaded on boats that made the short trip down Aquia Creek to the Potomac River and on to Maryland. Principio owned two forges on the Chesapeake Bay where pigs from the various furnaces were shipped for processing into bars.

Coal Landing seems the most logical shipping point for goods coming and going from Accokeek. It is quite likely, however, that Accokeek used at least two and possibly three shipping points, the second one being what we know today as Thorny Point and the third at Crow's Nest. Whether all of these landings were used simultaneously or whether Thorny Point replaced Coal Landing is unclear. In 1779 a governor's committee studying the possibility of reopening the mines at Accokeek mentioned using "the most convenient water on Aquia or Poto. Creeks or Poto. Run, for a Landing and Pasture for transporting Oyster shells[61] &c." (Fredericksburg in Revolutionary Days 90).

Thorny Point appears to have been a commercial shipping point from the earliest days of the county. Pigs outbound from Accokeek could have been hauled east on Courthouse Road and taken directly to Thorny Point. William Byrd stated that Accokeek used a landing about six miles from the furnace. By way of the old road, this landing is almost exactly that appears marked on the 1864 Gilmer map of Stafford County (Gilmer Map).

Historical interpretation in Stafford is always filled with discrepancies, however. Clouding this issue is the recent recovery from Crow's Nest of part of an iron pig. During the 18th century, Marlborough Point and Crow's Nest both boasted commercial shipping points. However, the poor roads leading to these points as well as the lengthy distance from Accokeek cast doubt as to their being used by furnace management. In fact, one would tend to automatically discount any serious consideration of the matter. The presence of the pig at Crow's Nest presents the researcher with a quandary. Discernible on the surface is what appears to be a date, "175_," the pig being broken after the "5." There was never a forge at Accokeek to process pigs into bars. There was, therefore, no reason for pigs to be coming *into* Stafford; one would expect all Stafford-made pigs to be outbound.

One possibility is that furnace management simply utilized whatever landing had a ship ready to sail. Recovery of the rest of the pig might reveal the name of the furnace that produced it and could help settle the question of Accokeek's shipping points. The shipping question is particularly frustrating because the company records make so few references to the landings used. On Jan. 29, 1749 one ledger entry records that Accokeek charged £1.3 "Cash paid for bringing a boat from Potomack Creek." Goods were

[60] From Robinson's quarry there were actually two roads leading to the furnace; the most direct road incorporated a steep hill, and a longer route that followed more level ground.

[61] Accokeek imported its oyster shells from Maryland. An entry in the Principio ledger dated Feb. 18, 1749 records that Accokeek paid £6.8.4 for shells.

unquestionably being delivered somewhere on Aquia Creek as on June 26, 1749 North East Forge charged Accokeek "for Barr Iron sent per Schooner Vulcan to Acquia with 500 bushels corn." On Mar. 11, 1745 North East charged Accokeek £131 for two months for a "Skipper in their Business from the first of May to the first of July." Each of the Principio-owned furnaces and forges had their own boats as well. On Apr. 10, 1750 Accokeek paid Kingsbury £16 for a flat with oars. That same day they paid Charles Waller (17_-1749) of Stafford "for a New flat with oars, masts anchors & cable."

By 1727 Accokeek was in blast and Principio had over £800 invested in their new venture. Company ledgers detail John England's expenses for setting up the Accokeek furnace, including food, crockery, livestock, and tools (made by Benjamin Turley at Principio's North East Forge). A year later company ledgers record a total investment of £824.8.6 in Accokeek and, by 1730 the furnace was in full production.[62] In 1730 276 tons of Accokeek pigs were shipped to England. Accokeek seems to have produced quality iron as evidenced by an advertisement appearing in the *Pennsylvania Gazette* newspaper. According to the advertisement, bars sold in Philadelphia were said to be "Mixed of Virginia Accokeek and Principio Iron, which makes it somewhat harder than Principio alone" (*Pennsylvania Gazette*, May 11, 1738). By 1730 Principio, under England's management, had two operational furnaces, most of the iron from which was used by the Royal Navy for all manner of ships' hardware.

Shipping the iron to England was the primary challenge facing colonial furnace operators. Company-owned sloops hauled pig and bar iron from Principio's various operations down to ports along the James and York rivers. There it was consigned to merchants who re-loaded onto tobacco ships bound either for the West Indies or England. Due to the weight, shipping pigs as freight was expensive, often costing more than their selling price in England. Principio partner, Joshua Gee, is credited with suggesting that the iron be sent as ballast on board the tobacco fleet, the pigs being packed between tobacco casks in the hold (Heite, "Pioneer Phase," p. 139). Early on, workers simply heaved the pigs into the ships' holds, this careless treatment resulting in broken pigs and a loss of profit for iron makers. Once the ship workers learned how to handle the iron, losses due to breakage were minimized and American-made pig iron was thus shipped successfully for over half a century. Most of the Accokeek iron exported to England began its transatlantic voyage from ports on the James River. Ledgers reveal that by the 1750s the company was regularly shipping pigs and bars from Maryland to England by way of the James and York rivers.

Nathaniel Chapman, Manager

John England oversaw the construction of the Accokeek furnace but company business soon called him back to Maryland. Ralph Falkner (died 1787) was appointed clerk/manager at Accokeek in 1725 but he soon moved on to John Tayloe's Bristol Iron Works. He was replaced by Nathaniel Chapman[63] (1709-

[62] Company ledgers also abound with Accokeek's accounts with the four Maryland furnaces and foundries. By Mar. 25, 1745 Accokeek owed North East a balance of £768.13.7 ½ for a variety of goods and services. That amount peaked by Sept. 25, 1748 when the Stafford furnace owed North East £4,006.17.5 ½..

[63] Nathaniel was the son of Jonathan Chapman (died 1749), a mason, and Jane Taylor. In 1737 Jonathan received a patent on land in Fairfax County, but his interest in iron working resulted in his living at various times in Virginia, Maryland, and Pennsylvania. Jonathan was buried at his home Summer Hill, now the site of Reagan National Airport. After her husband's death, Jane removed to Philadelphia. Nathaniel was his father's only living heir. Jonathan and Jane Chapman had issue:

--Sarah (1708-1767)—married John Gordon of Spotsylvania
--Nathaniel (1709-1760)—married Constantia Pearson
--John (born 1712)
--Taylor (1715-1749) married 1739 Margaret Markham of Stafford
--George (died 1747)—unmarried

Around 1732 Nathaniel Chapman married his cousin, Constantia Pearson (c.1712-1788), the daughter of Capt. Simon Pearson (died 1733) of Stafford. They had issue:

--Elizabeth (1733-1783)—married Dr. John Hunter (c.1721-1764) of County Ayr, Scotland
--Amelia (1735-1785)—married William Lock Weems (1735-1785), merchant of Anne Arundel and Prince George's counties, Maryland, and Georgetown, DC.

1760) who may have been either a clerk or assistant iron master under John England. Nathaniel appeared repeatedly in the company's early Maryland ledgers before being named manager of Accokeek. According to these ledgers, Nathaniel was a Principio employee by 1728 (Robbins, Principio, 52). At England's death in 1734, Principio's owners dispatched John Ruston to America as company manager. Two years later, however, Chapman assumed management of Principio's North American operations and continued in an administrative position until his death in 1761, some years after Accokeek ceased operations. He proved to be a capable and energetic manager. Upon assuming his duties at Accokeek Nathaniel Chapman received an annual salary of £100 but this was soon increased to £120. As furnace manager he assumed the responsibility of woodcutting, coaling, obtaining provisions, managing the furnace, keeping accounts, and shipping the pigs produced at Accokeek. During Chapman's tenure, Principio made Accokeek the company headquarters, probably due to the Washington family's residence in Stafford. The accompanying ledger contains numerous entries for payments to Nathaniel Chapman for trips taken on company business. His most frequent destinations were the courthouses at Stafford, Prince William, and Spotsylvania as well as to Williamsburg and Philadelphia. From the company store he purchased quantities of luxury items such as scarlet coat buttons, lace, silk, and a great quantity of chocolate.

In 1750 Nathaniel purchased a tract of land in Charles County, Maryland called Grimes' Ditch, though through the years the property has been known as Pomonkey, Chapman's Landing, Chapman's Fishing Pier, and Mt. Aventine. The tract has over two miles of Potomac River frontage and lies between the communities of Bryan's Road and Indian Head. Originally patented by Luke Gardiner, Nathaniel acquired it from his business partner, Edward Neale. At the time of Chapman's purchase, the tract was described as containing 580 acres including Cranery Island. A later survey determined that it actually consisted of 870 acres. The Chapman family purchased various adjoining tracts and Grimes' Ditch eventually became part of 2,225 acres known as the Chapman Forest tract.

It might well be argued that Nathaniel Chapman and the Principio Iron Company set the stage for the Industrial Revolution. Under Nathaniel's able management, colonial iron working expanded from small bloomery operations into an international business. Principio also grew into a major conglomerate and, prior to the Revolution, was the largest producer of iron in the colonies. One of Nathaniel's most important roles in this growth was the closing of Accokeek and Principio furnaces in the early 1750s and the purchase and construction of Lancashire and Kingsbury furnaces near Baltimore. The closing of Principio furnace (the forge remained in operation) was likely the result of lack of iron ore, a problem that John England had noted on his initial inspection of the area in 1723. For some years prior to its closure, iron ore for the Principio furnace had been shipped from the Whetstone Point mines on the Patapsco River. The cost of shipping reduced the company's profits and encouraged Chapman to seek a solution. The Principio partners purchased Lancashire, already in production, and built Kingsbury on land that they had held for some years. Both Kingsbury and Lancashire were close to this ore supply and were within two miles of each other.

Throughout the 18th century there was a lively interest in the "back country," that land lying west of the settled regions. The general population looked upon the area as a source of free or cheap land. Investors, industrialists, and land speculators envisioned unlimited development potential. Settlement west of the Blue Ridge began relatively early, the village of Opecan near Winchester dating from around 1735. Rich iron deposits, vast tracts of timber, and abundant water guaranteed ironmasters' attention to this wilderness area. Small furnaces and forges were erected throughout the back country, some of them profitable ventures and some blowing for only a season or two. Iron was essential to settlement and, thus, considered a wise investment. On Mar. 14, 1760 George Washington wrote in his diary that he and John

---

--Nathaniel (1740-1762)—drowned in New York Bay
--Louisa (1743-1763)—married Samuel Washington (1734-1781) as his third wife. She died in childbirth.
--Pearson (1745-1784)—married his cousin Susanna Pearson Alexander (1744-1815)
--George (1749-1814)—married Amelia Macrae

Nathaniel lived much of his life in Charles County, Maryland and in Loudoun County Virginia. Upon assuming management of Accokeek, he wrote to William Chetwynd assuring him that he would give the job his total attention. Dated June 25, 1736, this letter is on file at the Alderman Library, University of Virginia. In 1745 Nathaniel served as a justice of the peace in Stafford.

Carlyle had discussed "a Scheme of setting up an Iron Work on Colo. Fairfax's Land on Shannondoah.' They planned on enlisting the advice of their friend, Nathaniel Chapman, and making him a partner in the venture, he "being a perfect Judge of these matters." Nathaniel was asked "to go up and view the Conveniences and determine the Scheme." It is unknown if Chapman actually made the trip or not; he died unexpectedly later that year. Washington declined further involvement, though Carlyle and his brother-in-law, George William Fairfax, went ahead with the project (Jackson, vol. 1, p. 253).

Nathaniel left a sizable estate in Charles County, Maryland as reflected by his lengthy estate inventory. In addition to his ten-room house, Nathaniel left 31 slaves including Port "Aged about 45 years a Founder and forgeman" valued at £200. Port had been employed at Accokeek as had Tom Wood who was by then 60 years old. Nathaniel's estate also included one dozen black walnut chairs, an 8-day clock, a map of the world, a spinnett, "1 Dozen Perspective Sea Peices in Gilt Frames," a floor cloth, a substantial amount of silver, a "Shagreen Tea Chest Mounted with Silver," 8 beds, most with arrateen hangings, 291 barrels of Indian corn, 100 bushels wheat, 1 "New Chaise and Harness for two Horses," one old chaise and harness, 1 "Four foot Cart and Tumbling Body," 1 seine and ropes, farming equipment, 8 horses, 2 oxen, 44 cattle, 28 sheep, 30 hogs, 826 yards of Irish linen, 12 bushels salt, 1 chocolate pot, 1 silver watch, 1 pair "Temple Spectacles," and books on a variety of subjects (Robbins, Principio, 291-298).

Following Chapman' death, Thomas Russell II[64] (died 1786) was made manager of Principio's American operations. He arrived in Maryland in 1764 and, prior to the Revolution, became a citizen of that colony.

John England[65] was the backbone of Principio's American operations and, so long as England was present, Augustine Washington was able to carry on in Westmoreland County with his other interests. However, upon England's death in 1734, Augustine felt a need to pay more attention to activities at Accokeek, even with the business in the capable hands of Nathaniel Chapman. In 1735 the Washington family moved from their Pope's Creek plantation to another farm on Little Hunting Creek, now known as Mount Vernon, so that Augustine could be closer to the furnace. In late 1738 he moved his family to the Strother tract, also called the Ferry Farm, just across the Rappahannock River from Fredericksburg. It was here that young George Washington grew up. The primary reason for this last move was its proximity to Accokeek. Pope's Creek had been 48 miles from the furnace, Mt. Vernon over 40 miles. Ferry Farm was just 12 miles distant from Accokeek (Hudson, "Augustine Washington," 2617).

In April 1737 Augustine traveled to England to negotiate a larger share in the company. Successful in this endeavor, he was made resident manager of Principio's North American operations. By this time Principio was the largest producer of iron in all of the colonies and one of the largest in the entire British Empire (Heite "Pioneer Phase," 149).[66] Demand for iron continued and Washington's profits grew. At the time of Augustine's death in 1743, his shares in Principio were worth £1,200, in addition to which he left Wakefield, Mount Vernon, Ferry Farm, and several lots in Fredericksburg. Augustine's interest in Principio passed to his son, Lawrence, who built upon his inherited fortune by his continued involvement with the company.

By 1736 Accokeek was a thriving enterprise consisting of at least one mine, one furnace, a store, gristmill, stable, blacksmith's shop, and plantation. The company store provided all manner of goods for the workers and their families as well as for other people living near the furnace or doing business there.

---

[64] After the Revolution, Thomas Russell and William Augustine Washington (1757-1810), Augustine's descendant and heir) were the only two American shareholders in Principio. As a result, their interest in the company was not confiscated in 1781 when Maryland took over and sold the company's assets.

[65] For some time, company ledgers had listed John England as of "White Clay Creek, Delaware," suggesting that he might have retired from active management some time prior to his death. He left a small estate in King George, including an old horse, two yearlings, a cow and calf, a young steer, and a heifer. Household furnishings there included one bed and bedstead (King George Will Book, 1721-1744, p. 200). His holdings in Delaware were extensive. He may have left a son in Stafford as there are several John Englands listed in later county records. In 1803 one of these was appointed by the Stafford court to inventory the estate of Patrick Home of Rappahannock Forge in 1803.

[66] In 1781 when the company's property was seized as a British possession, it was written that a "certain Mr. Washington, a subject of the State of Virginia, is entitled to one undivided twelfth part thereof."

The company gristmill ground grain for the workers, as well as for local farmers and flour and meal were available for purchase from the company store.

Vitally important to Accokeek's success was its blacksmith's shop. Based upon the accompanying ledger, blacksmithing and repair work made up a substantial part of Accokeek's income. Unlike our modern "throw-away society," farmers found it cheaper to have broken or worn tools repaired rather than to buy new ones. The blacksmiths at Accokeek put new points on plows and coulters, laid new iron or steel edges on hoes and axes, and repaired broken ironwork on chains, carts, and small hand tools. Based upon the accompanying ledger, the many people who utilized Accokeek's blacksmithing capabilities represented the counties of Fairfax, Fauquier, Prince William, Stafford, Spotsylvania, and King George as well as the towns of Alexandria, Dumfries, Falmouth, and Fredericksburg. Accokeek's reputation for blacksmith work must have been outstanding as some Fairfax residents bypassed Occoquan and Neabsco furnaces in order to bring their work to Accokeek. It should also be remembered that at this time, many large plantations employed their own smiths and most villages had a blacksmith as well; yet these individuals still chose to bring their repairs to Stafford.

Accokeek's ore banks were located about 1 ½ miles northwest of the furnace. A road ran north from the furnace for a short distance, and then intersected with what is still called the Furnace Road. This road ran east and west between Moonack and Long branches of Potomac Run and White Ridge Road (part of which is now Courthouse Road—State Route 630) and crossed Mountain View Road (State Route 627) in the vicinity of Guy Lane. The Furnace Road was clearly marked and identified on an 1864 map of Stafford County and remains known to local residents as "the old mine road" (Gilmer Map). The area around the old mine banks is still very rough and only thinly populated.

Edward Heite theorized that Spotswood's Tubal mines might also have supplied iron to Accakeek, Neabsco, and Bristol, his land being outside the Northern Neck Proprietary and thus exempt from Lord Fairfax's mineral royalty (Heite, "Pioneer Phase," 144). The Principio/Accakeek ledgers do not include any purchases of ore form Spotswood. Occoquan and Neabsco were far removed from the Tubal area and shipped their ore from Maryland in order to avoid the royalty. If the Tubal mines supplied anyone with ore, it was most likely Rappahannock Forge that was located just a few miles down the river.

Pig Iron Production

Chesapeake iron production was based on the English "indirect" method rather than on the German "direct" method. The former required two facilities for making iron, a furnace in which to smelt iron from ore and a forge to process the crude pigs into usable wrought iron. Although this method was more involved and more expensive (considering the number of buildings and workers required to carry it out), it yielded a greater quantity of iron that the German method as well as a better quality finished product.

Accokeek furnace was capable of producing about 20 tons of pig per week (Robbins 56). According to company ledgers, in 1729 an additional sloop was hired at Annapolis to carry 200 tons of ore from Maryland mines to Accokeek (Middleton 170). There is no indication that the Accokeek mines ever ran short of ore. The importation of the Maryland ore probably reflects the company's reaction to Lord Fairfax's recent imposition of a one-third royalty on all minerals found on Northern Neck property. Because of this, Occoquan and Neabsco furnaces routinely brought ore from their mines on the Patapsco River in Maryland where Lord Baltimore did not claim such a royalty (Middleton 169). John Tayloe, a member of the Virginia Council and owner of both Neabsco and Occoquan furnaces, was instrumental in introducing an act for the benefit of himself and other ironmasters "praying leave to import from Maryland some of the Iron Oar of that province for the more easy fluxing of the Oar in their Mines without being obliged to pay the Port Duties or other Fees chargeable on Vessels importing Goods & Merchandize. It is accordingly Ordered that the Naval Officers of the several Districts into which any Oar shall be Imported do not require the Port duties or other ffees (except for a Permit only) for any Vessels importing Iron Oar from Maryland for the use of the Iron Works here so as such Vessels do not carry any other Goods & Merchandize" ("Journals of the Council," vol. 14, p. 236). Unfortunately, most of the records pertaining to Accokeek have been lost and that furnace's production figures were routinely combined with those of other Principio furnaces. In 1750 Accokeek sent 410 tons of pig iron to England, about 1/6 of Principio's combined export from Maryland and Virginia for that year. Production figures may be gleaned from the

1749-1760 ledger and from extant company correspondence. The following are the few known production figures for Accokeek:

1729—276 tons (partial shipment)
1749—404 tons
1750—410 tons
1752—169 tons

Accokeek produced not only pig iron but a variety of cast items as well. These included firebacks, latches, horseshoes, ladles, firedogs, compasses, bells, branding irons, iron heaters, and assorted small farming implements. Most of the pigs produced at Accokeek were shipped to the metal working centers of London, Bristol, Liverpool, or Birmingham or to Principio's forges in Maryland. The figures in the following chart are from material in the Virginia Colonial Records Project. The document from which these figures were gleaned was amongst the papers of the House of Lords and was titled "Account of the Quantity of Iron Imported from the Colonies in America from Christ. 1710 to Christ. 1749" (VCRP, House of Lords Papers). This report provided importation figures for iron produced in several different regions within the colonies. Unfortunately, figures for Virginia and Maryland were grouped together, though these two colonies consistently far out-produced all other regions. Iron was weighed using tons, hundredweights, quarters, and pounds. One quarter was equivalent to 28 pounds. Four quarters equaled one hundredweight or 112 pounds. One ton equaled 2,240 pounds or 20 hundredweight. An example from the chart below, 4.4.3.21 would be read, "4 tons, 4 hundredweight, 3 quarters, and 21 pounds."

YEAR	BAR IRON	PIG IRON
1718	3.7.0.0	0
1719	0.10.3.10	0
1720	4.4.3.21	0
1721	15.7.1.6	0
1722	0	0
1723	0	15.2.3.22
1724	7.0.0.0	202.9.0.0
1725	0	137.6.3.19
1726	0.3.0.0	263.0.3.3
1727	2.12.0.25	407.9.3.13
1728	0	643.6.0.24
1729	0	852.16.1.11
1730	0	1526.15.1.15
1731	0	2081.2.0.27
1732	0	2226.3.2.0
1733	0	2309.11.3.22
1734	0	2042.2.2.3
1735	44.9.0.21	2362.8.0.17
1736	0	2458.0.0.3
1737	0	2119.16.1.25
1738	0	2112.18.3.19
1739	0	2242.2.2.14
1740	5.0.0.0	2020.2.0.22
1741	5.0.0.0	3261.8.1.5
1742	0	1926.3.1.25
1743	0	2816.1.1.15
1744	57.0.0.0	1748.4.1.3
1745	4.5.2.14	2130.16.1.10
1746	193.8.3.12	1729.1.0.9

1747	82.11.2.11	2119.0.3.24
1748	4.0.0.0	2017.11.3.10
1749	0	1575.5.1.27

Fig. 1-7
A Finery/Chafery Forge (Diderot, Vol. IV, Forges, 4th Section, Plate VI).

Fig. 1-8
A forgeman using a water-powered trip hammer to convert pig iron to an ancony. The clanking of the trip hammers was deafening on site and could be heard for a mile or more beyond the forge (Diderot, Vol. IV, Forges, 4th Section, Plate VI).

Pig iron was a crude product that required further processing prior to being made into finished products. Pigs produced at Accokeek were shipped to Maryland where they were processed into bar iron at one of Principio's forges. This necessary step aligned the atom structure and drove out impurities remaining from the smelting process, both of which tended to make the iron brittle and difficult to work. The refining of the pig iron was a multi-staged process carried out at a forge. In the 18th century, a forge consisted of a finery and a chafery. The finery was a raised hearth measuring about 5'3" by 6'3" and was surmounted by a chimney. The sides of the finery hearth were lined with iron plates. A charcoal-fueled fire was used to heat the iron pits to a plastic consistency, the fire being regulated by a blast of air from a bellows. Once the fire was lit and the air blast introduced, the pig was gradually pushed forward over the fire. As the pig melted droplets trickled down to the bottom of the hearth. Using a long iron bar called a ringer, the finer stirred and worked the molten iron. Impurities separated from the iron forming a liquid slag that was periodically tapped off through a slag hole in the side of the hearth. In the bottom of the hearth the molten iron gradually congealed into a semi-solid mass.

In the second stage of refining, the semi-solid mass was broken into pieces. Using an iron crossbar called a furgeon, those pieces not sufficiently decarburised were raised up to the tuyere and reheated. In this last stage the iron was again melted. Pasty lumps formed in the bottom of the hearth and these were gathered together into a ball called a bloom or loop. The whole process of melting, refining, and balling took about one hour.

Using large tongs, the bloom was removed from the finery hearth and placed on an iron plate. It was beaten with a large hammer to remove the surface charcoal and slag. The bloom was then moved to a water powered tilt or trip hammer for further processing. The tilt hammer was comprised of a massive timber shaft on the end of which was affixed an iron hammer head measuring about 14 inches square and three feet high and weighing 500-600 pounds. A cumbersome water-powered horizontal drive shaft or drum was studded with projecting cams, or dogs, that raised the heavy hammer and allowed it to fall back under its own weight. Each hammer was paired with a correspondingly heavy anvil. As the drum turned, each cam raised the tilt hammer and let it drop under its own weight. The frequency of the hammer blows was regulated by the speed of the water wheel to which the drum was geared. Tilt hammers were notoriously noisy and the air around the working forge reverberated with the rhythmic thudding, which could be heard for several miles. This hammering forced out most of the remaining slag. The hammerman carefully shaped the iron into a 2-foot square block. This was returned to the finery and brought up to a welding heat in order to "sweat" out yet more impurities. The sweated block was returned to the tilt hammer and forged into a dumbbell-shaped bar called an ancony. On each end of the ancony was as a thick knot, one larger than the other, the bar between the two knows being about three feet long. By forging the center into a bar shape, the remaining slag was forced into the two knobs.

The next step in the refining process was carried out at the chafery. This was similar in design to the finery but was larger. In the chafery the smaller of the ancony knobs was heated about 15 minutes and forged under the tilt hammer into a bar shape. The thicker end, called the mocket head, required two heatings before being drawn out into a bar. Because a higher temperature was required in the chafery to sweat out carbonaceous particles in the iron, the chafery bellows were usually larger than those used in the finery. The hot iron was finally forged into the shape of a bar and the rough ends cut off with chisels.

Water-powered tilt hammers were used in forges in a variety of ways and the speed of the hammers varied according to the process being employed. Shaping or drawing blades required about 150 strokes per minute. In order to achieve rapid speeds, it was necessary to restrict the natural swing of the hammer. This was accomplished by mounting a stout wooden plank above the haft. This acted as a recoil spring to cushion and quickly return the rising hammerhead to the anvil. Another method, which accomplished the same end, was to set a recoil block in the floor of the forge beneath the tail end of the haft. The rapid striking of the hammers produced terrific vibrations that shook loose wedges and made constant repair and maintenance necessary. Someone, often a young boy, was employed to keep the hammerheads wet so that the wedges didn't work loose, allowing the huge heads fly off the ends of the timbers.

A tremendous quantity of bar iron was shipped from the Maryland operations back to Accokeek for resale or shipment to other destinations. The ledgers from Principio Furnace and North East Forge abound with entries of bar iron being sold and shipped back to Accokeek. The majority of this iron was consigned to merchants such as George Chapman (died 1747) of Spotsylvania and William McWilliams of Fredericksburg. Most of the company's exports were channeled through ports on the James and York rivers that included Warwick, Richmond, Petersburg, and Bermuda Hundred (near present day Hopewell). The following is a listing of iron shipments to Accokeek from Principio and North East, these figures being gleaned from the various Principio Company ledgers on file at the Historical Society of Delaware:

Date	Source	Quantity	Cost
__, 1729	Principio	117 bars	£36
June 12, 1729	Principio	17 tons pig	
Aug. 28, 1731	Principio	2 tons bars	
Jan. 1738		6 tons bars	205.14.0
Aug. 5, 1738	Principio	499 bars	243
Feb. 18, 1745	North East	12 tons bars	300
Mar. 10, 1745	North East	814 bars (11 tons)	290
Apr. 10, 1745	Principio	12 tons bars	240
Aug. 22, 1745	North East	8 tons bars	
Mar. 29, 1746	North East	12 tons bars	300
May 12, 1746	North East	12 tons bars	240
Sept. 10, 1746	North East	637 bars (10 tons)	
June 26, 1746	North East	166 bars (2 tons)	50
Apr. 19, 1747	North East	6 tons bars	120
	Principio	2 tons bars	40
Apr. 25, 1747	Principio	13 tons bars	260
May 20, 1747	North East	10 tons bars	200
May 28, 1747	Principio	12 tons bars	240
Oct. 12, 1747	Principio	10 tons bars	200
Mar. 2, 1749	North East	3 tons bars	75
Aug. 31, 1749	North East		
May 7, 1750	North East	857 bars (12 tons)	240
		38 bars	7
May 28, 1750	North East	370 bars (5 tons)	125
June 19, 1750	North East	236 bars (3 tons)	60
		134 bars (2 tons)	40
June 30, 1750	North East	525 bars (6 tons)	178
Dec. 1, 1750	Principio	4 tons bars	80
Dec. 31, 1750		202 tons pigs	
May 1, 1751	North East	727 bars (10 tons)	200
May 18, 1751	*no source listed*	484 bars (7 tons)	140
Feb. 1, 1752	Principio	374 bars (4 tons)	80
	North East	38 bars	10
Feb. 20, 1752	North East	235 bars	67.11.4
		290 bars	75.0.8
		170 bars	45
		477 bars	160
		56 bars	24
Mar. 14, 1752	Principio	786 bars (10 tons)	200
		157 "round" (2 tons)	41.4.0
July 7, 1752	"from the Furnace"	169 tons pigs 12 tuns hammers & anvils	845
Sept. 6, 1752	Kingsbury	390 tons pigs	
July 1, 1754	North east	60 tons pigs	
		20 tons blooms	
		14 tons bars	
July 20, 1754	North East	744 bars	288

Aug. 15, 1757	Principio	5 tons	
Aug. 29, 1757	North East	6 tons "for John Watson"	

Accokeek also sold a large quantity of hammers, anvils, and other castings. Some of these were cast on site, others were ordered from Principio and North East:

Date	Source	Quantity	Price
April 1731	Principio	4 tons "Pie metal" & 4.8 tons hammers	£63
July 25, 1737	Principio	4 ½ tons hammers & anvils	28
Jan. 4, 1738	Principio	1 cable	13.3.6
Mar. 24, 1745	North East	"for drawing 6 large & 24 small shovels and 4 ringers	1.2.0
Apr. 18, 1746	North East	Castings	8.1.8
Apr. 21, 1746	North East	12 hammers & 3 anvils (4 tons)	360
Mar. 11, 1749	North East	"a cable"	38.1.7
Mar. 24, 1749	North East	plating 2 dozen shovels	0.12.0
June 18, 1750	North East	15 tons hammers & anvils 1 ton forge plates	
Dec. 31, 1750		16 ¾ tons castings	

While Accokeek lacked the capability to make steel, other Principio operations did produce it. Essentially, steel is iron with most of the carbon removed. In order to be considered steel, the substance must have from 0.5 to less than 2.5% carbon. A carbon content exceeding 2.5% is cast iron. Steel can be shaped either by being melted and cast into molds or by being forged. Steel had a multitude of uses and was kept on hand at Accokeek where it was used for repairing tools and, occasionally, sold to customers. The ledger contains numerous references to "steeling an ax," a process by which a narrow steel edge was forge welded to an iron ax head to make it stronger.

Only small quantities of steel were produced in colonial America, the process being expensive, time consuming, and there being few people here who knew how to accomplish it. William Byrd wrote of his interview with Charles Chiswell (1677-1737) regarding the making of steel, "He told me a strange thing about steel, that the making of the best remains at this day a profound secret in the breast of a very few, and therefore is in danger of being lost, as the art of staining glass and many others have been. He could only tell me they used beechwood in the making of it in Europe, and burn it a considerable time in powder of charcoal; but the mystery lies in the liquor they quench it in" (Wright 351).

Date	Source	Quantity	Price
June 3, 1728	Principio	21 ½ lbs. blister steel	
Apr. 21, 1746	North East	6 bars (64 lbs.) blister steel	£1.14.11 ¼
		2 bars German steel (6 ½ lbs.)	0.4.10 ½
June 24, 1749	"Mr. Stevenson"	120 lb. Steel	2.8.0
July 6, 1752	North East	37 lbs. blister steel	0.15.5
		11 ¾ lbs. German steel	0.9.9 ½

Accokeek, North East, Kingsbury, and Principio sent quantities of goods back and forth to one another, these exchanges being carefully recorded in the various company ledgers. On July 25, 1737 Principio sent Accokeek 3 large broad axes, 3 2-foot rules, one 1 ½" auger, one ¾" auger, and one ½" auger. On Jan. 25, 1745 North East sent Accokeek 1 cask of 10-penny nails (£8.10.8) and 1 cask of 20-penny nails (£7.12.0). On Apr. 20, 1750 Accokeek paid North East for plating (making) 2 ½ dozen shovels. On Apr. 10, 1750 North East sent "3 set harnas for 6 horses" and on June 12, 1750 Accokeek paid North East for 8 barrels of tar and 18 pounds of cordage.

The Mercantile Connection

The Virginia tobacco fleet with its network of merchants and ship captains played a vital role in the Chesapeake iron industry by transporting American-made pig and bar iron to England and the West Indies. Although some iron was occasionally shipped to other colonies, this practice was not common. In 1750 an observer in Portsmouth, Rhode Island wrote that iron for shipwork and other uses came primarily from Maryland, Virginia, and Pennsylvania. More commonly, pig and bar iron were shipped to England and Scotland where it competed easily with foreign imports on which there were heavy tariffs.

The amount of iron carried on ships varied with the size of the ship and other considerations. The average ship clearance of the period listed 600-700 hogsheads of tobacco, several thousands staves, and 10 to 60 tons of iron. On average 34 pigs or 70 bars constituted one ton.

Principio Company iron was sold through a network of merchants in Maryland, Virginia, England, and Scotland. Understanding the system of trade during the colonial period is essential to appreciating the scope of operations at Principio and Accokeek.

During the 17th century, New England merchants dominated trade. By the early 18th century, enterprising Virginians along the York and James rivers began building sloops and schooners suitable for West Indies trade. Profitable trade was soon established between Virginia, the West Indies, and Madeira. Initially Norfolk and Yorktown were the centers of this new Virginia trade. While Yorktown was the more prominent port, Norfolk's proximity to Tidewater grain and Southside and Carolina meat and grain products soon pushed it to the forefront. Norfolk, a mere hamlet at the beginning of the 18th century, became a center for merchants interested in West Indies trade. By mid-century these merchants had made Norfolk Virginia's largest town.

Most of the 18th century merchants were from Scotland or England. They provided the ever-growing number of stores with a wide variety of essential items as well as with luxury merchandise. Storekeepers tried carrying lines of luxury fabrics and notions in order to attract customers, but found the practice risky. While the customers enjoyed viewing such items, they could afford to purchase them only in times of exceptional prosperity. Cheap cloth and ordinary farm and household goods were the best selling lines. Salt, rum and other liquors, and haberdashery were the mainstay of most stores and much of this was imported and delivered by the merchant/importers.

During this period, there was very little cash in the colony. Customers expected and received credit. Over the course of a year, the average customer made several purchases at any given store, but his bill rarely totaled more than a few pounds a year. An average store might serve 200-300 such customers in a year's time. Only a handful of customers might accumulate purchases totaling £25 or more. The great number of small debtors made debt collection difficult. It became customary in Virginia for merchants and planters to meet on the monthly court days in order to settle their debts, a system recorded numerous times in the Accokeek ledger.

Because the customers were short of cash, so were the shopkeepers and the merchant/importers who ordered the goods from England. English stores were forced to extend credit and often provided operating capital for the Virginia merchants. As a result, the Virginians either became partners with the English merchants or made credit arrangements with them. Thus, Virginia's economy was completely tied to England's and the English merchants felt the effects of economic problems in Virginia.

Virginia merchants were expected to pay their English suppliers within one year from the purchase time, but were sometimes granted extensions. Most merchants demanded that planters settle their accounts yearly. This was most often done in a combination of cash, tobacco, and produce or other goods. Sometimes planters managed to extend their debts two to three years before finally being cajoled into executing bonds or paying interest. In difficult cases, merchants resorted to debt collectors. When all else failed, merchants could take debtors to court. The merchants had the authority to sue for payment in either the county in which the debtor resided or in Williamsburg. This flexibility allowed the merchants to bypass county courts that seemed reluctant to press debtors for payment. This official reluctance was often spurred by the justices' own personal indebtedness to the very merchants bringing suits in their courts.

The most important court for debtor suits was the Williamsburg Hustings court. In 1736 the Assembly gave the Williamsburg court jurisdiction over debt cases regardless of where they arose. From then on, most merchant contracts called for payment in Williamsburg while the General Court was in session. This is reflected in the accompanying ledger which contains numerous references to Nathaniel

Chapman's trips to Williamsburg to collect debts. This law was repealed in 1770 thereby restricting merchants' suits to the county courts for debt collection. Accokeek's last blast occurred in the winter of 1751/52, though the ledger continues through June 1760. This reflects the system of debt payment and collection typical of the period, well over 100 pages being used to record these payments.

Carried on the Principio books was an account for merchant Nicholas Carew. In 1754 Nicholas's account with the Principio Company totaled nearly £20,000. Of that, £7001.17.11 was due to Accokeek. Whether this reflects sales of capital goods from the closing of Accokeek or was for previous iron shipments was not specified in the records.

In his study of the Principio Company, Michael Robbins found that during the 1750s to 1770s, much of Principio's iron was not shipped directly to England from Maryland ports such as Annapolis, but was carried down the Chesapeake Bay to Virginia. Once the iron arrived in Virginia it was consigned to merchants and chief shipping agents John Hoomes (died 1769) of Cumberland on the York River and John Hylton of Bermuda Hundred near the confluence of the James and Appomattox rivers. They were responsible for the transshipping of the iron to London, Liverpool, and Bristol (Robbins 228-229). From the available records, it is not possible to determine why the company chose to go to the additional time, effort, and expense to transship through Virginia. Some historians have suggested that there were more England-bound ships leaving from Virginia than from Maryland and, hence, more shipping opportunities there. The Accokeek ledger contains many references to the receipt of bar iron from the Maryland facilities and its transfer to shipping points on the James and York rivers.

There do not seem to be any surviving records from Accokeek's early period. However, an "Account of Piggs at Principio Furnace, August 1727" provides information on the production and selling cost of pig iron. This document reveals that it cost £4.5.9 to produce one ton of pig iron at Principio and that the iron sold for £10 per ton. The production cost included 11 shillings per ton for freight and import duty. Principio, then, realized a profit of approximately £5 per ton of iron produced at its Maryland plant. It may be assumed that this is representative of costs at Accokeek (Hudson, "Augustine Washington," 2614).

The following is a list of 112 Virginia-based merchants and 36 English/Scottish/Irish firms, the names of which were gleaned from the accompanying ledger. These merchants and companies bought, sold, and shipped Accokeek/Principio iron and delivered supplies to Accokeek and the other Principio operations. Towns and counties following each name indicate the areas in which these men lived and worked.

Merchants of Virginia

Gerard Alexander (c.1712-1761)—Fairfax County
Richard Ambler (died 1766)—Yorktown
Anthony Bacon (c.1717-1786)—York County, Williamsburg (?)
Alexander Baine (died 1806)—Norfolk, Louisa, and Hanover counties—Scottish
Richard Barnes
John Belfield (1725-1803)—Richmond County
William Bolden—Norfolk—Loyalist
Col. John Boling (1700-1757)—Henrico
Col. George Braxton (Brackstone) (c.1677-1748)—King and Queen, York, and Orange cos.
Francis Brown
Archibald Buchanan (1737-1800)—Nansemond and Prince Edward counties—Loyalist
James Buchanan (c.1737-c.1780)—Falmouth
Samuel Buckner (died 1764)—Gloucester and Caroline counties
Nicholas Carew
Maj. John Carlyle (1720-1780)—Alexandria
Robert Carmichael—Nansemond County
Christopher Chamney—towns of Richmond and Warwick
Col. John Champe (c.1700-1763)—Spotsylvania
George Chapman (died 1747)—Spotsylvania and Richmond(?)
James Clark—in business with Thomas Knox

Robert Hendley Courts (died 1775)
Patrick Coutts (1726-1777)—James City County, Richmond, Port Royal, and Williamsburg
James Cross (died 1787)—Prince Edward County
William Cunningham (born 1727)—Richmond, Petersburg, Dumfries, Falmouth
John Dalton (c.1722-1777)—Alexandria
Capt. Thomas Dansie (died 1768)—New Kent County
James Deans (died 1762)—Chesterfield County
John Dixon
Robert Donald—Warwick & Hanover Co. and Glasgow
Thomas Donald—Warwick in Chesterfield Co. and Glasgow
John Driver (died 1788)—Suffolk and Isle of Wight counties
Thomas Ferguson (born 1725)—Williamsburg
William Fowler (died 1774)—town of Richmond and Chesterfield Co.
Robert Franklin
Samuel Gist (1723-1815)—Hanover & Goochland counties—Loyalist
John Gordon—Richmond County (?)
Samuel Gordon (1716-1771)—Blandford (now Petersburg) and Prince George County
James Graham—Isle of Wight County (town of Smithfield)
John Graham—(1711-1787)—Dumfries
John Graham—Prince Edward & Northumberland counties
John Greenhow (1724-1787)—Williamsburg and Osbourn's Warehouse
John Hamilton (died 1816)—Nansemond and Halifax counties and North Carolina—Loyalist
Samuel Hanson
John Harmer—Amherst County, Williamsburg, and Bristol—Loyalist
Robert Hastie—Prince Edward County and Petersburg—Loyalist
Humprey Hill (1706-1775)—King and Queen and New Kent counties
William Holt (1730s-1791)—Williamsburg and James City County
John Hoomes (died 1769)—Cumberland (York River)
Col. John Hunter (died 1778)—Elizabeth City and Hampton—Loyalist
Capt. John Hutchings (1691-1768)—Norfolk and Princess Anne counties
Col. John Hylton (died 1773)—Henrico County (Bermuda Hundred)
Robert Jackson (died 1764)—Fredericksburg
John James—kept Marsh Store
David Jameson (died 1793)—York, Norfolk, and Princess Anne counties and Yorktown
George Jameson—Princess Anne County
Neill Jameson—Norfolk and Princess Anne counties—Loyalist
Francis Jerdone (1720-c.1771)—Louisa County
Peter Johnston—Henrico, Amelia, and Prince Edward counties
John Knox (in business with James Tarpley)—Williamsburg
Robert Langley (died 1792)—Petersburg—Quaker
Joseph and William Lewis
John Lidderdale
Richard Littlepage (1710-c.1767)—Henrico and New Kent counties
Alexander Mackie (1726-c.1766)—Prince Edward and Henrico counties, Upper Warwick on the James River
Thomas Mackie (born 1704)
Alexander McCall (McCaul)—Richmond, Manchester, and Charlottesville—Loyalist
William McWilliams, Jr.—Fredericksburg
William McWilliams, Sr.—Fredericksburg
David Mead (1710-1757)—Nansemond and Isle of Wight
Reese Meredith (1705-1777)—Philadelphia—Quaker
Hugh Miller (died c.1762)—Norfolk, Petersburg, and Prince George County
William Montgomery—York County ?
Thomas Moore (died c.1769)—King William, King and Queen, New Kent, and Hanover counties
James Morton

William Nelson (1711-1772)—Yorktown
Mann Page (1719-1781)—Spotsylvania County
John Pleasants (died 1767)—Henrico County—Quaker
Thomas Price—Norfolk and Middlesex
Robert Rae (1723-1753)—Falmouth
William Ramsay (1716-1785)—Fairfax County
Peter Randolph (1713-1767)—Henrico County
Col. Richard Randolph (1690-1748)—Henrico County
William Randolph (1681-1742)—Henrico County
John Robinson
Andrew Rosse (died 1836)—probably Falmouth
Robert Ruffin (died c.1777)—Dinwiddie and King William counties
Cuthbert Sandys--Fredericksburg
John Saunders—Princess Anne County—Loyalist
James Scott—Nansemond County
Joseph Scott—Nansemond County
Robert Sheddon (died 1826)—Portsmouth—Loyalist
Samuel Skinker (1677-1752)—King George & Port Conway
Sir William Skipwith (1707-1764)—Prince George County
George Sparling—Suffolk County—Loyalist
John Sparling—town of Norfolk and Suffolk Co. and Liverpool—Loyalist
Alexander Speirs (1714-1782)—Lunenburg, Mecklenburg, Prince Edward, Bedford counties and Petersburg, Portsmouth, Norfolk, and Richmond
Andrew Sprowle (c.1710-1776)—Norfolk and Portsmouth—Loyalist
Edward Stabler (c.1730-1785)—Henrico—Quaker
James Stark—Petersburg (?)
Charles Steuart (died 1797)—Portsmouth and Norfolk
Amos Strettel (1720-1791)—Philadelphia—Quaker
Robert Strettel (1693-1761)—Philadelphia—Quaker
Thomas Tabb (died 1769)—Amelia County
James Tarpley—Williamsburg
George Thomas—Hanover County
Robert Todd—Norfolk
Col. Robert Tucker (1701-1769)—Amelia County
Charles Turnbull—Henrico County, Prince George, and Dinwiddie counties
Anthony Walke (1692-1768)—Princess Anne County
John Watson—Henrico County
George Webb (1729-1786)—New Kent County
Thomas Yuille (1723-1792)—Halifax & Prince Edward counties—Loyalist

English/Scottish/Irish Merchants/Companies

Latham Arnold—London
Thomas Asselby (died 1783)—Bristol
Anthony Bacon (c.1717-1786)—London
Matthew, Patrick, and William Bogle and Company—Glasgow
Francis and Richard Brown—Bristol
Archibald Buchanan (1737-1800) and Alexander Speirs and Co.—Glasgow
James Buchanan and Co.—London
James Donald Jr. and Sr.—Glasgow
James and Robert Donald and Company—Greenock and Glasgow, Scotland
Norman Durwood—London
Joseph Fisher

[Thomas] Flowerdew and Norton—London
William Gale—Whitehaven
Charles Gore—London and Liverpool
___ Griffiths—Bristol
Peter Hone—Whitehaven
Edward Hunt—London
George Kippen and Co.—Glasgow
Thomas Knox—Bristol
John Lidderdale, John Harmer, ___ Ferrell—Bristol and London
Lyonel (1682-1737:42) and Samuel Lyde (died 1806)—London
John Maynard—London
John (died 1759) and George Murdock and Company--Glasgow
Neilson & Boyd
John Perks
John Price—London
George Randall—Seren, Cork, Ireland
Anthony Rhodes (died 1761)
John Sparling—Liverpool
John Sturman
Thomas, Griffiths, and Thomas—Bristol
Morgan Thomas and Co.—Bristol
Watson & Cairns--Edinburgh
Francis Wild
Williams, Evans and Co.—Bristol
Wilson and Nixon—Whitehaven

The Washington Family and Accokeek Furnace

Around 1718 Augustine Washington (1694-1743), a typical middle class farmer, purchased a small plantation in Westmoreland County on Pope's Creek, a tidal tributary of the Potomac River. A short distance to the east of Augustine's farm was Stratford Hall, home of the wealthy and influential Lee family with whom the Washingtons had little in common. But Augustine was a highly motivated man with an excellent sense of business. About 18 miles southwest of his plantation and on the Rappahannock River near Leedstown was a deposit of bog ore, a type of iron ore well-suited to smelting in a charcoal blast furnace. The colonists suffered from a perennial shortage of iron products and the discovery of iron ore deposits signaled a bright business opportunity. Augustine, then a very young man, availed himself of this opportunity, becoming involved in the newly-formed Bristol Iron Company. Working with Col. John Tayloe (1721-1779) of Mount Airy, Richmond County, Augustine either helped with or was manager of the Bristol Iron Works that straddled the present boundary between Westmoreland and King George counties (Hudson, "Augustine Washington," 2612). His dealings with Bristol undoubtedly provided him contact with the Principio Company, members of which were building their first structures in Maryland at about the same time Bristol was building on the Rappahannock. Aware of Principio's desire to expand and locate new ore deposits, Augustine assembled several parcels in Stafford County that eventually would become the site of Accokeek Furnace and headquarters for the Principio Company.

Accokeek was a profitable segment of the Principio conglomerate, making wealthy men of Augustine Washington and the other shareholders and managers. Augustine's profits from Accokeek propelled him up the social ladder and enabled him to educate his sons, Lawrence (c.1716-1752) and Augustine, Jr. (c.1718-1762) in England and also paved the way for son George's marriage to the heiress Martha (Dandridge) Custis.

At first content to allow John England to manage the works at Accokeek, Augustine remained in residence on Pope's Creek. England's death in 1734 forced Augustine to reconsider his level of involvement at Accokeek. He negotiated with the English partners for a larger share in the company and became manager of their North American operations. Determining that he needed to be closer to Accokeek, he moved his family nearer to the works and the surviving Principio and Accokeek ledgers indicate that he was highly involved and frequently on site until his death. It seems likely that Augustine's proximity to Accokeek was a significant factor in that works becoming the company's headquarters. Augustine Washington and Nathaniel Chapman, another brilliant and tireless businessman, directed Principio's efforts and established the company as the largest American manufacturer of iron products prior to the Revolution. Together they managed an international industrial conglomerate and set the stage for the Industrial Revolution to follow.

Augustine Washington died in 1743, leaving Mt. Vernon and his 1/6 interest in Principio and "all the right Title and Interest I have to in or out of the Iron works in which I am Concerned in Virginia & Maryland," to his son Lawrence (Ford 4). Lawrence shared his father's business acumen and was instrumental in the company's expansion. He and Nathaniel Chapman seem to have worked well together and continued to guide the company ably and profitably. In 1751 Lawrence signed on behalf of the company for their purchase of Lancashire Furnace[67] in Baltimore County, Maryland. This acquisition provided the company with a furnace that was close to a major deposit of quality ore, a lack of ore near the Principio Furnace having been a long-standing concern.

Unfortunately, Lawrence was consumptive and died in 1752, leaving his daughter Sarah "all my Stock, Interest and Estate in the Principio, Accokeek, Kingsbury, Lanconshire, and North Eastern iron works, in Virginia and Maryland, reserving one-third of the profits of the said works to be paid my wife as hereafter mentioned" (Ford 75). Lawrence's death was likely a contributing factor in the decision to close Accokeek and consolidate the company's interests in Maryland. It is certainly not a coincidence that the year of Lawrence's death was also the last season of blasting at Accokeek.

Sarah died shortly after her father and without issue and the interest in the company passed to Lawrence's brother, Col. Augustine Washington. Augustine also took an active interest in the Principio Company, though by the time he became a partner all smelting and manufacturing were being conducted in Maryland. In the Accokeek ledger is an entry dated June 14, 1755 in which Col. Augustine Washington was charged "to account of insolvents for his part of old debts lost £34.10.0." At the meeting of the Council of Virginia held on Nov. 5, 1757 Augustine Washington, "in behalf of himself and other adventurers in iron who praying leave to import from Maryland bar and pig iron of that province without being obliged to pay the port duties and other fees chargeable on vessels importing goods and merchandize" (Hillman, vol. 4, p. 71). As a result of his petition, the fees were rescinded and he was able to import Maryland iron into Virginia without additional charges.

At some point half of Augustine Sr.'s 1/6 interest in Principio seems to have been sold for by 1764 a 1/12 interest came to be vested in Mrs. Anne (Aylett) Washington (died 1774) of Wakefield, the widow of Augustine, Jr. and the only surviving American partner. Perhaps in need of money or perhaps simply having no need of or interest in the iron company, Anne wished to sell the share. On May 4, 1764 she wrote to her London agent, Gilbert Francklyn, asking him to examine the company's books in order to determine a valuation of her inherited share. John Price, manager of the company's London office, told Francklyn that company by-laws forbade anyone with less than a 1/10 share from looking at the books (Francklyn, Gilbert to Anne Washington, letter). The matter didn't die, however, and on Feb. 27, 1771 the Principio partners wrote to Mrs. Washington agreeing to let her agent, William Lee, obtain copies of the quarterly reports for her (Lee, William to John Augustine Washington, letter). The writer suggested that as Thomas Russell would soon be returning to America, he could bring them. Once she had an opportunity to go over the reports, they could negotiate for the sale of her share (Francklyn, Gilbert to Anne Washington, letter). How events played out thereafter is unknown. Anne died in 1774 never having disposed of the share. Shortly thereafter, the Revolution erupted and not until long after the war did the Maryland legislature settle the affair with the next heir, William Augustine Washington[68] (1757-1810).

---

[67] Lancashire stood on Green Berry Point on the west side of North East Creek, very near the Penn Central railroad tracks, and east of General Pulaski Highway (U. S. Route 40). The property was purchased in 1751 for £2,675 and the furnace was within two miles of Kingsbury Furnace and close to navigable water.

[68] William was the son of Augustine Washington, Jr. (c.1718-1762) and Anne Aylett. He was born in Westmoreland County at Wakefield, which he inherited from his father. William married first Jane Washington (died 1791), secondly Mary Lee, and thirdly Sally Tayloe of Mt. Airy. He died in Georgetown, Washington, DC.

Closing Down

Accokeek Furnace closed during or following the 1752 season and workers and equipment were transferred to Principio's other plants in Maryland. No direct documentary evidence survives explaining the closure and over the years two primary theories have emerged. Some researchers have thought that the furnace was closed because the supply of iron ore ran low. Others have suggested that there was not enough timber at Accokeek and the furnace ran out of charcoal. Neither of these theories is correct. The closing of Accokeek seems merely to have been an attempt on the part of Principio management to consolidate their efforts in Maryland. By 1752 Principio owned five furnaces and forges, all but Accokeek being in Maryland. Accokeek had become the company headquarters around 1727 when it first opened and there had been talk among shareholders about adding a second furnace to the complex. John England believed that the supply of ore in Stafford was such to justify a second furnace and the Principio Furnace then in operation in Maryland did not have an adequate ore bank. For whatever reasons this expansion never materialized, but Augustine Washington's residence so near the furnace and Nathaniel Chapman's move from Maryland to Accokeek probably were the determining factors in the furnace being designated as the company headquarters. Perhaps there had been discussions about consolidation prior to Lawrence Washington's death. Tuberculosis was a long and arduous illness that always led to a predictable end. The cessation of blasting at Accokeek followed too closely on Lawrence's death to have been merely a coincidence. As Lawrence had no sons, his share in the company passed to his daughter. Nathaniel Chapman may have felt that this was the perfect opportunity to consolidate the company's manufacturing efforts on the other side of the Chesapeake Bay. Had it not been for Lawrence, this move might have taken place upon Augustine's death in 1743.

In the 19th century the old Principio works in Maryland were purchased and renovated by the McCullough Iron Company. Henry Whitely wrote a history of the Principio Company for the benefit of the McCullough Company. He said of the closing of Accokeek, "About the time of Lawrence Washington's ill health and subsequent death, the supply of ore at Accokeek failed and the furnace was necessarily abandoned. In 1753 the moveable effects were distributed among the other works, slaves and store goods, horses, cattle and wagons were sold, and the business in Virginia as far as related to iron making was gradually closed up, some of the real estate being sold in 1767" (Whitely 195). Company ledgers reveal that the slaves and store goods from Accokeek were moved to Kingsbury Furnace and North East Forge. An Accokeek ledger entry dated May 28, 1753 includes a list of goods sent from the Accokeek store to Kingsbury. Tools and surplus iron and steel were transferred to that plant, also. Later legal wrangling over the property indicated that the Accokeek furnace was dismantled sometime thereafter. Although iron making ceased, the office remained open until 1760 paying and collecting old debts. This was a slow and not always successful undertaking and by 1776 the company books still carried £606.2.9 in old debts form Accokeek, though nothing remained there but the real estate. After the furnace closed, the gristmill continued under least to William Kendley.

Henry Whitely believed Accokeek's closure was the result of depleted ore deposits, yet 20 years after Principio pulled out James Hunter of Rappahannock Forge reopened the site to supply his own furnace near Falmouth. Hunter utilized the ore banks until his death in 1784. As far as closing the furnace due to a shortage of charcoal, Accokeek leased land in addition to the acreage assembled by Augustine Washington, John England, and Nathaniel Chapman. There is no indication that lack of timber or ore was responsible for the decision to close the works.

Although the state allowed Hunter to purchase 200 acres of the Accokeek tract, the old Principio management continued to hold much of the furnace property. In 1800 William Russell (1740-1818) of Maryland, "Agent for the Mine Co.," paid land taxes on 5,161 acres in Stafford. By 1801 the furnace acreage had decreased to 4,761. Two years later the tract was divided and sold. William Russell last appeared in the tax records of 1803 when Col. Enoch Mason (c.1769-1828) paid taxes on 400 acres "for William Russell." Enoch lived at neighboring Clover Hill and this 400-acre tract was part of Accokeek's outlying acreage. The remaining furnace acreage was probably divided and sold at about this time though there are no surviving deeds for the conveyance.

Closing Accokeek was not as straight forward an affair as it might seem. In 18th century America, few people paid cash for goods purchased from stores and other businesses. Instead they paid their accounts once or twice a year, cash flow permitting. It was not unusual for a merchant to wait several years to collect money owed if the patron fell on hard times or chose to ignore the debt. The various furnace, forges, and stores owned by the Principio Company operated the same way. An entry dated July 1, 1754 in the North East Forge ledger reveals that Accokeek owed the forge a balance of £6,869.17.4 ½ for goods plus £241.19.11 ½ for cash advances. Collecting past debts was often a lengthy process for all concerned. In the case of Accokeek, many people still owed the company money long after the business had ceased operating. An entry dated Dec. 31, 1757 in the North East Forge ledger records that on that date Accokeek sent the forge £712.11.11 ¼ for past accounts. On June 26, 1758 Accokeek sent North East Forge £1150.1 from "Sundry Accounts at Accokeek." Those listed in this particular entry were all merchants to whom bar iron had been consigned. On Aug. 28, 1758 North East received another installment of £129.10. 14 ¼. Company ledgers show sporadic payments received from Accokeek over a five-year period. When the personal property at Kingsbury Furnace was inventoried in 1781, the following were listed: 118 Grubbing hoes, 7 Coopers Adzes, 12 Narrow Axes, 11 Coopers Axes, 13 Narrow hoes, 1 Carpenters adz, 5 Scrubbing Chissels, 6 Pearing Chissels, 1 ½ inch Auger, 1 1 ¼ do. To this was added the notation, "Came from Accokeek all Damaged and good for little" (Kingsbury Furnace inventory, 1781).

As the situation between the Americans and British worsened it became increasingly difficult for English firms to operate in the colonies. Although Accokeek had been long closed, Principio continued to operate its other four furnaces and forges in Maryland. Principio's English owners had little say over any of their American property. William Russell (1740-1818), then company manager, threw his fortunes with the rebellious colonies and became a Maryland citizen. Despite his change in citizenship, he continued to operate the Principio works, providing the fledgling American government with a major supply of iron and weapons. In 1780 the Maryland General Assembly passed an act authorizing the seizure and confiscation of all British-owned property in the state. Three years later, Principio's property was seized and much of it was sold for the benefit of the state. As a result of Russell's timely change of citizenship, his interest in the business and real estate (and that of the Washington family) were spared, though it took years of legal wrangling to settle claims on the property.

Accokeek Reopened

For over 20 years, the Accokeek property sat abandoned. Intensifying hostilities between England and America, however, focused attention on the colonies' need for an abundant supply of iron. In an effort to spur iron manufacturing, the Virginia General Assembly passed several acts for the encouragement of iron works one of which authorized James Hunter (1721-1784) of Rappahannock Forge to obtain ore from the Accokeek tract. In part due to the closing of Accokeek, Hunter had opened his own furnace and forge c.1759 just above Falmouth. By 1760 Hunter's forge was manufacturing a wide variety of iron products and domestic consumer goods. Initially, Hunter's smelting was limited and he purchased most of his pig and bar iron from Principio's Maryland facilities, from Neabsco and Occoquan furnaces, from Baltimore Iron Works, and may well have bought or leased Alexander Spotswood's old mines. As relations with England deteriorated and shipping became treacherous due to English blockades and attacks on American ships, James sought a nearby source of ore. A petition to the Virginia Assembly resulted in Principio's management being given 30 days to either re-open the site or risk losing it. On May 1, 1777 the Assembly passed an act stating:

> Whereas, the discovery and manufacturing of iron requisite for the fabricating the various implements of husbandry, small arms, intrenching tools, anchors, and other things necessary for the army and navy, is at this time essential to the welfare and existence of this State, as the usual supplies of pig and bar iron from foreign States is rendered difficult and uncertain, and James Hunter, near Fredericksburg, hath erected and is now carrying on, at considerable expense and labour, many extensive factories, slitting, plating, and wire mills, and is greatly retarded through the want of pig and bar iron; and whereas, there is a certain tract of land in the county of Stafford, called or known by the name of Accokeek furnace tract, on which a furnace for the making of pig iron was formerly erected and carried on, which has been since discontinued... James Hunter is therefore authorized to enter upon two hundred acres of the Accokeek tract, including the old furnace, if its owners or agents should fail in one month to begin and within six

months to erect thereon a furnace equal to or larger than the former one, and prosecute the same for the making of pig iron and other castings (Hening, vol. 9, pp. 303-304).

This act authorized James Hunter to take over the old furnace seat and dam "and pay a sum for the property as determined by the court." Based upon later correspondence, it appears that Hunter may have suggested rebuilding Accokeek Furnace, though he never did so.

This act was followed by a "Report of the Commissioners Appointed to Locate Land for Iron Works" and dated Nov. 15, 1777:

> Resolv'd as the Opinion of this Committee that the Memorialist ought to be allow'd to locate two Hundred acres of Land of the Accokeek Tract, lying in the County of Stafford including the old Furnace Seat & dam, & if a sufficient Body of Iron ore is not discovered therein that he be at liberty to explore and open any other unimproved Lands belonging to the said Furnace tract, and upon discovering a sufficient Body of ore, to locate ten acres thereof (in case the proprietors or their agents shall not within a reasonable time open them) paying to the proprietors such valuation as well of the two Hundred acres as of the Ten acres, as shall be made by a July of twelve good & Lawfull Freeholders upon Oath (Executive Communications, Report).

The committee further provided that "if a Body of Iron Ore is not discovered on the Accokeek Tract, the Memorialist ought to be allowed to explore & open for the discovery of Iron ore, any other unimproved Lands within the circuit of Thirty Miles." The report also noted that "these [lands] being situated on the same direction & vein of Ore with Mr. Spotswoods & Mr. Chissels[69] is likely to be of same quality and though improper for Bars is yet exceeding fit for various other manufactures set on foot here and if opened may be instantly rendered serviceable from their vacinity to Hunters works in want of Pig Mettal and at present not procurable from the neighbouring States untill these and other Lands can be explored for Ore and convenient Furnace Seats fixed on His slitting, plating & wire Mills being on a scale large enough to supply this State provided he can secure the Pig Mettal without which it is impossible for him to furnish the Country with Bar Iron for Plating and many Utensils, the Army & Navy with Arms, Entrenching Tools, Anchors & all sorts of shop work which they have hitherto depended on him for besides his steel Furnace now in operation with the Publick and Private Factorys for Arms" (Executive Communications, Report).

On Nov. 1, 1777 a jury consisting of James Withers[70] (1717-1784), Francis Stern, Thomas Arrasmith[71] (c.1748-after 1820), Enoch Benson (born 1729), Joel Reddish[72] (born 1748), Thomas

---

69 Charles Chiswell's Fredericksville Furnace in Spotsylvania County.

70 James was the son of James Withers (1680-1746) and Elizabeth Keene (died 1769), the only child of Matthew and Bridget Keene. Young James married first, before Nov. 8, 1748, Catherine, the daughter of Thomas Barbee (1690-1752). He married secondly Jemima Garner and died in Fauquier County. James Withers served as foreman for the jury that allowed James Hunter to take over the old Accokeek mines. Interestingly, from 1797-1799 a Thomas Withers was employed as overseer of Stanstead, James Hunter's plantation. Family researchers have not determined which Thomas Withers this was.

71 In 1782 James Arrasmith married Lavina Elizabeth Hefflin (c.1765-1845). According to his Revolutionary War pension application submitted by his widow, James enlisted in either Stafford or Fauquier County. In 1820 he was 72 years old and a resident of Fauquier (Revolutionary War Pension Application).

72 Joel Reddish was the son of Joseph Reddish (died c.1793) and Sarah Reddish (died after 1794) of St. Paul's Parish (now King George County). Joel left the Church of Virginia and joined Chopawamsic Baptist Church which later sponsored Hartwood Baptist Church (established Mar. 26, 1771). Joel's name appears on page 6 of these church records stating that he was excommunicated and restored to membership. On page 12 it was noted that on Dec. 13, 1783 he was again excommunicated for excessive drinking. In 1787 Joel was granted 930 acres in Bourbon County, Virginia (later Mason County, Kentucky). Joel sold his Stafford land in 1812, apparently in preparation for his removal west. By Aug. 19, 1814, he was residing in Bourbon County.

Edrington (1743-c.1796), William Edrington (c.1741-1794), Zachariah Benson,[73] Thomas Stephens, Joseph Reddish[74] (died c.1793), and Cossom Horton[75] (died c.1820) was called for the purpose of determining the fair value of 200 acres of the old Accokeek Furnace tract (Stafford Deed Book S, page 1). Prior to the hearing, Hunter was directed to write to Principio's administrators inquiring as to their plans for Accokeek. At the hearing Hunter produced a letter dated July 10, 1777 from Principio's manager, Thomas Russell, who stated, "I have no thoughts of enlarging our Concern in Iron Works at this time especially at so remote a distance and where the Prospect is not very inviting. You are to use your own discretion in this matter and I doubt [not] but our Gentlemen will receive sufficient [compensation] for any of their Lands that may be appropriate for ye building of Iron Works and making of Iron" (Executive Communications, Letter, Thomas Russell). That same month, Travers Daniel[76] (1741-1824) surveyed the parcel in the presence of the commissioners and jury. The 200-acre tract included the old mill seat, pond, and dam as well as the ore bank. It was described as:

> two hundred acres of land at Accokeek old Iron Works Beginning at the Chestnut oaks on a millside at A near the Head of the pond & run from thence So. 25 d east 350 po. to B a stake near the head of a small branch, thence so. 63 east 80 poles to C near an appletree & in the Companys lower line from thence North East 90 poles to D two small white Oaks & a Hickory in the s'd Line thence N. W. 116 poles to E, a white Oak thence No. east 70 poles to F, a large gum near a spring, thence N. W. 154 poles to G the old Dam, thence up the side of the pond N. 64 W. 48 poles to H, from thence across the Head of the pond to the Beginning (Executive Communications, Memorial).

The jury determined "the value of the 200 acres of land of the Accokeek Iron Works…to be £500 current Virginia money."

George Washington was also aware of Hunter's need and desire to utilize the Accokeek site. In mid-1778 James wrote to Washington apparently asking to purchase Mary Ball Washington's old mine tract on the north side of Potomac Run. James' letter has been lost, but a copy of George's answer remains on file amongst his papers:

> Valley Forge June 15th 1778
>
> Sir;
>
> Your favor of the 12th last did not come to my hands till yesterday. The land therein mentioned hath not been legally conveyed, or properly secured to me by my Mother—this reason if no other would prevent me from selling either the land, or the wood that grows on it; but I have other reasons against it, equally forceable; one is, that I have had an interest, which my present situation & absence have been the only Bar to the execution of building a saw mill, for the purpose of sawing up the Pines which I am told the land abounds in, and which constitute the chief value of it provided its bowels have been stripped of all the ore and which is denied by some.
>
> If no disadvantage on acct. of Roads into the land, and the consequent destruction of the wood & Timber by the Miners, or their followers was to result I shd

---

[73] Zachariah was the son of Robert Benson (1685-1757) and Frances Prou (born 1689) of King George County. He was the brother of Enoch Benson who also served on this jury.

[74] The Reddish family was established in Virginia in 1623 when John Reddish was listed as a headright for Capt. William Perry. The Reddishes came from the village of Reddish near Stockport, Cheshire, England (now part of the city of Manchester).

The parents of Joseph Reddish of Stafford are unknown, but a James Reddish and Elizabeth Massey were married in St. Paul's Parish (now King George County) on Aug. 19, 1726. Joseph may well have been a son of this marriage. In Stafford, the Reddish family lived at Reddish Hill on the west side of Mountain View Road (State Route 627) near Moore's Corner.

[75] Cossom lived on the old Hope Patent on Aquia Creek.

[76] Travers resided at Crow's Nest, the peninsula between Accokeek and Potomac creeks. He served as county surveyor from 1777 to at least 1795 and as a justice from 1765 through at least 1774.

have no objection so far as the matter depended upon me. The thing at the same time, appearing absolutely necessary for the well being of your Works, to part with the ore upon terms which shall be judged reasonable between Man and Man. Wishing you success in your undertaking, I am with great Esteem and Regard

Sir,
Your mo. obt. Humble servt.,
G. Washington

(American Memory, Letter, George Washington)

There is no documentary evidence proving that James Hunter utilized Mary Washington's mine banks; however, there is circumstantial evidence suggesting that he did so. For at least ten years prior to his acquisition of the land at Accokeek, James had owned 1,176 acres in the vicinity of and probably adjoining the Accokeek mine tract on Potomac Run (Vogt, vol. 1, p. 68). Additionally, his estate inventory listed ten slaves at Accokeek, presumably digging ore and cutting timber (Stafford Deed Book S, p. 283). These workers were doing something constructive or Hunter would not have kept them there. Finally, as late as 1864 the Woodcutting Road (now Mountain View Road—State Route 627) made a nearly direct line south from Accokeek to Rappahannock Forge (Gilmer Map). Only in later years did the lower portion of the road cease to be used.

James Hunter also managed to acquire timber rights to land around Accokeek, though there are no known records that explain exactly how he did so. In 1803 some 5,400 acres of his remaining property was sold. While some of this land was located northwest of the forge along U. S. Route 17 (Warrenton Road), other parcels were located in the vicinity of Accokeek and Potomac runs. Because of the loss of records, it is unknown when, how, or from whom he acquired much of this property. Hunter used much of this property for charcoal production, cutting large quantities of timber from the area and hauling it to the forge via "The Woodcutting Road," now Mountain View Road (State Route 628).

In 1779, just two years after he was granted the property, Hunter's claim at Accokeek was challenged. A committee appointed by the governor of Virginia was assigned to study Hunter's need for and claim to the tract. The committee wrote, "The Accokeek Iron Mines in Stafford County, belonging to a company in England whose property they still remain, were worked but discontinued upwards twenty years ago because they had larger and richer Banks of ore, with greater conveniencys of Wood and Water in Maryland, where all their Hands, stocks and utensils were removed" (Executive Communications, Oct. 4 to Dec. 24, 1779). With regards to the condition of the land around Accokeek they wrote, "the Lands are said to have been offered for sale by the company's agent and probably from their being very broken without Timber and the soil excessive poor, have not been sold." The committee also reported that Hunter "ought to be allow'd to locate two Hundred acres of Land of the Accokeek Tract...including the old Furnace Seat and dam, and if a sufficient Body of Iron ore is not discovered therein that he be at liberty to explore and open any other unimproved Lands belonging to the said Furnace tract" (Executive Communications, Oct. 4 to Dec. 24, 1779). For the second time Hunter's claim was upheld.

Ownership of the 200 acres was contested for years, later generations of the Washington family seeking to regain what had been lost. In 1780 Maryland sequestered all English-owned property in that province and sold the tracts for the benefit of the state.[77] Most English companies and Loyalist

[77] Beginning on Aug. 9, 1781 and continuing for several weeks the *Maryland Gazette* newspaper carried a notice of the sale of Principio Iron Works in Maryland. The notice read:

> Office for the preservation and sale of forfeited estates, Annapolis, July 26, 1781.
> Pursuant to an act of assembly, will be sold, at public auction, on the 11th day of September next, at the Lancashire furnace, in Baltimore county, Between thirteen and fourteen thousand acres of valuable land, lying near to Baltimore-town, late the property of the Principio company; on which are erected two convenient furnaces, and two grist mills. The land will be chiefly parcelled out into small and convenient farms. That part on which the furnaces are erected, will be first sold, in order that any person or persons, who may incline to purchase, with a view to carry on the iron works, may have an opportunity of securing such other parts of the land as they may think necessary. At the

businessmen had long before returned to England. Principio's owners, nearly all of whom had remained residents of that country, lost what little control they had of their Maryland interests. Rightful ownership of Accokeek, held jointly by English and American investors, was unclear. Since Principio had leased the Accokeek land from the Washingtons, that family believed they held legal title. In March 1780 a meeting was held at Garrard's Ordinary[78] to determine ownership of the property. Charles Carter (1738-1796), escheator for Stafford County, met with a jury consisting of Travers Daniel (1741-1824), John Rowzee Peyton (1754-1798), John Gregg, James Kenyon, John Ralls (1748-c.1783), William Routt[79] (1735-1823), Francis Stern, William Waller (1740-1817), William Mountjoy (1737-1820), and Charles Porter. They were charged with determining whether "a certain Tract and five Plantations thereon containing 200 acres more or less known as the old Mine tract" could be claimed by James Hunter according to the laws of escheat. "They do say that the said land, Messuage and Plantations, are and were ___ day of April 1775 the property and fee simple estate of several British Subjects; that ___ Washington of the County of Westmoreland in this Commonwealth, claims as an estate in fee of 1/12 part of the said Lands, Messuages and Plantations called the Mine tract and that the said lands and premises (subject to the said Interest or estate of ___ Washington as claimed aforesaid, and to Peter Byran who claims as heir at Law of Sarah Byran one h___ acres of said Mine tract, under a deed of Intail (Deeds produced) from Alice Fritter to Sarah Byran, and a claim of Thomas Russell of the state of Maryland for one half of the said Tract and that the said Lands, Messuages and ___ and become forfeit and vested in the commonwealth subject to the claims above recited (Stafford Deed Book S, p. 1). While only the first part of this document survives, James Hunter's claim was upheld.

In 1783 Thomas Russell and William Augustine Washington petitioned the Virginia Legislature citing the fact that they were not British subjects and that Maryland had confiscated only company shares owned by British subjects (Heite, Historic, 25). Russell and Washington claimed that Hunter had never built anything on the 200 acres he had escheated in 1777 and other tenants on the Accokeek tract had been refusing to pay their rents because of the uncertainty caused by Hunter's disputed claim.[80] Russell and Washington asked the Assembly to revoke Hunter's grant and allot a 1/12 interest to Washington and 1/8 interest to Russell in proportion to the shares they had owned in the company. They also suggested that the remaining interest should revert to the state as loyalist property (Virginia Legislative Petitions, Petition of William Augustine Washington). The official response to this petition is unknown. In the meantime, Russell continued to operate North East Forge in Maryland.

> same time will be sold; the utensils and stock, of every kind belonging to the said works; among which are about one hundred valuable slaves, of different ages and sizes; sundry of which are excellent tradesmen, such as founders, colliers, blacksmiths, &c. The money to be paid down, if agreeable to the purchasers; if not, they may give bond with security, to pay one third of the sum bid on the first day of September 1782, another third on the first of September 1783, and the remaining third on the first of September 1784, with interest, in gold and silver, or the new bills of credit to be emitted, in pursuance of an act of the last session, at their actual value at the time of payment.
>
> By order Jo[seph] Baxter, clk.

[78] Garrard's Ordinary stood just northeast of the courthouse near the site of the present school board offices.

[79] The Routt family was long seated in Northumberland County. In 1753 William Routt of Stafford married Winifred Bryan. In 1784 William and his son, William Routt (born 1756), sold 250 acres lying on both sides of Accokeek Run to John James (c.1732-c.1794) (Stafford Deed Book S-176). This land had been part of Thomas Norman's (c.1625-c.1709) patent from the 1690s and probably adjoined the Accokeek Furnace tract. William Route (1735-1823) was the son of Peter Routt (died 1765) and Elizabeth Norman.

John James who purchased the Routt land was the son of George James (c.1702-1753) and Mary Wheeler. John married in 1763 Anne Strother and was a justice for Stafford County 1766-69, 1772-82 and possibly other years as well.

[80] A detailed account of this complaint is contained in the papers of Thomas Ruston, "Principio Iron Works Law Suit Legal Copy Book," Library of Congress.

On May 18, 1794 George Washington wrote to his nephew, Robert Lewis, to whom he wished to give his mother's Accokeek land. Considering Washington's letter to James Hunter in 1778, George seemed surprisingly unaware of the events involving his family's Stafford property. He wrote, "I have been told, that some person in Falmouth (whose name I do not recollect) had pillaged the Land of the most valuable Pines thereon; and that either he, or some other, talked of escheating it; but I never supposed injustice would prompt any one to such a measure" (Fitzpatrick, vol. 33, pp. 370-372). Assuming that Washington was referring to James Hunter, it is unclear how he could not know who had been using the property. He had written to James in June 1778 approving of Hunter's use of the land. Several of Washington's diary entries also pertain to James Hunter. On Aug. 3, 1770 he paid Hunter £10.5 for "Mill spindles Gudgeons &ca" to be used in his new mill. Apparently, this was a balance due for parts as about six weeks earlier he had sent him £15 on account of the mill. On Mar. 17, 1775 George recorded that he had dinner with James Hunter, having been "detained by Wind" on his way to the Second Virginia Convention held at Henrico Parish Church (Jackson, vol. 3, p. 314).

Ownership of the Accokeek tract as well as Mary Ball Washington's property continued to be disputed in the courts for years, tough the Washingtons never regained control of the property. Thomas Russell II died in 1786, but his son, Thomas Russell III (died 1806), continued with William Augustine Washington (1757-1810) to press their claim for the Accokeek land. In 1796 Russell wrote to Washington suggesting that papers in the Northern Neck land office might contain evidence that would clarify their claims to the land. However, the issue was never resolved.

Most of James Hunter's real estate was sold at public auction in 1803. There are no surviving deeds for any of these conveyances though a list of buyers and the acreage they purchased was included in the list of alienations in the land tax records of 1804. Although not included on this lengthy list, Francis Foushee[81] acquired the 200-acre furnace tract at about this time. From 1804-1810 William was listed in the tax records with the original 200 acres. In 1811 he added an adjoining 150 acres, both designated as "Old Mines." On Apr. 11, 1804 he conveyed the property to his son, William (1749-1824), this transaction being revealed in a later deed of trust dated Feb. 13, 1826 between Dr. William Foushee[82] and R. C. L. Moncure (1805-1882), trustee. According to this instrument, William was indebted to Robert Dunbar (c.1745-1831) of Falmouth for $250.93 (Stafford Deed Book GG, p. 82). By 1826 Dunbar was in severe financial straits of his own and was calling in all his earlier loans. Foushee deeded to R. C. L. "The Old Furnace containing about Two hundred acres, being the same tract of land conveyed to the said William Foushee by his Father, the late Francis Foushee, deceased" along with two other tracts.

William was unable to pay his debt in a timely manner and in 1826 the property was sold in front of Thornton Alexander's tavern.[83] According to the deed, a *William* Dunbar of Falmouth was the highest bidder, offering $265 (Stafford Deed Book GG, p. 186). This was probably a transcription error, the name *William* being confused with *Robert* Dunbar; there is no record of a William Dunbar of Stafford or Falmouth. On July 1, 1826 Robert Dunbar conveyed the tract to James Alexander (Thornton Alexander's brother) for $265 (Stafford Deed Book GG, p. 187).

In order to pay for his purchase, James Alexander executed a deed of trust to John Moncure Conway (1779-1864) and Capt. John Macrae[84] (1792-1830). He defaulted on the loan and Conway announced the upcoming sale of the land. The newspaper notice read:

---

[81] On Feb. 6, 1790 Francis received a grant for 1,000 acres in Mason County, Kentucky. On July 29, 1795 he received 7,813 more acres in the same county.

[82] William was born into a Huguenot family in the Northern Neck and educated in Edinburgh, Scotland. He first practiced medicine in Norfolk, then moved to Richmond in 1777. William lost an eye gouged out by a "low fellow" who, without provocation, attacked him in a billiard parlor. During the Revolution, William served in Thomas Hall's militia unit and was a surgeon with the Virginia State troops. He was granted half pay as a pension. In 1782 Richmond was incorporated into a town and William Foushee was elected mayor.

[83] Thornton Alexander (c.1788-1840) kept a tavern that was located roughly in the vicinity of Lewis Insurance Agency on the corner of U. S. Route 1 and Courthouse Road (State Route 630).

[84] John Macrae lived in Fauquier but "breathed his last at Mrs. Nelson's Boarding House" in Fredericksburg. According to his obituary, John "was a gallant and distinguished officer during the last war, and a lawyer of brilliant prominence—a member of the late State Convention of Virginia." He died at the age of 38 (*Virginia Herald*, Jan. 23, 1830).

Trust Sale. By virtue of a deed of trust executed by James Alexander to me and duly recorded in the Clerk's Office of the County Court of Stafford, for the purpose of securing certain sums of money therein mentioned, I shall proceed to sell on the premises—Monday the 24th day of December next, to the highest bidder, at public auction for cash, the tract of land whereon the said Alexander now resides, containing two hundred Acres, more or less—this land is situated within three miles of Stafford Court House, is principally low grounds and very fertile, the Houses are comfortable and in good repair—also, the following property to wit: Three negroes, a man, woman and small girl, all the stock consisting of a mule, two or three horses, fourteen head of cattle, and thirty or forty head of hogs, the crop of corn, farming utensils and Household and Kitchen Furniture. This sale will be subject to a previous lien upon the land and part of the personal property executed to John Macrae Esq. but should Capt. Macrae consent, which is probable, to join in the sale which will consummate the entire title to the purchaser, timely notice will be given. Acting as trustee I shall only convey such right as is thereby vested in me.

J. M. Conway, Trustee

(*Fredericksburg Political Arena*, Oct. 24, 1828)

John Macrae did decide to join in the sale and on Dec. 16, 1828 he announced in the same paper:

Trust Sale. I shall offer for sale on the second Monday in January next, before the front door of Col. Thornton Alexander's Tavern near Stafford Court-House, a Tract of Land called the "Furnace Tract" whereon James Alexander now resides, estimated to contain two hundred acres, and a slave named Giles. I shall make the sale under a deed of Trust from said James Alexander to me, bearing date the 20th day of October, 1827, and duly recorded in the Clerk's Office of Stafford County Court.

John Macrae

(*Fredericksburg Political Arena*, Dec. 16, 1828)

It is unclear who bought the property at this sale, but by 1837 the old furnace tract belonged to R. C. L. Moncure (1805-1882). In 1848 he conveyed the property to his brother, John Moncure (1793-1876), and it became part of John's 699 ¼-acre Woodbourne tract. Interestingly, the land tax records of 1858 refer to Woodbourne as the "Furnace."

Accokeek Furnace is the second oldest iron works identified in Virginia, predated only by Tubal Furnace, and is one of the earliest blast furnaces in the country. Visible remains in the immediate vicinity of the furnace include the pond site, two mine pits, a raceway, the furnace site on the southwest bank of Accokeek Creek, mill wheel pit, a retaining wall, a terraced hillside above the furnace, a slag dump on the northeast bank of the creek, and a cemetery above and to the north of the furnace foundation. One mine pit is on the southwest side of the creek and one is on the northeast side. The area around the furnace site and in and along the run is littered with sandstone, bricks, and pieces of dark green glassy slag, remnants of the smelting operation.[85] Time and forest litter have obliterated many other remnants of the operation. The writer was unable to find any oyster shell or charcoal around the furnace, materials that are easily discovered at the Rappahannock Forge site near Falmouth. According to Edward Heite in his report on Accokeek, "In assessing the Accokeek site archeologically, the industrial buildings can be expected adjacent to the furnace, but the residences should be sought either on the bluff above or in the flatlands below. A mill, possibly attached to the furnace raceway, might be located within a half-mile of the furnace

[85] The hillside behind the furnace foundation is terraced. If the design of the Accokeek complex mirrored that of other 18th century iron furnaces, then upon these terraces stood the coal and ore houses that sheltered the raw materials for smelting. On one of the terraces, just to the west of the furnace foundation, is a cemetery containing an unknown number of graves. At least twenty of these are marked with pieces of field stone; none are inscribed. A number of depressions in the ground indicate other unmarked burials on this terrace.

tself. Coal, ore, and flux storage should be expected on the hillside, above the level of the top of the stack, ır thirty feet above the stream" (Heite, "Historic Site Report, p. 38).

In 1984 the Virginia Historical Landmarks Commission placed Accokeek Furnace on the Virginia ,andmarks Register though nothing beyond a preliminary archeological survey has yet been done. The \ccokeek Ironworks site contains unexposed, unaltered ruins that have remained largely undisturbed since he furnace was dismantled in 1755. These will provide a wealth of information if and when they are tudied. Three acres (including the furnace) now belong to the George Washington Boyhood Home 'oundation and the site is now under the management of the Kenmore Association.

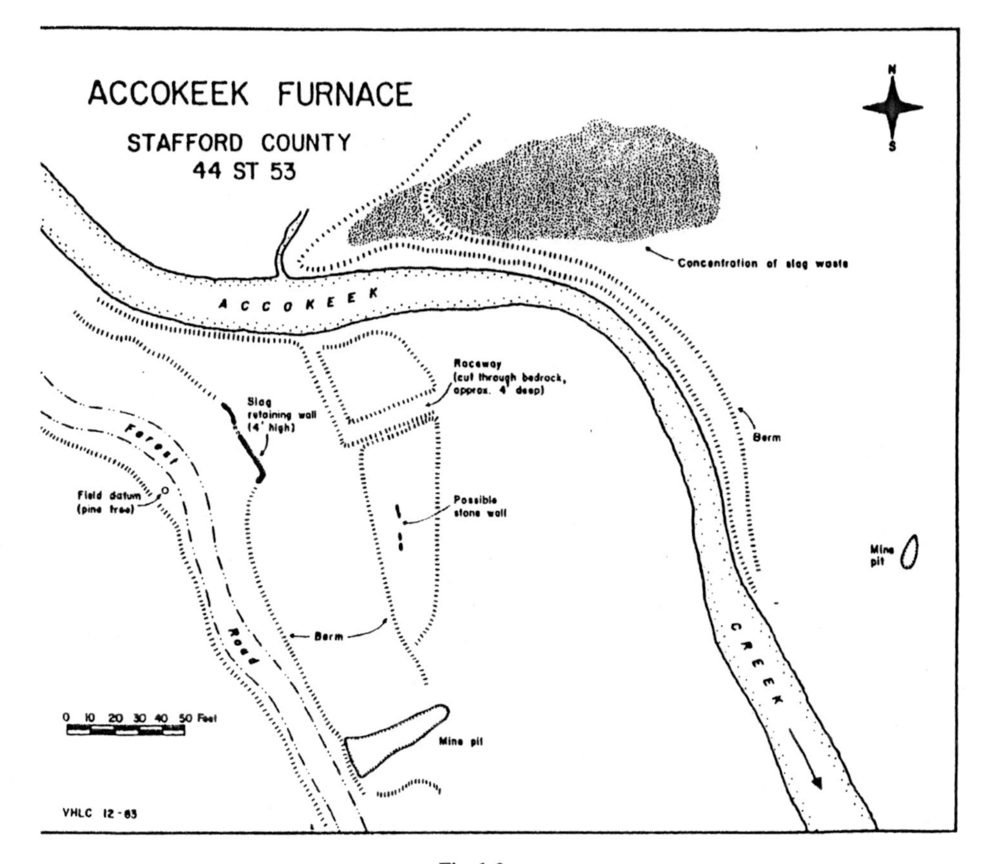

Fig. 1-9
The Accokeek Furnace Archeological Site
Courtesy of the Department of Historic Resources

The Ohio Company

In 1748 King George II was presented with a petition from John Hanbury (died 1758), London merchant, on behalf of himself and Thomas Lee (1690-1750), Thomas Nelson (1716-1782), Col. Thomas Cresap[86] (1694-1790), Col. William Thornton,[87] William Nimmo (died c.1749), Daniel Cresap[88] (1728-1798), John Carlyle (1720-1780), Lawrence Washington (c.1716-1752), Augustine Washington (c.1718-1762), George Fairfax (c.1724-1787), Jacob Giles,[89] Nathaniel Chapman (1709-1760), and Joseph Woodrup who sought to settle the "ohio Country" and extend British trade beyond the western frontier of Virginia. They also sought to halt French claims to this same land (Mulkearn vii). The king ordered the governor of Virginia to grant the petitioners half a million acres, 200,000 of which were to be located immediately. This portion was to be held rent-free for ten years provided the company seated 100 families there within seven years and built a fort to protect them. Originally considered to be within the bounds of Virginia, this enormous tract included land on both sides of the Ohio River, between the mouths of the Monongahela and Kentucky rivers as well as parts of what are today Kentucky, West Virginia, Ohio, and Pennsylvania.

The purpose of the Ohio Company was to secure a share of the lucrative Indian trade west of the Alleghenies that had been profitably exploited by the Pennsylvanians and French. The members also planned to build forts, establish settlements, construct roads, and otherwise develop the country. England's concern over French settlement and control in the Ohio Valley forced Britain to adopt an energetic western frontier policy.

Leadership of the Ohio Company was comprised of the regions wealthiest and most prominent citizens of London and Virginia. Other members who joined the company after the initial 1748 petition included Gov. Robert Dinwiddie (died 1770), George Washington (1732-1799), George Mason (1725-1792), John Mercer (1704-1768), George Mercer (1733-1784), James Mercer (1737-1793), John Francis Mercer (1759-1821), Richard Lee[90] (1726-1795) of Lee Hall, Thomas Ludwell Lee (1730-1778), Philip Ludwell Lee (1727-1775), Robert Carter[91] (1728-1804) of Nomini, Col. John Tayloe (1721-1779) of Mt. Airy, Gawin Corbin (died 1760) of Pecatone, Rev. James Scott (1699-1782), Presley Thornton[92] (1721-1769), Arthur Dobbs[93] (1689-1765), and Lunsford Lomax[94] (1705-1771).

---

[86] Col. Thomas Cresap was one of Maryland's most prominent frontiersmen and land speculators. The *Maryland Gazette* of Apr. 29, 1756 reported that Thomas and Daniel Cresap, sons of Col. Cresap, "with others, dressed as Indians and attacked the women and children at the Indian towns while their warriors were out doing the same to our settlements."

[87] Possibly William Thornton (born 1717), the son of Francis Thornton (1692-1737) of Gloucester County. William removed from Gloucester to Brunswick County, which he represented in the House of Burgesses in 1756-1759, 1761-1762, 1765, and 1767-1768. He married Jane Clack (born 1721).

[88] Daniel Cresap was born in Havre de Grace, Maryland and was the son of Col. Thomas Cresap and Hannah Johnson. Daniel married first c.1749 Martha (Flint) French, widow of John French. He married secondly in 1750 Ruth Swearingen.

[89] According to a notice placed in the *Maryland Gazette* of Dec. 13, 1759, Jacob was an ironmaster at Susquehanna Iron Works in Maryland. In 1750 he patented land in Maryland that later became the Bush River Furnace.

[90] Richard Lee was also known as "Squire Lee." He represented Westmoreland County in the House of Burgesses.

[91] Robert Carter was the son of Robert Carter (1704-1732) and Priscilla Churchill (1705-1757) of Westmoreland County. Young Robert married Frances A. Tasker (died 1787) of Maryland.

[92] Presley was a partner with John Tayloe (1721-1779) in the Occoquan Iron Works in Prince William County.

[93] Arthur Dobbs was the son of Richard Dobbs, High Sheriff of County Antrim, Scotland and Mayor of Carrickfergus and Mary Stewart. Appointed governor of North Carolina, Arthur left Scotland c.1755 on the ship the *Garland*. He made his voyage in the company of his son, Edward, and his nephews Richard Spaight and Cornet McManus, and another relative, the Rev. Alec Stewart.

[94] Lunsford Lomax was the son of John Lomax (1675-1729) and Elizabeth Wormeley of Essex County. He and his father were two of the original 18 magistrates appointed for Caroline County. Lunsford served as a

From July 1763 the Ohio Company met regularly at Stafford Courthouse. George Mason, a member from 1752 until his death in 1792, was very active in the company, receiving and forwarding supplies for settlers, attending to the surveying of the land, and calling meetings of members for business purposes. A number of these individuals were involved with Accokeek Furnace and their names appear on the accompanying business ledger.

The company surveyed the Ohio River Valley, traded with the Indians, built forts, and established some early settlements. However, the French and Indian War (1754-1763) blocked efforts to settle the west and the French destroyed the company's strongholds in 1754. Further legal wrangling over land titles resulted in the company's economic failure in 1792.

A second Ohio Company, officially called the Ohio Company of Associates, was organized in Boston on Mar. 1, 1786. This company successfully established settlements in the Northwest Territory and remained in business until 1832.

---

burgess for that county from 1742-1751 and from 1752-1755. He was the fourth generation of his family to live at Port Tobago, about 5 miles south of Port Royal. In 1729 Lunsford married first Mary Edwards, the daughter of William Edwards of Williamsburg and Surry County. He married secondly in 1742 Judith Micou of Essex County.

Bibliography

Alcock, John P. Five Generations of the Family of Burr Harrison of Virginia, 1650-1800. Bowie, MD: Heritage Books, 1991.

American Memory, http:// lcweb2.loc.gov, George Washington Papers at the Library of Congress, 1741-1799, Series 4, Letter, George Washington to James Hunter, June 15, 1778, Image 10.

Armstrong, Lyn. Woodcolliers and Charcoal Burning. Horsham, Sussex, England: Horsham House, Ltd., 1978.

Bailey, Kenneth P. The Ohio Company of Virginia and the Westward Movement, 1748-1792: a Chapter in the History of the Colonial Frontier. Glendale, CA: Arthur H. Clark Co., 1939.

Bining, Arthur C. British Regulation of the Colonial Iron Industry. Philadelphia: University of Pennsylvania Press, 1933.

Bining, Arthur C., Pennsylvania's Iron and Steel Industry. Gettysburg, PA: The Pennsylvania Historical Association, 1954.

Bruce, Dr. Kathleen. Virginia Iron Manufacture in the Slave Era. NY: The Century Co., 1931.

Brydon, G. MacLaren. "The Bristol Iron Works in King George County." *The Virginia Magazine of History and Biography*, vol. 42, no. 2, (April 1934), pp. 97-102.

Chapman, Nathaniel. Letter, June 25, 1736. Alderman Library, University of Virginia, Charlottesville, VA.

Chetwynd, William. Letter to John England. Sept. 19, 1725. MS 669, Maryland Historical Society, Baltimore, Maryland.

Chetwynd, William. Letter to John England. Nov. 9, 1726, Emmett Collection, New York Public Library, New York.

Condit, William W. "Virginia's Early Iron Age," *The Iron Worker*, (Summer 1959), vol. 23, no. 3, pp. 1-7.

Coulter, Calvin. "The Virginia Merchant." Diss. Princeton University, 1949.

Crumrin, Timothy. "Fuel for the Fires: Charcoal Making in the Nineteenth Century." *Chronicle of the Early American Industries Association*, (June 1994).

Darter, Oscar H. Colonial Fredericksburg and Neighborhood in Perspective. NY: Twayne Publishers, 1957.

Dibert, James A. Iron, Independence and Inheritance: the Story of Curttis and Peter Grubb. Cornwall, PA: Cornwall Iron Furnace Associates, 2000.

Diderot, Denis. Recueil de Planches sur les Sciences, les Arts Liberaux, et les Arts Mechaniques. Paris, 1763.

Dobson, David. Scots on the Chesapeake, 1607-1830. Baltimore, MD: Genealogical Publishing Company, 1992.

Eaton, David W. Historical Atlas of Westmoreland County, Virginia. Richmond, VA: Dietz Press, 1942.

England, E. Walter. The England Family Genealogy, self-published Sept. 1977. Historical Society of Delaware, Wilmington, DE.

England, James B. "Genealogical Record of the England Family in America Descended from John England and Love his Wife of Staffordshire, England. Also a Short Record of Some of the Bouldin Family Descendants of James Bouldin, of Pencader." Manuscript. Historical Society of Delaware.

Executive Communications. Correspondence, 1774-1920. Accession #36912, Library of Virginia, Richmond, VA:

Memorial of James Hunter to George Wythe. May 31, 1777.

Report of the Governor's Committee. Nov. 15, 1777.

Report of the Governor's Committee. Oct. 4 to Dec. 24, 1779.

Fall, Ralph E., ed. The Diary of Robert Rose. Verona, VA: McClure Press, 1977.

Felder, Paula S. Fielding Lewis and the Washington Family. Fredericksburg, VA: The American History Co., 1998.

Felder, Paula S. Forgotten Companions: the First Settlers of Spotsylvania County and Fredericksburgh Town. Fredericksburg, VA: American History Co., 2000.

Fitzpatrick, John C., ed. The Writings of George Washington. Washington, DC: U. S. Government

Printing Office, 1939.
Ford, Worthington Chauncey. *Wills of George Washington and his Immediate Ancestors*. Historical Printing Club, 1891.
Francklyn, Gilbert. Letter to Anne Washington. May 4, 1764. MS 877, Washington Papers. Maryland Historical Society, Baltimore, Maryland.
Fredericksburg Circuit Court Records:
Dejarnett vs. Suddoth. SC/H/1807/90-18
Lawson vs Tayloe. CR/SC/H/1786/171-5
Fredericksburg Political Arena. Dec. 16, 1828. Advertisement for sale of the Furnace Tract and a slave
Fredericksburg Political Arena. Oct. 24, 1828. Advertisement for the sale of the Furnace Tract belonging to James Alexander
Futhey, J. Smith and Cope, Gilbert. *History of Chester County, Pennsylvania*. Philadelphia: Louis H. Everts, 1881.
Gale, W. K. V. *Ironworking*. Buckinghamshire, England: Shire Publications, 1998.
Gardner, Col. Robert E. *Small Arms Makers: a Directory of Fabricators of Firearms, Edged Weapons, Crossbows, and Polearms*. NY: Crown Publishers, Inc., 1963.
Gillespie, Charles C., ed. *A Diderot Pictorial Encyclopedia of Trade and Industry*. NY: Dover Publications, Inc. 1959.
Gilmer, Maj. Gen. Jeremy. Map of Stafford County, 1863. Virginia Historical Society, Richmond, VA.
Gordon, Robert B. *American Iron: 1607-1900*. Baltimore, MD: Johns Hopkins University Press,1996.
Gordon, Robert B. *A Landscape Transformed: the Ironmaking District of Salisbury, Connecticut*. NY: Oxford University Press, 2001.
Gray, Gertrude E. *Virginia Northern Neck Land Grants*. Vol. 1, 1694-1742. Baltimore: Genealogical Publishing Co., 1993.
Hayden, Rev. Horace E. *Virginia Genealogies: a Genealogy of the Glassell Family of Scotland and Virginia*. Baltimore: Genealogical Publishing Co., 1979.
Heite, Edward F. "Excavation of the Fredericksville Furnace Site." *Archeological Society of Virginia Quarterly Bulletin*, vol. 25, no. 2, (December 1970), pp. 61-97.
Heite, Edward F. "Historic Site Report: Accokeek Furnace Property, Stafford County, Virginia." Camden, DE: Smithsonian Institution, 1981.
Heite, Edward F. "The Pioneer Phase of the Chesapeake Iron Industry: Naturalization of a Technology." *Archeological Society of Virginia Quarterly Bulletin*, vol. 38, no. 3, (September 1983), pp. 133-181.
Heite, Edward F. "Unearthing Business History: Can Archeology Provide Evidence for Interpreting Management Styles?" *Journal of the Society for Independent Archeology*, vol. 18, nos. 1 and 2 (1992).
Hempstead, Joshua. "Cecil County in 1749." *Maryland Historical Magazine*, vol. 49, Baltimore, MD: Maryland Historical Society, (1954), p. 346-350.
Hiden, M. W. "The Hiden Family," *Tyler's Quarterly Historical and Genealogical Magazine*, vol. 24, (1943), pp. 125-147.
Hillman, Benjamin J., ed. *Executive Journals of the Council of Colonial Virginia*, vol. VI, Richmond, VA: Virginia State Library, 1966.
Hudson, J. Paul. "Augustine Washington and the Iron Works." *The Iron Worker*, vol. 25, no. 3, (Summer 1961), Lynchburg, VA.
Hudson, J. Paul. "Iron Manufacturing During the Eighteenth Century," *The Iron Worker*, vol. 21, no. 4, (Autumn 1957), pp. 9-13, Lynchburg, VA.
Hudson, J. Paul. "The Story of Iron at Jamestown, Virginia—Where Iron Objects Were Wrought by Englishmen Almost 350 Years Ago." *The Iron Worker*, vol. 20, no. 3, pp. 2-14, (Summer 1956), Lynchburg, VA.
Hudson, J. Paul. "Augustine Washington: First Stafford Industrialist." *Northern Neck of Virginia Historical Magazine*, vol. 24, no. 1, (Dec. 1974), pp. 2611-2618.
Hunter, Louis. "Heavy Industries Before 1860," in *The Growth of the American Economy*. NY: Prentice Hall, 1949.
Ince, Laurence. *The Knight Family and the British Iron Industry*. Merton, England: Ferric Publications,

1991.
Jackson, Donald, ed. The Diaries of George Washington. Charlottesville, VA: University Press of Virginia.
Jefferson, Thomas. Notes on the State of Virginia. NC: University of North Carolina Press, 1955.
Johnston, George. The History of Cecil County, Maryland. Baltimore: Genealogical Publishing Co., 1998.
"Journals of the Council of Virginia in Executive Sessions, 1737-1763," *Virginia Magazine of History and Biography*, vol. 14, (1907), p. 236.
Kemper, Jackson. "American Charcoal Making in the Era of the Cold-Blast Furnace." United States National Park Service, 1941.
King George County Court Records:
WB 1721-1744, p. 200—Will of John England
King, George H. S. The Register of Overwharton Parish, Stafford County, Virginia, 1723-1758. Fredericksburg, VA, 1961.
Lease, Augustine Washington to Principio Iron Company, July 24, 1726, Maryland Historical Society.
Lee, William. Letter to John Augustine Washington. Dec. 15, 1774, MS 877, Washington Papers, Maryland Historical Society, Baltimore, MD.
Lewis, Ronald L. Coal, Iron, and Slaves: Industrial Slavery in Maryland and Virginia, 1715-1865. Westport, CT: Greenwood Press, 1979.
Lewis, Ronald L. "Slavery in the Chesapeake Iron Industry, 1716-1865." Diss. The University of Akron, 1974.
Little, Lewis P. Imprisoned Preachers and Religious Liberty in Virginia. Lynchburg, VA: J. P. Bell Co 1938.
MacMasters, Thomas E. The History and Technology of Governor Spotswood's Iron Furnaces in Spotsylvania County, Virginia, 1721-1792. Los Angeles, CA: University of California, 1974.
Maryland Gazette, May 27, 1746, William Williams reported a runaway servant
Maryland Gazette, Sept. 16, 1746, notice of Ralph Falkner for bar iron and a billiard table
Maryland Gazette, Aug. 9, 1781, notice of the sale of Principio Iron Works
May, Earl C. "Principio to Wheeling, 1715-1945," A Pageant of Iron and Steel. NY: Harper and Brother, 1945.
McIlwaine, H. R., ed. Journals of the House of Burgesses of Virginia, 1727-1740, petition of the inhabitants of Opekkon, Richmond, VA: Virginia State Library, 1905, p. 330.
Meima, Ralph C. Spotswood's Iron: the Story of the Birth of the Industrial Revolution in the New World. Fredericksburg, VA: Chancellor's Village Press, 1993.
Middleton, Arthur P. Tobacco Coast: a Maritime History of the Chesapeake Bay in the Colonial Era. Newport News, VA: Mariner's Museum, 1953.
Moller, George D. American Military Shoulder Arms, vol. 1: Colonial and Revolutionary War Arms. 1993.
Northern Neck Land Grants, Library of Virginia, Richmond, Virginia:
Grant Book 3, p. 266, July 21, 1710, patent of Thomas Leachman, 632 acres
Grant Book 4, p. 97, June 30, 1712, patent of John Spry, 278 acres
Grant Book 4, p. 103, July 30, 1712, patent of Michael Dermot, 248 acres
Grant Book A, p. 143, May 10, 1725, patent of Augustine Washington, 349 acres
Grant Book B, p. 49, Mar. 3, 1726, patent of John England, 430 acres
Grant Book A, p. 218, Aug. 12, 1726, patent of Col. John Lomax, 200 acres
Grant Book A, p. 219, patents of Augustine Washington, 396 and 4,360 acres
Grant Book B, p. 91, July 12, 1727, patent of John England, 197 acres
Grant Book B, p. 108, Jan. 26, 1728, patent of John England, 333 acres
Grant Book B, p. 124, May 23, 1728, patent of John England, 156 acres
Grant Book B, p. 122, May 29, 1728, patent of John England, 128 acres
Grant Book C, p. 52, Mar. 25, 1730, patent of John England, 428 acres
Grant Book E, p. 394, Jan. 11, 1741, patent of Nathaniel Chapman, 120 acres
Survey, Feb. 1, 1726, to John England for 430 acres
Survey, Feb. 22, 1728, to John England for 156 acres
O'Dell, Cecil. Pioneers of Old Frederick County, Virginia. Marceline, Missouri: Walsworth Publishing

Company, 1995.
Overman, Frederick. The Manufacture of Steel. Philadelphia: A. Hart Publishing Co., 1851.
Overman, Frederick. Mechanics for the Millwright, Machinist, Engineer, Civil Engineer, Architect, and Student. Philadelphia: Lippincott, Grambo and Co., 1851.
Pease, John B. A Concise History of the Iron Manufactures of the American Colonies up to the Revolution and of Pennsylvania Until the Present Time. NY: Burt Franklin, 1878.
Peden, Henry C. Inhabitants of Cecil County, Maryland, 1649-1774. Lovettsville, VA: Willow Bend Books, 1993.
Pennsylvania Gazette, May 11, 1738, advertisement for Accokeek iron
Principio Collection, Joseph S. Wilson Collection, 1724-1903, Historical Society of Delaware:
Accokeek Furnace Journal/Day Book, 1748/9-1760
Ledger 1728, "This Ledger for Company in England"
North East Forge Journal, 1745-1746, 1748-1750
North East Waste Book, 1754-1755, 1772-1773
Kingsbury Furnace Journal, 1768-1770
Kingsbury Furnace Inventory, 1781
Principio Ledgers—1728, 1731, 1734, 1737-1738, 1739-1740, 1754-1755.
Principio Company Records Concerning Augustine Washington, 1723-1769. Accession #29453, Business Records Collection, The Library of Virginia, Richmond, Virginia.
Principio Partners. Letter to John England. Sept. 15, 1725. MS 669, Maryland Historical Society, Baltimore, MD.
Principio Partners. Letter to Nathaniel Chapman. Sept. 11, 1758. Accession #LC 29600, Library of Congress, Washington, DC.
Principio Partners. Letter to Mrs. Anne Washington. Feb. 27, 1771. MS 877, Washington Papers, Maryland Historical Society, Baltimore, MD.
Principio Partners. Undated Statement. Accession #29453, Alderman Library, University of Virginia, Charlottesville, VA.
Revolutionary War Pension Applications, National Archives and Records Administration: James Arrowsmith, W5643
Reynolds, John. Windmills and Watermills. NY: Praeger Publishers, 1970.
Richmond County Court Records:
DB 8-488-491—May 7, 1729—agreement—Lyonel Lyde, John King et al to John Tayloe
Richards, William A. Forging of Iron and Steel. NY: Van Nostrand Co., Inc., 1915.
Robbins, Michael W. "Maryland's Iron Industry During the Revolutionary War Era." Report prepared for the Maryland Bicentennial Commission, June 1973.
Robbins, Michael W. The Principio Company, Iron Making in Colonial Maryland, 1720-1781. NY: Garland Publishers, 1986.
Ruston, Thomas. Papers, Principio Iron Works Law Suit Legal Copy Book, "Lands Taken up for the Account of the Accokeek Company with the Persons Names of Whom Bought or Leased," Library of Congress.
Ruston, Thomas. Papers. Deed from Allen and Joseph England to John Price. Dec. 31, 1741. Library of Congress, Washington, DC.
Schallenberg, Richard H. "Evolution, Adaptation, and Survival: the Very Slow Death of the American Charcoal Iron Industry," *Annuls of Science*, vol. 32, (1975), pp. 341-358.
Scribner, Robert L. and Tarter, Brent. Revolutionary Virginia: the Road to Independence. Charlottesville, VA: University Press of Virginia, 1981.
Smith, Cyril S., ed. Sources for the History and Science of Steel, 1532-1786. Cambridge, Massachusetts: Society for the History of Technology, 1968.
Spoede, Robert W. "William Allason: Merchant in an Emerging Nation." Diss. College of William and Mary, 1973.
Stafford County Deeds and Wills:
DB Z-130—Jan. 9, 1701/2—deed—Thomas Norman to John Adams
DB Z-129—Jan. 24, 1701/2—deed—Thomas Norman to John Russell
DB Z-235—July 13, 1704—deed—Thomas Norman to John Adams
DB J-284—July 11/12, 1726—lease—Richard Brooks to John Tayloe

WB O-89—Jan. 31, 1750—will of George Williams
WB O-114—July 10, 1750—inventory of George Williams
DB S-1—Nov. 1, 1777—report of the escheat hearing concerning the Accokeek Furnace tract
DB S-176--___, 12, 1784—deed between William Routt, Sr., William Routt, Jr. and John James
DB S-283—Apr. __, 1785—inventory of the estate of James Hunter
DB S-384—May 8, 1786—deed—George Brent, Esq. to Robert Stuart
DB GG-82—Feb. 13, 1826—deed of trust—William Foushee to R. C. L. Moncure
DB GG-186—June __, 1826—deed—R. C. L. Moncure to William (Robert?) Dunbar
DB GG-187—July 1, 1826—deed—William (Robert?) Dunbar to James Alexander
Scheme Book Court Orders, 1790-1793, Oct. 10, 1791, p. 89, will of George Williams

Stafford County Land Tax Records, 1782-1861, 1867 to present. Stafford Courthouse, Stafford, VA.

Stell, Geoffrey P. and Hay, Geoffrey D. Bonawe Iron Furnace. Edinburgh, Scotland, 2000.

Swank, James M. History of the Manufacture of Iron in All Ages and Particularly in the United States from Colonial Times to 1891; also a Short History of Early Coal Mining in the United States and a Full Account of the Influence Which Long Delayed the Development of all American Manufacturing Industries. Philadelphia: The American Iron and Steel Assn., 1892.

Thomson, Robert P. "The Merchant in Virginia, 1700-1775." Diss. University of Wisconsin, 1955.

United States Department of the Interior, National Park Service. "Hunter's Iron Works." 1971.

Virginia Colonial Records Project, Library of Virginia, Richmond, Virginia:

House of Lords Papers, MS Lists, 1766-1774,. "Account of the Quantity of Iron Imported from the Colonies in America from Christ. 1710 to Christ. 1749." Survey Report #3101, Reel #597

Maps from the Colonial Office Library Concerning Virginia, 1698-1791, "The Courses of the Rivers Rappahannock and Potowmack in Virginia, as Surveyed According to Order in the Years 1736 and 1737." Survey Report #0908, Reel #497

Virginia Gazette, June 13, 1766, Samuel Duval's notice of the sale of mineral coal.

Virginia Herald, Jan. 23, 1830, obituary of John Macrae.

Virginia Historic Landmarks Commission. "Accokeek Iron Works: National Register of Historic Places Inventory," 1983.

Virginia Legislative Petitions, Petition of William Augustine Washington and Thomas Russell, Dec. 7, 1783, Library of Virginia, Richmond, Virginia.

Vogt, John and Kethley, T. William. Stafford County Tithables: Quit Rents, Personal Property Taxes and Related Lists and Petitions, 1723-1790. Athens, GA: Iberian Publishing Co., 1990.

Ward, Harry M. and Greer, Harold E. Richmond During the Revolution, 1775-1783. Charlottesville, VA: University Press of Virginia, 1977.

Watkins, C. Malcolm. The Cultural History of Marlborough, Virginia. Washington, DC: Smithsonian Institution Press, 1968.

Watson, Aldren A. The Blacksmith: Ironworker and Farrier. NY: W. W. Norton and Co., 1990.

Wayland, John W. The Washingtons and their Homes. Staunton, VA: McClure Printing Co., 1944.

Whitely, Henry. "The Principio Company: a Historical Sketch of the First Iron Works in Maryland." *Pennsylvania Magazine of History and Biography*, vol. 11 (1888), pp. 63-68, 190-198, 288-294.

Wightwick, John. Letter to John England. Oct. 2, 1730. MS 669, Maryland Historical Society, Baltimore, MD.

Wilson, Samuel M. The Ohio Company of Virginia, 1748-1798. Lexington, KY: 1926.

Withers, Robert E. Withers Family of the County Lancaster, England and of Stafford County, Virginia Establishing the Ancestry of Robert Edwin Withers, III. Richmond, VA: Dietz Printing Co., 1947.

Wood, Ray. "The Accokeek Furnace Site: An Assessment." Washington, DC: Engineering Science, Inc. July, 1989.

Wright, Louis B., ed. The Prose Works of William Byrd of Westover: Narratives of a Colonial Virginian. "A Progress to the Mines." Cambridge, Mass.: Belknap Press, 1966, 339-378.

Contributors:

Steven T. Bashore
Carmen Baxter
Elmer Biles
Turner and Estherleen Blackburn
James Brothers
James Courtney
Ruth Daiger
Karen Dale
Ralph and Janet England
Terri Furguiele
Rick Gamble
Francis Gill
George L. Gordon
Dr. Warren R. Hofstra
Dr. Peter King
Barbara Kirby
Alaric R. MacGregor
Patrick McCurdy
Barry McGhee
James Moncure
Mary Catherine O. Moncure
Elgin and Susan Perry
Tom Peterson
Stephen L. Ritchie
Ben Ritter
Dr. Michael W. Robbins
Walter V. "Pete" Roberts
William G. Scroggins
John A. Washington

Chapter 2

Rappahannock Forge

Historians have largely overlooked and underestimated the scope and importance of James Hunter's Rappahannock Forge, an 18th century manufacturing center located just above Falmouth on the Rappahannock River. This is likely due not only to the loss of most 18th century records in both King George and Stafford counties, but also to the loss of many early Virginia records of that period. The few brief articles that have been written about Rappahannock Forge have been based primarily on local oral tradition combined with a smattering of references from sources such as the minutes of the House of Burgesses and the Council of the State of Virginia. With the assistance of Walter V. "Pete" Roberts, a noted firearms authority, the author has located a wealth of previously untapped resources that prove Rappahannock Forge to have been a far more extensive operation than previously believed. These records also prove that, contrary to oral tradition, Rappahannock Forge continued operating long after Hunter's death.

Some of the author's theories about the construction of and production at Rappahannock Forge are admittedly speculative, the loss of many 18th century records making it difficult to document all facets of Hunter's business. Oral tradition has long held that James Hunter personally financed the construction of every building at his forge as well as tons of firearms, swords, naval hardware and stores, meal, flour, wool and cotton cloth and many other supplies utilized by the military during the Revolution. It is the author's contention that in 1776 and 1777 the state of Virginia invested substantial sums of money to enlarge and improve the physical plant at Rappahannock. The purpose behind this investment was to increase the production capabilities of Hunter's diverse manufacturing center. With the outbreak of the Revolution, Virginia authorities were desperate to capitalize on this already established and productive manufactory while simultaneously building new factories focused solely on the making and repairing of firearms. Contemporary accounts also state that prior to and during the Revolution Rappahannock Forge was one of the largest manufacturing centers in North America; items made there ranged from domestic goods such as wool combs and farm tools to military supplies such as firearms and anchors. Sadly, the fate of the forge records is unknown. Additionally, many of the state records pertaining to orders, payments, and other communications between the forge and the Virginia government are missing. On Feb. 7, 1781 Thomas Jefferson admitted to Hunter that there were problems with record keeping at the state level. He wrote, "By the loss of many of our papers we find ourselves unable to say how stand our orders with you for camp-kettles" (Boyd, vol. 4, p. 551). There were obviously many orders for supplies placed with Rappahannock Forge, yet only a handful may be documented (See Appendix A). Based upon the sheer size, capabilities, and resources available at Rappahannock, there must have been a great deal of correspondence between forge management and the Virginia authorities, yet only a fraction of these documents survive. How is it then that so few people are aware of Rappahannock and so little research has been done on it? Rappahannock Forge faded from memory not because of insignificance, but because many easily located references to it have been lost or destroyed since the plant's closing nearly 200 years ago. This, coupled with extensive Union vandalism that removed most traces of the buildings at Rappahannock Forge, has resulted in a general ignorance among modern historians with regards to the role the forge played in colonial Virginia and the American Revolution and in the area's overall economy.

The Revolutionary period teemed with hundreds of unsung heroes whose names are forever lost and their patriotic sacrifices long forgotten. Many of these men willingly relinquished all they owned for the cause of freedom, yet no markers stand to commemorate their selfless acts. Their names are not recorded in schoolbooks to be memorized by younger generations. For many of these men, their staunch convictions resulted not in glory and fame, but in financial ruin and disgrace. James Hunter was one such individual who has been largely forgotten. Faced with a government unable to fully pay for essential war materials, Hunter invested his own fortune and borrowed additional money to keep his manufactory afloat and productive. He provided the Continental Line with all manner of supplies from camp kettles to muskets, wool uniforms to swords. His unselfish contributions to the war effort helped ensure the liberty we take for granted today.

Rappahannock Forge opened c.1759 largely in response to the closing of Accokeek and the reduction in production at Tubal furnace. Hunter's initial efforts were directed to meeting local domestic need, there being no other forge in the immediate area.[95] Long before open hostilities erupted between America and England, Hunter had established a major manufacturing center capable of producing a wide array of essential consumer goods. With the outbreak of the Revolution, it was a relatively simple matter to shift from the manufacture of domestic goods to the production of war materiel.

The loss of many official records and all of the forge buildings accounts for much of the misinformation that has been published by otherwise scholarly authors. From the time of its opening, this complex was generally known as "Hunter's Works." The name "Rappahannock Forge" came into use during the Revolution and was stamped on firearms produced here. These interchangeable names have led some to believe that Hunter's Works and Rappahannock Forge were two separate facilities. This, of course, was not so. There are also several highly respected books dealing with Revolutionary arms manufacture that confuse Rappahannock Forge with the Fredericksburg Manufactory, an armory that operated concurrently with Hunter's Works. There is confusion about the two James Hunters who were related and lived within a few miles of one another.[96] There is also confusion about the variety of products manufactured at Rappahannock. Finally, there have been few attempts made to determine the names and backgrounds of the hundreds of people employed at the forge. Due to the magnitude of the facility, Rappahannock had a profound effect on the region's economy and employed a considerable percentage of the local population.

Today the Rappahannock Forge site is heavily timbered. Several steep ravines break the high ridge that parallels the north side of the river. This ridge rises sharply very near the riverbank, leaving little ground suitable for building. A narrow strip of relatively level terrain follows the river's north bank and it was here that Hunter built his remarkable forge complex. Its rugged beauty would strike most visitors who wander up the river to this site. The hawk screams overhead. Water rushes over Embrey Dam and there is a serenity here that can be found only along a river. The visitor here will not likely notice the numerous barely visible building foundations now secluded by vines and leaves. He might not even recognize the remains of Hunter's canal, thinking it only an old roadbed if he pondered it at all.

How different was the scene 225 years ago when the forge was at its peak. The air was filled with the shouts of men, trying to talk above the constant rush of water and the splashing, groaning, and creaking of the great water wheels. The ground shook with the rapid, incessant pounding of trip hammers in the forge, plating, and fulling mills. The ringing and thudding of these hammers could be heard three or more miles away and was deafening on site. When the iron furnace was in blast, the great double bellows forced a continual stream of air into the base of the massive stone structure, resulting in a constant, guttural roar. Horses screamed and mules brayed as they stood in harness, slaves loading wagons with heavy wooden crates destined for all corners of the colonies. Gunshots rang out as inspectors checked firearms manufactured in the armory. Everywhere was heard the sound of hammering as the blacksmith forged tools, the wagon maker nailed planks to a flatbed and the millwright made repairs to a leaky flume. The ringing of the stonemasons' chisels was heard from the quarries that dotted the river bank. Blocks of granite cut from these quarries were loaded on batteaus and floated down Hunter's canal for use in new forge buildings. The few acres upon which were clustered more than 30 industrial buildings fairly teemed with humanity as several hundred people carried out their daily responsibilities. Then, there were few if any trees on the steep hillside that rose above the forge. Along the top of the ridge hundreds of tiny, primitive cabins overlooked the river and provided simple shelter for the many people who worked in the forge shops and on the plantation. And over all of this was the stench. The air was filled with smoke from cooking fires and

---

[95] Sometime prior to 1764, a bloomery/foundry was established in the lower part of Fredericksburg. Although it employed some 100 people and was valued at between £25,000 and £50,000, it operated only a few years. In 1772 it was converted to a brewery under the management of William Fitzhugh (1741-1809).

[96] For many years there were two James Hunters living in the Stafford/King George/Spotsylvania area. James Hunter (1721-1784), builder of the iron works, was the uncle of the second James Hunter (1746-1788). In an effort to distinguish between the two, local officials usually styled the elder James as "James Hunter, Sr." or "old James." The younger James Hunter, son of James, Sr.'s uncle William Hunter (died 1753), usually appeared in the records as "James Hunter, Jr." or "Jamie Hunter." For the sake of clarity, in this text the elder James Hunter will be referred to by his given name. The younger James will be referred to as Jamie.

soot from charcoal making, the iron furnace, forge, steel mill, and blacksmith shop. From the tanyard emanated the rancid odor of curing hides. The smell of sweat was everywhere as black slaves and white servants labored without pause at their various industries. On cool damp days, when the air was still and the smoke filled the valley of the Rappahannock, visibility was nearly zero. A continual stream of effluents flowed into the river, raw sewage from the scores of workers and livestock and run-off from the tanyard.

On the high ground above the river, the land was level and suited to agriculture. This was James Hunter's Stanstead. Unlike today, there were very few trees on the forge property and many hundreds of acres were in cultivation. Black workers plowed the fields, harvested the crops, sheared sheep and butchered hogs in order to feed and clothe the people laboring at the forge and working on the plantation. The sounds of African chants and songs wafted through the air. Cattle lowed to one another as they grazed in the fields; hogs grunted as they rooted for acorns in the damp woods. Along the far edges of the plantation could be heard the sounds of axes and crashing timber as scores of men cut trees destined to become charcoal to power the furnace, forge, steel mill, brass foundry, and blacksmith shops. Carts and wagons, men, mules, oxen, and horses were in constant motion, transporting wood, vegetables, stone, charcoal, and a myriad of other supplies from one place to another.

Only at dusk did a hush fall over the forge and plantation as weary workers found their way to their tiny cabins. Only then did stillness reign. The soothing notes of a dulcimer drifted through the air as exhausted men spent a few precious hours with their wives and children. Oil lamps or candles palely illuminated a few cabins. In the summer the songs of whip-poor-wills and night insects filled the night and brought a sense of peace. At first light, it would all begin anew.

The builder of this great complex was James Hunter (1721-1784). He was the son of James Hunter[97] (1693-1747), merchant of Duns, Scotland and Helen Simpson. In his late teens or early twenties, James began making trips to Virginia, helping manage the family business. Like many 18th century Scots families, the Hunters were engaged in the mercantile business. They shipped tobacco from the colonies and rum, sugar, and molasses from the islands. They sold and traded English-made goods for raw materials produced in the colonies.

James' uncle, William Hunter[98] (died 1753), established himself in Fredericksburg as early as 1736, one of the earliest Scottish merchants to arrive in the area. In 1741 William purchased from the estate of Henry Willis (1692-1740) a river front lot in Fredericksburg adjacent to the public warehouses of John Allen (died 1750) and Nathaniel Chapman (1709-1760) of Accokeek Furnace. On this property he kept his home, a tavern, and ferry, and utilized two existing warehouses for his business. At the same time he purchased a 30-acre meadow well behind his waterfront lot that was later sold by his heirs (Felder, Companions 111, 169; Spotsylvania Will Book B, p. 185). James Hunter settled permanently in Virginia around 1746 with the intention of helping his ailing uncle manage the mercantile firm in Fredericksburg. James proved to be an able businessman and increased his fortune handsomely.

During the late 1740s and early 1750s, James invested in large tracts of land in the counties surrounding Fredericksburg including King George, Stafford, Spotsylvania, Culpeper, and Fauquier. On the north side of the Rappahannock he bought Stanstead, Charles Carter's (1707-1764) 3,900-acre plantation, part of which is now Servicetown Truck Stop. This tract was part of some 13,000 acres assembled and patented by Robert "King" Carter (1663-1732) and given to his son Charles. Until the boundary change of 1777, this land was part of King George County (Hening, vol. 9, 244-245). Charles lived on the farm for a number of years before moving further down in King George where, by 1750, he settled on a new plantation he called Cleve.

Although Charles Carter had built a comfortable home on Stanstead, James Hunter probably had to build his own dwelling after purchasing the farm. William Waller Hening wrote of Charles Carter and his house:

> He was distinguished by the appellation of Blaze Charles Carter from a remarkable redness in his face and was more frequently called Old Blaze than by any other name. He died in Fredericksburg about three or four years ago, totally insolvent. That no person

[97] This James Hunter did not settle in Virginia, but remained in Scotland as head of his mercantile house.
[98] In 1744 William married Martha Taliaferro, the daughter of Col. John Taliaferro (1687-1744) and Mary Catlett of Spotsylvania. Martha's brother, William (1726-1798) was in business with James Hunter, Sr.

since his death has undertaken the management of his affairs is easily accounted for, because he left nothing to manage. He had once possessed a very considerable fortune but from the moment he came to the enjoyment of his estate he began to spend it in various chimerial [sic] projects. Among these may be reckoned his building a very costly house on one estate and in a short time after it was completed taking it down and carrying it to another. His affairs became so involved that about the peace he conveyed his whole estate in trust to Gavin Lawson[99] of Stafford and others for the purpose of paying his debts...So great has been the reverse of fortune on him that for some years before his death he had been compelled to take in boarders at an academy at Fredericksburg in order to provide the means of support ("British Mercantile Claims," vol. 28, 224-225).

There are no surviving deeds among the King George County records documenting Hunter's purchase of Stanstead; however, the King George quit rent rolls of 1771 list James as paying rent for an 870-acre tract plus 3,900 acres purchased "of Carter." Based upon a mid-19th century drawing,[100] the Stanstead house was a large rectangular 2 ½-story home with a chimney on each end and three dormers across the front of the gabled roof. From the drawing it is not possible to determine whether the house was of brick or frame construction, but it seems to have been destroyed during the Union occupation of 1862-63.

James Hunter first appeared in the Stafford County records in 1768 when he paid quit rents on 1,176 acres purchased from James Baxter.[101] Exactly when or from whom he acquired the tract is unknown. There is no surviving deed for this transaction, but the land tax records reveal that this property was on the northeast side of what is now Abel Lake and adjoined the old Accokeek iron mines. By 1782, when the county began collecting land taxes instead of quit rents, he was listed as owning the 200-acre Rappahannock Forge tract (assessed at an astounding £25 per acre) and 1,870 acres assessed at £13 per acre. A comparison of Hunter's property values with those of his neighbors reveals that his was, indeed, prime real estate; most assessments in the vicinity of the forge averaged £4 to £8 per acre. In 1782 Hunter also owned two lots in Falmouth and his brother Adam owned one.[102] James was listed twice on the 1783 Stafford personal property tax rolls. The first entry listed his personal slaves and livestock and included 104 slaves, 20 cattle, 71 horses, and a four-wheeled vehicle. This was one of only ten four-wheeled carriages owned by county residents that year. His forge holdings were listed separately and included 156 Negroes, 23 horses, and 10 cattle. During his lifetime, James was one of Virginia's largest slave holders.

During the mid-1750s, James also purchased over 1,200 acres in Culpeper County. On June 26, 1754 he paid William Russell[103] of Culpeper £44.18.1 for 388 acres in the great fork of the Rappahannock, this land described as being bounded by the lands of Mutton Lewis, Samuel Ball,[104] and Joseph Cooper

[99] Gavin Lawson (c.1745-1805) was a Scottish merchant from Lanarkshire. He arrived in Virginia sometime prior to 1776 and settled initially in Culpeper County. Gavin married in King George County Susanna Rose (1749-1825), the daughter of the Rev. Robert Rose (1704-1751) and his second wife Anne Fitzhugh (born 1721). He lived in Stafford before moving to Geneva, New York. His name appears frequently in the early records of most of the surrounding counties. He bankrupted shortly before his death.

[100] Drawing in the collection of D. P. Newton, White Oak Museum, Stafford, Virginia.

[101] James Baxter was the son of William Baxter (died 1749) whose will is on file in Stafford County. James was employed at Accokeek Furnace and traveled frequently to Williamsburg, Philadelphia, and various places in Maryland conducting business for the Principio Iron Company.

[102] From 1782-1800 James and Adam Hunter paid land taxes on lots in Fredericksburg. In 1782 James owned 3 ½ lots valued at £155. Adam owned a 1/5 interest in one lot that interest valued at £8. Adam's last entry was in 1783, though the tax records skip from that year to 1787. From 1789-1800 James' estate owned lots 274 and 275, which were rented out at various times to Mrs. Mary Sullivan, Robert Crookshank, Mrs. John Baggott, Jacob Grotz, and John Atkinson. The latter had been Alexander Hanewinkle's partner in the hemp business in Fredericksburg. On June 20, 1800 Patrick Home, James' last surviving executor, sold the lots to Mary Sullivan.

[103] Col. William Russell (c.1699-1757) of Culpeper.

[104] Samuel Ball (1686-1751) was the son of Capt. William Ball (1641-1694) of Lancaster County. Samuel moved to Culpeper where he was one of the vestry of St. Marks' Parish. In 1717 he married Ann Catherine Tayloe, the daughter of Col. William Tayloe of Richmond County.

(Culpeper Deed Book B, p.146). On Oct. 16, of that year he paid George Hume £28.10.4 for 200 acres bounded by the lands of John Spotswood's heirs and William Stanton (Culpeper Deed Book B, p.306). Three years later he purchased 642 acres in the great fork of the Rappahannock on the north side of Mountain Run and adjoining Robert Spotswood and Cabin Run. He paid £1,636.16.10 for this tract (Culpeper Deed Book C, p. 74).

Iron production had long been an important element in the Virginia economy. As an industry, it had witnessed a phenomenal growth from the earliest production attempts at Jamestown and Falling Creek to the many pre-Revolutionary furnaces that dotted the colonial Virginia landscape. Michael D. Thompson wrote, "On the eve of the American Revolution, the iron industry of the thirteen colonies was producing annually in excess of 30,000 tons of pig iron. The size of the industry in the North American colonies far exceeded anything the British could muster" (Thompson 18) and by 1775 the American colonies were the world's third largest producer of iron, following behind Russia and Sweden (Bining, British 122). Virginia and Maryland jointly were the largest producers in the colonies, regularly out-producing the other colonies by a ratio of ten to one. In 1771 the American colonies produced some 30,000 tons of iron, but only 7,000 tons were sent to England. The remaining 23 tons were used domestically in finished goods specifically forbidden by English law. Bining estimated that at the outbreak of the Revolution there were 257 furnaces in the colonies, one-fourth of these in Maryland and Virginia and another one-fourth in Pennsylvania, though the latter colony had no works that could compare in production with Principio or Baltimore. As America plunged forward into the Revolution, her iron producing capabilities exceeded those of England, though she fell far short of the Mother County in terms of manufacturing. Thompson wrote of the American iron industry, "The growth of the North American iron industry in the period 1715-1775 rates with the great economic miracles of history up until that time. It placed the colonies in a position of being at least potentially the equal to Great Britain in the size, quantity, and quality of arms that both armies could field during the American Revolution" (Thompson 18).

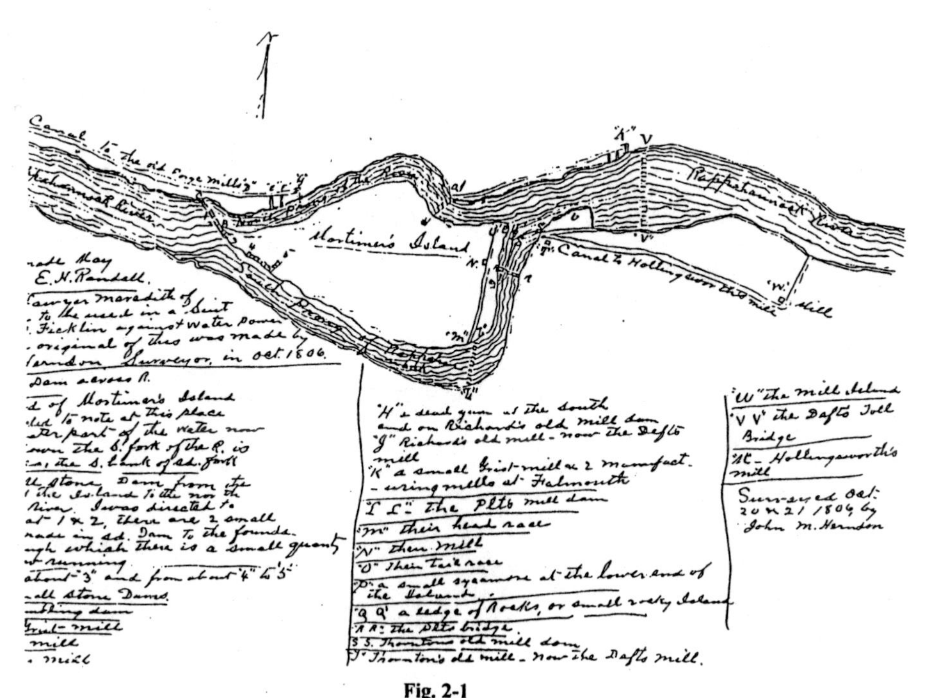

Fig. 2-1
Survey of the old forge site dated Oct. 20 and 21, 1806. Note canal and three raceways at E, F, and G (Stafford County Loose Surveys).

James Hunter was a merchant and businessman, not an iron master and he learned about iron manufacturing through trial and error. Shortly after establishing his iron works on the Rappahannock, he became involved with another iron furnace in Frederick County, Maryland. This second operation was called Fielderia Furnace and will be discussed later. James' early experiences at Fielderia were neither positive nor profitable; the furnace never produced enough iron to pay expenses. His financial rewards from Rappahannock were also limited despite the fact that operations there were far more diversified. One is left to wonder if Rappahannock would have been economically successful had it not been for the unpaid government contracts dating from the Revolution that bankrupted so many other arms producers.

Based upon a cursory examination of the Rappahannock Forge site, there appear to have been two furnaces there, though one was likely a replacement for the other. Foundation #X13 on the accompanying site map appears to be the early furnace site. A few oyster shells and bits of charcoal remain around the site, as does the sand casting bed and a considerable quantity of dross, a waste product of smelting. To the immediate west of this site is a small branch that probably powered the bellows. Today, this branch carries very little water, probably due to changes in drainage patterns brought about by development along Route 17. For some reason, a second furnace (Foundation #X7 on the site map) was built c.1777 (Shelley 417). This site shows a great deal more activity and is heavily littered with oyster shells, charcoal, cinders, dross, and iron scraps. The black sand floor used for casting is also still present as is the branch to the immediate west that likely provided head for the bellows.

The furnace indicated by #X7 on the site map shows evidence of heavy use. A road, which is still visible, ran almost due south from the furnace to the river. Traces of a second dam remain on the north side of the river as well as on the edge of Vicaris'/Lauck's/Hunter's/Beck's/Mortimer's Island.[105] The furnace road led to this dam and, at some point, was used as a pubic crossing to the island.

All iron furnaces produced a large quantity of dark green, glassy waste material known as slag. This was tapped from the furnace, allowed to harden, then broken up, carted away, and dumped. Curiously, the Accokeek site is littered with slag while the substance is conspicuously absent from Rappahannock. An average-sized furnace produced tons of slag each season. The slag dump at Accokeek is now a large hill with trees growing on it. The Rappahannock furnace should have produced no less slag than Accokeek, yet not a trace may be seen on site. Because the sharp-edged slag was hard on the feet of the horses and oxen, Hunter's workmen gathered nearly every piece of it and hauled it away. A diligent surface search of the entire area has produced not a single piece, though smooth-faced, river-washed specimens may be picked up at leisure along the river's edge below the forge. Great quantities may also be found beneath the Falmouth Bridge. As a search of Mortimer's Island revealed no traces of slag, it appears that Hunter's workmen carted the material down to the river, dumped it in, and simply waited for the inevitable floods to carry it downstream and out of the way.

There has long been debate as to when Hunter actually began his iron works. Without question, the opening of Rappahannock was in response to the closures of the other works in the area, especially Accokeek. Hunter seems to have begun construction of his dam and canal c.1758-1759 and was probably experimenting with the first of two blast furnaces built on the site. During the first half of the 18th century, the three leading producers of iron in the Fredericksburg region were Alexander Spotswood's Tubal Furnace west of Falmouth, Spotswood's Massaponax Furnace south of Fredericksburg, and Principio's Accokeek Furnace in central Stafford. Tubal produced iron products from c.1716 to c.1767. This furnace's later years were marked by a steady decline in production and it eventually closed for want of money. Thomas Fox (c.1710-1792) leased Tubal from Spotswood's heirs in 1772, reopened the works, and ran it until his death, though no production figures survive. Massaponax[106] (the New World's first air furnace) ceased production in 1758 and Accokeek was in blast from 1727 to 1752. Rappahannock Forge filled a void created by the closings of these three major producers and Hunter probably saw iron production as an opportunity to increase the fortune he had already made in the mercantile and shipping business.

When Hunter began building Rappahannock, the Virginia economy was suffering the effects of several years of poor tobacco harvests. James was deeply involved at the time in the shipping and slave

[105] This dam appears on an 1806 survey of the site (Fig. 2-1).

[106] The Massaponax Furnace was built near New Post between 1730 and 1732. William Byrd described it as a double air furnace fired with mineral coal. James Hunter later built an air furnace at Rappahannock and used it for the production of steel.

trade, which should have turned a good profit. The purchasers of his slaves, however, were under financial stress as well and they put off paying what they owed Hunter. Consequently, he was forced to borrow money to pay for construction at the forge. A letter[107] from John Mercer[108] (1704-1768) to his son, George Mercer[109] (1733-1784) details some of the financial problems Hunter faced during the early years of construction:

> As to Mr. Hunter you must partly know, that his negro consignments obliged him to be punctual in his remittances whether he rec'ed the money or not from the purchasers. They fell so short, that he was obliged to draw on his bro. William[110] for more thousands than he could pay & was therefore obliged to return his bills, therefore to maintain his credit, as he had many thousand pounds due on account of negroes, abot. £5000 advanced for Jno. Campbell Esqr. (now in Jamaica to raise the money) about £3000 for Spotswood's estate[111], & above £6000 an Ironwork in Maryland in partnership with one Gantt, & as he was erecting a forge with four hammers above the falls, with a merchant mill &c to make the race of wch he was obliged to employ near 200 men, & for which Fras. Thornton hath brought an action against him in Spotsylvania court[112] for ten

---

[107] This letter was undated, but Mercer historians believe it to have been written between Dec. 22, 1767 and Jan. 28, 1768.

[108] In 1730 Mercer bought the failed village of Marlborough and invested great sums to bring it back to life. He had a fine brick home built there and the refurbished town boasted a racetrack, ordinary, shops, and all the amenities of an 18th century Virginia settlement.

[109] George Mercer was the son of John Mercer of Marlborough and his first wife, Catherine Mason (1707-1750), sister of George Mason III (1690-1735) and only daughter of George Mason II (1660-1716). George Mercer was born at Marlborough. On May 27, 1765 he was appointed official Stamp Distributor for Virginia, a position that no doubt troubled his patriotic father. Both George and his father were involved with the Ohio Company (see article Chapter 1). George was sent to London as an agent of the Ohio Company where, in 1767, he married Mary Neville (died 1768).

Mary Mercer submitted a Loyalist Claim "on behalf of George Mercer, late Lieutenant Colonel in the Virginia Regiment but now insane, and her daughter Martha, a cripple," for the loss of 6,479 acres on the Shenandoah River in Frederick County with stock, tobacco, hemp, wheat, timber, mills; for 3,170 acres on Little River and Bull River in Prince William County with timber, mill, crops; and for 180 slaves, worth in all £30,362.18.0. Her deposition contains a long account of Col. Mercer's activities including his serving as chief distributor of stamps for Virginia, Maryland, and North Carolina, and during the Revolution. In 1782 the £400 pound per annum pension granted him in 1776 was reduced to £300. Mary Mercer was buried in Richmond, Virginia.

[110] In a letter from William Hunter to Mrs. Marianna Hunter dated Feb. 16, 1769 the former quotes from a letter he had written to his brother, James, Sr., "The Balance of your private accot to me exceeds £6000 sterg. [James] Mills is upwards of £1000, and both your African Engagements to me are considerably more...in future I may possibly be an Exile from the British Dominions if you neglect much longer to send my Effects to me...I cannot cease without testifying my astonishment that you have entirely forgot Jamie Hunter. I request you directly to remit him wherewithal to enable his return to Virginia, to take possession of his Estate."

[111] In 1759 James Hunter bought 1,750 acres in Culpeper County from Spotswood's heirs. Whether this tract included any of Spotswood's mines is unknown, the deed having been lost.

[112] Around 1758 Hunter's workmen built a dam across the Rappahannock to route the water through his own canal running along the north side of the river. This effectively left Francis Thornton (died 1784) with no water with which to operate his grist mill on the south side of the Rappahannock. Papers relating to this case, which was heard in the Spotsylvania court, have been lost. Mention of this case occurs in the court records of Stafford County dated July 14, 1783. "Your Petitioner Francis Thornton humbly sheweth that his Grandfather Francis Thornton [died 1795], upwards of sixty years ago, erected a Grist Mill on the south side of Rappahannock River in County of Spotsilvania opposite to lands of James Hunter and John Richards Gent; above the Town of Falmouth, which mill hath from that time been kept up, and remained in the possession of your petitioner's said Grandfather, his Father, and himself, and now is kept up by and in possession of your Petitioner. That at and from the time of erecting the mill, the main current of the River

thousand pounds damage for stopping the water to his mill, he was obliged to borrow money of Colo. Corbin[113]: I looked upon it that Mr. Hunter had picked up a valuable friend, who had chosen an opportunity to assist, with his large fortune, a valuable member of society, by enabling him to bring his works to perfection & retrieve his credit, but how much was I deceived, when upon a visit from our worthy friend [Hunter] about a month ago, he informed me that he had been obliged, to secure Corbin by a bill of sale of his whole estate, which I had never heard of before, & that [Hunter] had urgently pressed him ever since April to advertise it for sale, I suppose to become the purchaser of it, and as he was then sensible that it was almost brought to perfection & could not cost Hunter so little as £200 a month.

I can now inform you with great pleasure that his four hammers are at work, so that I am in hopes he will be able to get quit of all the Harpies that waited to prey upon him, but he assured me that tho one of his hammers began to work, in September he had not yet received five pounds for Iron tho [it is] the best article in the country to produce cash (Mulkearn 206).

Mercer continued by lamenting the difficult economic conditions plaguing the colonies. He wrote, "for my own part, I can affirm that for the last two years, I never was master of five pounds for five days, I may almost say, for five hours."[114] Economic problems would continue to plague Hunter until his death.

---

ran on the south side, so as to supply the same at all times with a sufficient quantity of water without your Petitioner's extending his dam across the River, until the said James Hunter some few years ago, by extending a dam across the River a small distance above you petitioner's mill turning the water out of its usual courses, and from your petitioner's mill, thereby rendering it of but little value and making it necessary that your Petitioner should run his dam across the river and join the same to the bank on the opposite side, that for the reason above set forth, no application was made at the time your petitioner's mill was erected for the condemnation of an acre of land on the opposite side of the River. Nor hath such an application been made at any time since your Petitioner therefore prays relief in the premises, and that your Worships will please to appoint two of your body to view and value one acre of land lying upon the opposite side of the River for your Petitioner's mill" (Stafford Deed Book S, p. 78, July 14, 1783).

[113] Col. Richard Corbin (1714-1790) of Laneville, King and Queen County was the son of Col. Gawin Corbin (1659-1744) and his second wife, Jane Lane (died after 1715). From 1748-1790 Richard represented Middlesex County in the House of Burgesses. He was a member of the Council from 1748-1783 and Receiver General from 1754-1776. He served as County Lieutenant of Essex County 1752 and beyond. In 1737 he married Elizabeth Tayloe (1729-1784), the daughter of John Tayloe (1687-1747) of Richmond County. A known Loyalist throughout the Revolution, Richard wisely remained neutral. His high moral character and affable personality made him admired even by those who disagreed with his political views. Throughout the period he conducted himself with such discretion that he was permitted to retain his property after the war.

The amount of Hunter's debt to Corbin is unknown but by the time he died, Corbin still had not collected all that he had loaned. Richard's will reads, "I give bequeath and assign unto my son Thomas Corbin James Hunter's bond and Mortgage for two thousand five hundred and twenty nine pounds seven shillings and eleven pence" ("The Corbin Family" 303-370; Fleet 160-162).

[114] Around 1766 Mercer decided to supplement his greatly reduced income by building a brewery at Marlborough. This necessitated assuming even greater debt in order to construct a brewhouse and malthouse, hiring a master brewer, and buying 40 slaves to make the grain necessary to the process. These cost some £8000 and he prevailed upon his friend, James Hunter, to sign as security. The money was loaned by another friend, John Robinson who was treasurer of Virginia. The brewery was a failure. John Robinson died in 1766 at which time it was discovered that he had nearly bankrupted the colony by making loans, most unsecured, to financially strapped friends (Mulkearn 190).

Fielderia Iron Furnace

About six years after opening his works on the Rappahannock River, James Hunter became involved with another iron furnace in Frederick County, Maryland. This small furnace was the brainchild of one Fielder Gantt (c.1730-1807), also from Frederick County. Hunter's involvement in Fielderia Furnace is of interest primarily because of the wealth of court records that reveal information about Hunter and 18th century ironworking. Fielderia was built c.1765 in Frederick County, Maryland, about three miles from Harper's Ferry on the Harper's Ferry Road. The furnace likely stood somewhere in the vicinity of present-day Sandy Hook, a tiny one-street village pressed against the railroad tracks on the north side of the Potomac River. By the mid-1760s, this was still a frontier area, though Harper's Ferry had been settled in 1733 and Opequon, just a few miles upstream, had seen its first white settlement c.1730. An abundance of iron ore, waterpower, timber, and promising new investment opportunities set the stage for numerous iron furnaces in western Maryland. Known as the "back country" or "the Western Lands," merchants and businessmen eyed these frontier communities as lucrative markets for everything from ribbon to salt.

James' business partner in this venture was Fielder Gantt, the son of Thomas Gantt (c.1680-1765) of Prince George's County, Maryland and Priscilla Brooke, the daughter of Thomas Brooke (1659-1731). He was the grandson of Edward Gantt (1660-c.1685) and the great grandson of the immigrant Thomas Gantt (c.1615-c.1691). Fielder also had a nephew by the same name[115] with whom he is often confused. Fielder the ironmaster had a daughter named Margareta (1770-1821) who married Nicholas King (1771-1812) and a son John (c.1775-1803) who predeceased his father and died without issue. Neither Fielder Gantt left descendants to continue the name.

Fielder began acquiring land in Frederick County, Maryland in 1756. He moved to Frederick in 1764 and from 1765-1766 represented that county in the lower house of the Maryland legislature. He served in that body again from 1779-1780. From 1779-1785 he served as a justice of the peace for Frederick and was overseer of the roads in 1788. At the time of his first election to the legislature, he owned 13,081 acres in Prince George's and Frederick counties (Papenfuse 340-341). Fielder married Susannah Bowie(?) about whom little is known and made his residence on a plantation called Come by Chance. He was buried at All Saints Episcopal Church in Frederick.

In early 1764 Fielder began acquiring land in Frederick County with the intention of building an iron furnace. On Mar 30, 1764 he purchased from Adam Ramsburgh land that included "all the iron ore and store on the two tracts" (Frederick County Deed Book J, p. 262). The furnace was constructed on a 150-acre tract called Fouts Delight purchased from Conrad Licklider for £350 (Frederick County Deed Book J, p. 116). Fielder eventually assembled 8,151 acres much of which was intended to supply the furnace with charcoal. The furnace property adjoined his own Come by Chance, the entire tract consisting of some 10,481 acres that spanned the area from South Mountain, across the Frederick/Washington County line, and through the Catoctin and Middletown valleys.

Much of what is known about this furnace is found in the records of the Maryland High Court of Chancery. The partnership between Gantt and Hunter was stormy and unsuccessful but the relationship continued for about 20 years. Following James Hunter's death, his executors sought to settle his estate, which included selling the 8,151-acre Fielderia tract. From about 1785 until at least 1826 the Maryland Court of Chancery was occupied with settling disputes over past debts and land titles related to Fielderia. The resulting records are voluminous and provide a fascinating view of the furnace and those involved with its operation. On July 23, 1804 Fielder Gantt provided a lengthy deposition that was included in an 1826 dispute over title to part of the Fielderia tract (Maryland High Court, "Hoffman, John and others"). In this document Gantt openly accused his former partner of fraud and deception. Other depositions from the Maryland chancery records, provided by men who knew both Gantt and Hunter and were familiar with their partnership, suggest that Fielder's statements were not altogether true. Other surviving records, including family letters, suggest that James was painfully frugal and rarely subject to feelings of pity or regret even when his own family members were in trouble. These same documents, however, also indicate that he was an honest man and anything but deceitful. By the time Fielder Gantt made his deposition in 1804, Hunter had been dead for twenty years and was unable to speak for himself. In order to gain some understanding of

[115] Fielder Gantt (1764-1824) of Prince George's County, Maryland.

Hunter's point of view, the court collected correspondence between the two, these letters constituting a considerable portion of the chancery records.

Fielder's 1804 deposition was the result of a court case instituted by Robert and Samuel Purviance. On June 28, 1785 Robert and Samuel obtained letters of administration from William Buchanan, Registrar of Wills for Baltimore County, giving them the authority to settle Hunter's affairs in Maryland. The following month the Purviances asked Fielder Gantt "to pay and discharge the principal and Interest due on the said Indentures of lease and release and well hoped he would have complied with such their reasonable request" (Maryland High Court, "Hoffman, John and others"). According the plaintiffs, Gantt owed Hunter's estate £7,839.14.2 plus interest that had been due since Dec. 13, 1771. The bill of complaint charged, "the said Fielder Gaunt not only refused and failed to pay the same or any part thereof but hath suffered the Interest thereof to raise greatly in arrear whereby the said principal and interest is greatly more than the value of the premises."

In answer to the charge Fielder explained how he and James had become acquainted and connected in business. In the early 1760s Gantt decided to establish an iron works in Frederick County and began purchasing land for that purpose. As he did not have enough capital to carry out the venture on his own, he sought several partners willing to contribute cash towards this expensive project. In 1763 he purchased Negroes at Annapolis jointly with the Hon. Benedict Calvert[116] (c.1724-1788) fully expecting Calvert to join him as a partner in the works. It was at this time that Fielder met James Hunter, also in Annapolis on business. Hunter told Gantt about 60 slaves that he wished to sell for £23.10 each. James informed Fielder, "they were a bargain and that he would join the defendant in the profit and loss on the sales and allow him five per cent for his trouble." Gantt sold the slaves at a profit, and returned to James with the money earned from the sale. Hunter was very pleased and "this sale was the foundation of all subsequent transactions between the said James and the defendant" (Maryland High Court, "Hoffman, John and others").

Over an unspecified period of time, Gantt continued trying to work out a partnership between himself, Benedict Calvert, and Gov. Horatio Sharpe[117] (1718-1790) in the proposed iron works. While in Baltimore for these negotiations, Gantt lodged with Hunter. The conversations between Gantt, Calvert, and Sharpe "becoming frequently the subject of conversation between them the said James proposed that the defendant should admit him the said James as a partner with the defendant in those works, to which proposition the said defendant unfortunately consented...not knowing that the said James had a difference with Mr. Calvert" over the sale of some slaves (Maryland High Court, "Hoffman, John and others"). When Gantt informed Calvert of Hunter's joining in the partnership, "the said Calvert utterly refused and declined being therein concerned alledging his bad opinion of the said James, by which means the defendant was not only deprived of the countenance, assistance and advances of an opulent partner, but also the said Calvert was induced to require security for monies due him for negroes purchased for the works by the defendant, and afterwards brought suit for the same against this defendant and securities, by means whereof the defendant was greatly injured." According to the deponent, Calvert also influenced Gov. Sharpe to decline the partnership, leaving Gantt with James Hunter as his only partner.

At the time of receiving Hunter as a partner, Gantt explained that he "had expended all his resources and exerted his Credit to the utmost in order to prosecute his intended works, and his object in receiving partners was by their advances of money." Hunter assured Gantt that he had the "ability to make all such necessary advances and gave him the most solemn promises to furnish him with, when required the

[116] Benedict was the acknowledged but illegitimate son of Charles Calvert (1699-1751), Fifth Lord Baltimore. In 1742 Charles sent Benedict to America and placed him in the care of Dr. George Steuart (c.1700-1784) of Annapolis. Benedict was Collector of the Customs for Patuxent District in the 1740s. He married Miss Elizabeth Calvert, "only surviving daughter of the late Hon. Charles Calvert, Esq. [c.1664-1733], deceased, formerly Governor of Maryland" (*Maryland Gazette*, Apr. 27, 1748). He lived at Mt Airy in Prince George's County, Maryland and was the father of George Calvert (1769-1838).

[117] Horatio Sharpe was born near Hull, Yorkshire, England. In 1753 he came to Maryland from the West Indies, having received the appointment of colonial governor. He carried out his official duties with great care and energy, effectively balancing the demands of the royal government, the proprietorship, and the needs of the Maryland colony. In 1760 he was ordered to determine the bounds of the Maryland province at which time an agreement was struck creating the Mason and Dixon's Line. In 1773 he was summoned to England to handle family affairs there and spent the Revolutionary period in London. While in Maryland he resided at Whitehall near Annapolis.

sums necessary for completing and prosecuting the said works." Gantt stated that he completed the works in 1764 "at which time the said Hunter had only advanced twenty five barrels of Pork amounting to one hundred and one pounds five shillings and one hundred and fifteen pounds Virginia currency. This was far short as "the advances of the defendant on their joint account were not less than six thousand pounds, altho' according to the agreement between them he should have advanced to an equal amount for all original advances as well as for the value of lands, works, and other property." Gantt charged that Hunter's estate owed him a large though unspecified amount plus interest and that he was not indebted to James' estate in any way.

James visited Fielderia in April 1765 "and appeared much pleased with what the defendant had done, the Furnace being then ready to go into blast[118]—that he the said James continued there several days, examined the books and directed the Clerk to enter several articles in his account without producing any vouchers or receipts for the money he said he had advanced to all which the defendant was unable to pay due attention from the great Hurry of business at that particular time." This accusation of fraud was countered in a deposition from William Murdock Beall[119] (c.1743-1823), former Fielderia clerk, who said that Hunter did not take credit in the books for anything for which he did not produce receipts (Maryland High Court of Chancery Records, Hunter, James of Virginia, Creditor of Fielder Gantt).

Gantt and Hunter decided to formalize their business relationship and engaged attorney Thomas Jennings[120] (c.1736-1796) to draft articles of partnership. During that trip Fielder told James that he would have to depend upon him for further advances "as his own fortune and credit had been already so strained that nothing farther could be expected from either." Hunter promised, upon returning to Virginia to furnish Gantt with 20 Negroes from his plantation and that in the fall he would advance £6,000 "with which sum he expected that they might be able to extend their business, as Iron Works were not the only object he had in view."

Jennings drafted the articles of partnership, but James found them "insufficient" and made amendments. Hunter claimed to have urgent business in Virginia and couldn't wait for Jennings to rewrite the articles but promised to return soon to sign the revised articles. William Beall commented on Hunter's dissatisfaction with Jennings' initial articles saying that one of his objections "was a clause vesting the said Fielder Gaunt with the Supreme management of the Works." James may have been concerned about investing large sums without having an equal voice in the management of the operation. By Jan. 22, 1766 no formalization of the partnership had occurred. Hunter wrote to Gantt proposing "as soon as the Ice hath broken away, and the Roads became passable, to meet you at the Furnace, as you are pleased to say you will comply with every engagement on your part on the like being done on mine, which is all I ever required, and I shall then be ready to fix my Estate on a fair equitable footing, whereby said Estate and its produce are not alienated from myself and subjected to your sole disposal" (Maryland Chancery Records, Hunter, Adam, Legatee). James continued, My Council confirm my own opinion touching my right to the Furnace, and its produce, and I will exert it against all attempts and indirect practises of yours to lessen it. While I have life or a penny left to support it, nor shall any owner ever be the manager, a thing very essential to prevent your effecting my ruin" (Maryland Chancery Records, Hunter, Adam, Legatee). The partnership was never formalized and a ledger entry dated Apr. 19, 1765 includes a debt of £4741 due Hunter for 14 slaves "on condition I became a partner in Fiedleria Furnace and lands (which was never effected)" (Maryland Chancery Records, Hunter, James, Late of Virginia).

In the meantime James drew up a bond in which Gantt conveyed to him title to a moiety in the works. Fielder claimed to have such trust in James that he readily signed the bond, believing that his partner would indeed return soon to sign the articles of partnership and advance the promised money (Maryland High Court of Chancery Records, "Hoffman, John and others"). One must wonder why Gantt

---

[118] The construction of the furnace and forge was overseen by Basil Beall (b.1744) (Maryland High Court of Chancery Records, Hunter, James of Virginia, Creditor of Fielder Gantt).

[119] William was the son of Nathaniel Beall (c.1715-1757) of Frederick County and Ann Murdock (c.1717-c.1804). Around 1770 William married Mary Tannyhill (c.1742-1810).

[120] Thomas Jennings was probably born in England. He settled in Maryland c.1759 and was a resident of Annapolis from at least 1772 until his death. In 1786 he was elected to the board of St. John's College and held this position until his death. He practiced law in Annapolis and the counties of Frederick, Anne Arundel, Baltimore, Charles and Prince George's. Thomas served in Maryland's Senate and Lower House, was attorney general from 1768-1778, and was mayor of Annapolis.

would have signed away a half interest in the works without having received something in return. According to records in the Frederick Courthouse, on Oct. 13, 1767 Hunter paid his partner £7,323.1.7 for which he received a receipt. Gantt apparently was unable to produce his share of the cash at that time and secured his debt to James by conveying his interest in nearly 9,000 acres plus "all the goods Chattles Utensils Implements Cattle Live and Dead Stock Slaves Servants and other things whatsoever mentioned in the Schedule hereto annexed...and also all and every other thing and things belonging to Fielderia Furnace aforesaid...Together with all Mines Oar Works Houses Tennaments Erections buildings Forges Ways Waters Water Courses Plantations Trees Woods" (Frederick County, Maryland Deed Book K, p. 1165). One is left to wonder why Fielder would have signed such a deed of trust had he not owed Hunter money. On the other hand it would have been out of character for James to invest money in a business venture without first obtaining adequate security.

Land purchases continued for several years as Gantt sought to acquire adequate land for mining and charcoaling. Beginning in June 1765 James Hunter's name began appearing on the deeds, perhaps marking the beginning of his official involvement with Fielderia (Frederick County, Maryland Deed Book K, page 145).

On one of his visits to Frederick County James and Fielder discussed the advantages of building a forge in which Fielderia pigs might be processed into bars. By all accounts Hunter approved of the idea and Fielder commenced working towards that end. Gantt contracted with workmen to do the construction and wrote to James reminding him to send money. Fielder claimed that when James returned to Fielderia, the two men rode to the place where the buildings were to be erected and James seemed satisfied but asked that the principal millwright visit his works near Fredericksburg "to view a mill seat he had there." The millwright rode down to Hunter's and, during his absence, Gantt had the dam built and the race cut "so that only the forge House and mill wright work remained to be done." When the millwright returned from visiting Hunter, Fielder was surprised to be told that Hunter "forbade his erecting the forge" on Catoctin Creek. Gantt had already invested £885.11.3 ½ in the project, none of which was contributed by Hunter.

By the time Fielder Gantt made this deposition, Hunter was long dead and unable to dispute his partner's version of the story. However, a letter dated Oct. 13, 1770 from Hunter to Gantt clearly stated James' thoughts. Hunter's objection was not with the forge itself, but with its location. He felt that "the money depended therein might have been laid out to so much better purpose on Manocasy which is just in the way of our Carriage and would render our whole Concerns more compleat, the other is dividing them and can never be extended where there is want of water" (Maryland High Court of Chancery Records, Hunter, James of Virginia, Creditor of Fielder Gantt, 1789/90). Based upon Fielders deposition, the men had decided to send the Fielderia iron to Georgetown for shipment to and sale in England rather than to sell it locally. They planned on carting the iron from the furnace east to a landing at the mouth of Monocacy Creek, and then floating it down the Potomac River on boats. Catoctin Creek, the location of Fielder's intended forge, was west of the furnace, thus requiring longer overland hauling of the iron. Hunter refused to contribute to the construction of a forge in what he viewed as an impractical location. On July 15, 1756 Hunter wrote to Gantt with regards to the Catoctin forge, "I am sorry you have made any preparation for erecting Forges at Ketoctan, as both yourself, Mr. Bond,[121] and other good judges disapproved of the stream, besides being inconvenient to our intercourse between the Furnace and George Town or Great Falls" (Maryland Chancery Records, Hunter, Adam, Legatee). Why Gantt changed his mind about building the forge on Catoctin Creek is unknown.

Thoroughly disgusted, Gantt sought to relinquish management of the works and engaged a qualified person for that purpose. When informed of this decision, Hunter refused the man and Gantt was forced to continue at Fielderia. Fielder's creditors began bringing suit against him and "in order to Extricate himself he made every Exertion to put the furnace again into blast and did actually put her into blast, but could not continue it for a defect of funds the said Hunter neglecting all the while to give him any assistance except by furnishing salt and some other articles." Thus, smelting was aborted and part way through its second season Fielderia's brief and unproductive life came to an end.

Working relations between Hunter and Gantt deteriorated and both sought to free themselves from their unsuccessful partnership. Fielder said that he repeatedly wrote to Hunter demanding that he settle the accounts, "when the said James to the utter surprise and astonishment of the defendant wrote him that as

---

[121] Possibly Richard Bond (1728-1819) of Cecil County. He served as a sheriff and justice in Cecil County and represented that county as a delegate. My 1776 he had established a gun factory in Maryland.

soon as his attorney had completed the proper deeds for establishing his the said James, a title to the lands &c he should sell to the first chap that offered." Of course, James was perfectly within his rights to sell the property. Fielder had conveyed his interest in the works to secure his still-unpaid debt to Hunter. Despite Fielder's claims that James owed him money, it appears that it was actually Gantt who had not satisfied the requirements of the deed of trust. Selling the property was the simplest means of settling the issue.

In his deposition Gantt claimed that during this time his financial situation was so precarious "that he could not without danger go abroad, being in constant apprehensions of confinement." Hunter "took a most cruel advantage and gave a final stab to his credit" by bringing suit on the defendant's bond. Gantt and Hunter agreed to turn the matter over to arbitrators[122] "who however after several meetings declined to act." This curious situation was explained by William M. Beall who said that the reason some of the arbitrators refused to act was because Gantt "had said something disrespectfull of some of them" (Maryland High Court of Chancery Records, Hunter, James of Virginia, Creditor of Fielder Gantt, 1789/90). The arbitrators declined to hear the dispute and the issue remained unresolved.

Gantt charged that James owed him for expenses "incurred in erecting the buildings for forges in Kettoctan which was began by his Express desire [and] that proper credit has not been given by the said James for pig Iron shipped by the said James in the years [left blank] to his brother William Hunter of which no account whatever has yet been rendered." Fielder claimed that despite many entreaties for James to pay his share of expenses at Fielderia, the latter consistently refused to do so, though he made many promises of forthcoming money. Gantt deposed that "in consequence of this conduct of the said James and the said James failing in all his promised advances, he the defendant experienced much distress in his circumstances in addition to which injuries great part of the Iron made this year 1765 was shipped for England by the said Hunter and never yet accounted for or carried to the credit of the defendant" (Maryland High Court of Chancery Records, "Hoffman, John and others"). Gantt further stated he had "no hopes that he can ever recover any sum to compensate for the loss of the labour of twenty four years of the defendant's life passed in most bitter distress and all occasioned by the fraud and delusion of the said James." Fielder believed that had he never met James Hunter "he should have lived in Ease and Credit and been at this day possessed of an opulent fortune."

Fielderia was in blast for about a season and a half. Considering that an average charcoal blast furnace produced about 800 tons per season, Fielderia's total yield of 292 tons was meager at best. A ledger included in the chancery records lists the following monthly production figures:

May 1768	58 tons
September	28
November	25
May 1769	82
December	87 and 12 ½

(Maryland High Court of Chancery Records, "Hunter, James of Virginia, Creditor of Fielder Gantt")

The exact reasons for this dismal production are uncertain; Gantt claimed that he was limited by lack of funds and Hunter believed it was due to poor management on Gantt's part.

Hunter and Gantt soon realized that they were unable to work together but were at a loss as to the best way to terminate the partnership. Each hoped the other would offer a buy-out agreement, but neither had the cash to do so. On Sept. 10, 1767 Hunter wrote to Gantt that he was "content to take for my part of the furnace &c the money I have advanced with Interest...payable in Pig iron, on Navigable Water...two hundred Tons this fall two hundred Tons August 1768 and the remainder before August 1769." He added, "I told you I was neither inclined nor able to buy [the furnace] but it was very agreeable to me to have the whole Stock sold."

At some point James informed his partner that he would take his share of the partnership in pig iron. Fielder deposed that James had contracted with John Ballendine and had agreed to "pay the said Ballendine for transporting the Iron by Water." Hunter then wrote to Gantt telling him to send the iron overland to their agreed upon shipping point. Fielder had 170 tons of iron carted to a landing near the

[122] According to the chancery records, one of the arbitrators was Charles Graham (c.1721-1779), a Scottish merchant who resided in Lower Marlboro, Calvert County, Maryland. Prior to the Revolution, Charles was the principle agent for James Russell's extensive mercantile business in Maryland and supervised Russell's interest in the Nottingham Iron Works in Baltimore County.

mouth of the Monocacy "and was afterwards compelled to engage Waggons to transport the same to George Town or loose it, the said Hunter neglecting to remove the same." Gantt claimed that Hunter failed to pay for the hauling of the iron to Georgetown, necessitating Fielder to pay the drivers from his own funds. This allegation was refuted, however, by the deposition of Robert Peter, Georgetown merchant, who stated that James Hunter had lodged money with him for that purpose. Peter further stated that on the few occasions when James had not left money with him ahead of time, Peter had paid the drivers and Hunter had reimbursed him (Maryland Chancery Records, Peter, Robert).

Fielder's deposition contains no claim that he ever satisfied Hunter's demands regarding the deed of trust or the specified quantity of iron required for the termination of the partnership. Gantt failed to satisfy the deed of trust and Hunter foreclosed. The Maryland court decided in Hunter's favor and James became the sole owner of the Fielderia property (Frederick County, Maryland Deed Book 10-137). In addition to the furnace and land James received two new wagons, a variety of hand tools, 14 wheel barrows, 24 falling axes, 6 cattle, 38 hogs, plates, dishes, basins, brass and copper coffee pots, a coffee roaster, kitchen pots and pans, knives, forks, household furniture, china cups and saucers, wine glasses and decanters, napkins, towels, blankets, rugs, 21 horses, two carts, one pair bellows, one anvil, one vice, tongs and hammers, wheat, rye, Indian corn, salt beef, salt pork, cordwood, and coal (Frederick County, Maryland Deed Book K, p. 1165). What he did with these items is unknown.

Exactly what transpired at Fielderia after Hunter foreclosed is also unclear. At least 200 tons of pig iron were shipped to Georgetown (Frederick County, Maryland Deed Book K, p. 1165). There is no indication that smelting ever resumed, yet Fielder Gantt seems not only to have remained in residence on the property, but to have overlooked the fact that he no longer owned it. In 1791 Abner Vernon wrote to Baker Johnson concerning a court action to evict Gantt, Vernon writing that he hoped Fielder had not damaged the property out of vindictiveness (Vernon, Letters). Following James' death, his executors had the Fielderia tract divided into 59 lots and auctioned them in 1792. Baker Johnson[123] (1747-1811), owner of the Catoctin Furnace, purchased the Fielderia Furnace and several additional parcels. There is no evidence that Johnson ever re-opened Fielderia Furnace.

By the time Gantt made his court deposition, James Hunter was long dead and unable to provide his own testimony. In place of this the court ordered all correspondence between Hunter and Gantt admitted as evidence. To say the least, Hunter's perspective, as reflected in these letters, differed greatly from that of his partner, and though their relationship soured, initially the two had been on cordial terms. More than a year before James' name began appearing on deeds with that of Fielder, he was actively involved in providing Gantt with supplies. He wrote, "If you could procure a good small craft to meet you at Caves Warehouse in Potomack Creek about 10 miles from my House between the said 15th & 22d May my own and other Waggons may be got to carry over [Port wine Salt] and fill up with Tar" (Frederick County, Maryland Deed Book K, p. 1165). Hunter's numerous letters to Gantt reflect the evolution of their relationship from cordial to antagonistic.

Although the two had discussed a formal partnership, no firm agreement was ever reached. By the fall of 1764 James was somewhat perplexed as to how he stood and sent a letter to Fielder by way of his brother, Richard Hunter. James wrote, "I have all a Long Considered myself a Partner with you in the works yet you have not fixed the share, whether a half or a third which will depend on the part you are able to hold yourself" (Maryland Chancery Records, letter, Sept. 26, 1764). The tone of this letter seems to be

[123] Baker Johnson was the son of Thomas Johnson (1702-1777) of St. Leonard's Creek, Calvert County, and Dorcas Sedgwick (1705-1770). Early in life he moved to Frederick County where he practiced law. In 1784 Baker married Catherine Worthington (1761-1814), the daughter of Nicholas Worthington (1734-1793). Baker was for some time a partner with his brothers, James, Roger, and Gov. Thomas Johnson (1732-1819), in furnaces, forges, and mills in western Maryland. In 1774 James Johnson and Company built Catoctin Furnace about 12 miles northwest of the town of Frederick. The following year they built a forge on Bush Creek along with the Johnson Furnace at the mouth of Monocacy Creek (Pearse 19). Baker eventually bought out his brothers' interests and was sole owner of these extensive works.

In 1774 Baker was a member of the Committee of Correspondence for Frederick County, this group appointing him in 1775 to solicit subscriptions to purchase arms and ammunition. He was a member of the Maryland Conventions of 1774, 1775, and 1776, and served as a colonel in the 34th Battalion during the Revolution. Baker died in the town of Frederick, Maryland "after fifteen months severe indisposition" (*Maryland Gazette*, July 3, 1811).

one of cooperation, Hunter only asking that some definitive arrangement be agreed upon. Hunter well understood that his role at Fielderia was as a source of capital and, by this time, he had supplied Gantt with a quantity of supplies, especially salt. By February of the following year nothing had been settled with regards to the partnership, but the concern had shifted to labor at Fielderia. Hunter proposed sending 20 of his own slaves to Frederick County including three carpenters and a blacksmith. He wrote, "The Principal Tradesmen must all come from Pensylvania," the source of his own skilled workers (Maryland Chancery Records, letter, Feb. 5, 1765).

With regards to their accepting other partners in the venture, again Hunter's opinion as expressed in his correspondence varies from that of Gantt. Fielder claimed in his deposition that he had initially sought to involve Benedict Calvert and Gov. Sharpe as partners. The matter of adding partners, however, seems to have been discussed after Hunter became involved rather than before. Of Calvert Hunter wrote, "You know when I agreed to join you in the Adventure there was but one man in all Maryland I objected to be concerned with." Hunter's difference with Benedict regarded lack of payment for a group of slaves. He encouraged Gantt to pursue getting Horatio Sharpe as a third partner, "for I have the highest opinion of his integrity & Honour."

As he stated in his deposition Gantt did press Hunter for cash on numerous occasions. It was a fact of the times that cash was nearly impossible to come by. The colonial economic system was based almost wholly upon barter and credit, most people having not so much as loose change in their pockets. Hunter was no different, stating to Gantt, "I have not five pounds paper in the House, nor can command it on so sudden emergency" (Maryland Chancery Records, letter, Feb. 26, 1765). Fielder's exact need at the time is unknown, though the correspondence reflects repeated requests to Hunter for cash.

By August 1765 what had begun as a cordial relationship had begun to deteriorate. Hunter had received word that business at Fielderia was not going well. The initial problem involved Gantt's treatment of his workmen. James wrote, "I am rather astounded any workman should serve you at all for beside the want of money I am informed your Behavior hath become so intolerable, that every Servant you have threaten to Leave you. Be not angry my dear Sir for when I represent to you it is the same as looking into myself, but be assured there is a reciprocal duty between Master and Servant and when either has passed beyond certain bounds whereby that Cement & Cordiality is broke the Interest of the former is sure to suffer" (Maryland Chancery Records, letter, Aug. 2, 1765). From this point on the relationship between the two men deteriorated, James gradually becoming aware that his partner was not what he had first appeared to be.

Fielder claimed that Hunter made repeated promises to send him money, though none was forthcoming. The validity of this claim is difficult to determine from the court records on hand. In the fall of 1765 Gantt sent Hunter word that he planned on sending iron to Georgetown and asked the latter to arrange for a ship to pick it up from there. Between the time he asked for the ship and the appointed arrival of the iron, Gantt decided not to send the iron as he had not received what he believed to be adequate cooperation from his partner. James did as he was bid, paying for a ship to pick up the pigs. There being no iron to load, the captain informed Hunter of the fact. Incensed, James wrote to Gantt that the charges for the ship "shall be debitted your private Account." He also demanded that Gantt find a capable manager for the furnace "as yourself is no way qualified for the Charge." Disgusted with the business arrangement, Hunter added, "it is high time for me to take measures for defeating your intentions and Securing my own property. The ruin begun and effected by yourself, and what you would wickedly charge to me is a meaness not worth refuting...I design as soon as my Attorney can perfect deeds and establish my Title to half the Lands Furnace & Appurtenances to dispose of the same to the first good Chap that offers" (Maryland Chancery Records, letter, Oct. 16, 1765).

With regards to the building of a forge, some discussion seems to have taken place, though Gantt's deposition made it appear that his partner had agreed to the proposition. One of Hunter's letters indicates that, so far as he was concerned, the subject never advanced beyond an initial conversation. He wrote, "You have wantonly begun a Forge without my knowledge or consent and I now give you notice that I disapprove of it and will not agree to its being erected at least for the present, till I am fully satisfied how my Lands are Situated & the quantity and that Sufficient Funds are provided for the execution thereof." Hunter's concerns were not only with the convoluted shipping required by locating the forge on Catoctin, but by the lack of water during dry periods. He finished the letter by saying, "If you had been removed from the management nine months after the works were begun we should have benefitted at least £2000 thereby" (Maryland Chancery Records, letter, Oct. 16, 1765).

The business relationship between Gantt and Hunter had ended by 1768. The court approved James' foreclosure on his partner and the Fielderia Furnace tract became vested solely in him. Several questions remain unanswered, however. Why did Hunter fail to take any legal measures to recover the money Gantt owed him? Why did he not sell Fielderia Furnace and its 8,000 acres? His own finances were tenuous due to his massive investment at Rappahannock; yet James seems to have done nothing with the Fielderia property. There are no records indicating that he had any further dealings in Frederick County, Maryland. Following James' death Adam Hunter and Abner Vernon, faced with meeting immense debts due by the estate, brought suit against Gantt. The court found in favor of the plaintiffs, determining that the latter owed the Hunter estate over £22,000 including interest, penalties, and court costs. Since the suit was heard several times thereafter, one is left to assume that Gantt was unwilling or unable to make the payment. The furnace was never used again. The fate of the forge is unknown, but it is unlikely that it was ever completed much less operational. Adam and Abner had the 8,151-acre tract divided into lots and sold and Fielderia Furnace was forgotten.

James had other business interests beyond ironmaking. According to papers filmed as part of the Virginia Colonial Records Project, James Hunter and Company was a partnership of several Scottish merchants. On September 1803 the Commissioners of Claims, considering memorials from various Loyalists who had lost property as a result of the Revolution, read the claim of the heirs and representatives of James Hunter, William Campbell, William Cunningham,[124] William Ballendine, Patrick Ballendine[125] (died 1770), and John Ballendine[126] (died 1782), merchants trading in Virginia under the firm of James Hunter & Company.[127] Mention is made in VCRP papers dated Jan. 10, 1767 that the company was based in Ayr, Scotland and was likely the same company headed by James' father. James of Rappahannock Forge seems to have continued his association with the company after his father's death. As a point of interest, John Ballendine was a part owner of Occoquan Iron Furnace in Prince William County. James and Adam Hunter (1739-1798) also maintained a long-term partnership focused on the shipping of tobacco and other goods between Virginia, England, Scotland, the West Indies, and western Europe. The brothers used at

---

[124] William Cunningham was a merchant of Glasgow and traded for many years as part of the firms of Cunningham, [Robert] Findlay and Company and Cunningham, Browne and Company. William came to Virginia in 1748 to establish his business and returned to Glasgow in 1768, leaving his brother Alexander (died 1772) to manage affairs here.

[125] Patrick was a merchant from Ayr, Scotland.

[126] John was a merchant from Ayr, Scotland. In 1757 he received a 20-acre patent in Prince William County on the Occoquan River adjoining the lands of the heirs of Charles Ewell and Valentine Peyton. Upon this he built the Occoquan flourmills, which later became the site of the town of Occoquan. He also built and resided in a large stone house he called Rockledge which still survives. In 1760 he took over management of the Occoquan Iron Works and described himself as "of the County of Prince William, Gent." After 1765 he appeared in the Fairfax records as being a resident of that county. John bankrupted in 1767 at which time a large number of Occoquan's skilled slaves were sold.

Having failed in the iron business, in 1773 John went to England to study canal building. He returned the following year and took up residence at Amsterdam near Little Falls in Fairfax County where he sought to interest important people in the concept of a canal on the falls of the Potomac. This project met with some difficulties and was not completed. In 1775 John removed to the James River where he began to cut a canal there at his own expense. That year he received a state subsidy for building the James River canal and an air furnace at Westham to which was added a cannon factory. Numerous cannon were produced here during the Revolution. In January 1781 Benedict Arnold's troops destroyed his property, including Westham. His iron contracts unfulfilled, the canal incomplete, and his property heavily mortgaged, he was financially ruined.

Genealogist George H. S. King wrote of John, "He was a brilliant man who conceived big manufacturing enterprises but could not always attend to the details to carry these projects over."

[127] James Hunter and Company kept a store somewhere in Westmoreland County. He was also involved with two other firms, James Ballendine and Company and John Ballendine and Company, both businesses operating in Virginia. In 1803 a Loyalist Claim was submitted by the "Heirs and representatives of James Hunter, William Campbell, William Cuningham, William Ballendine, Patrick Ballendine, and John Ballendine."

least two ships, the *America* and the *Friendship* (VCRP, Admiralty), the latter belonging to their father. The same VCRP papers also indicate that the two companies shipped considerable quantities of goods in and out of Accomack, then known as Metomkin.

Neither James nor Adam married and the two brothers were closely associated in business during the many years they lived in Virginia, yet relatively little is known of Adam. On July 9, 1765 George Mercer (1733-1784) was named distributor of tax stamps for Virginia and Adam was appointed official inspector for Maryland and Virginia (VCRP, Treasury Warrant). The author has been unable to locate any other reference to this appointment or Adam's performance as a tax official. His position as distributor of the hated tax stamps seems quite opposite to the political views espoused by his brother and quite possibly contributed to the authorities later suspecting him of Loyalist sympathies. In 1781 authorities seized his papers along with those of James Sommerville (1742-1798) and William Wyat, two other Scots merchants of Fredericksburg (Palmer, vol. 2, p. 279).

James maintained partnerships with a number of other merchants. From at least 1767-1774 James was also a partner with John Glassell,[128] a Scottish merchant who, prior to the Revolution, returned to Scotland and submitted various Loyalist claims for lost property. The Culpeper records contain a mortgage and a lease involving Hunter and Glassell, merchants of Fredericksburg (Culpeper Deed Book E, p. 415; Culpeper Deed Book H, p. 61). During this same period, James was a partner with James Mills[129] (1718-1782), a prosperous merchant of Urbanna and Tappahannock. On June 9, 1760 Alexander Wright (1723-1772), merchant of Fredericksburg, sold James Hunter and George Frazier[130] (died c.1765) a 2/3 interest in lots 23 and 24 in the town.

While building his manufacturing center on the Rappahannock James, with Adam's participation, continued his involvement in the mercantile and shipping business with which he was so familiar. A series of letters included in the Virginia Colonial records Project indicates that during the 1770s and 1780s the brothers owned one, and possibly two sailing vessels and were shipping tobacco, iron, and other products to the Caribbean and British Isles.[131] Though Adam was a partner in all his brother's business dealings, it is uncertain exactly how much direct involvement Adam had at Rappahannock during his brother's lifetime. Based upon extant records, his primary responsibility seems to have been with their shipping and mercantile interests. Some of the letters from the Colonial Records Project are addressed to "James and Adam Hunter, Merchants of Fredericksburg" and some to "Adam Hunter, Merchant." This probably explains why there is so little mention of Adam in the surviving records pertaining to the forge. Adam does seem to have filled in during periods in which James was absent. On July 31, 1781 Adam answered a letter from Gov. William Smallwood[132] of Maryland stating that James had been residing at his Culpeper plantation "ever since our late alarm of the enemy coming this way" (Pleasants, vol. 47, p. 376).

The brothers maintained their mercantile partnership until James' death. The Virginia Colonial Records Project includes an informative claim pertaining to the seizure of tobacco owned by the Hunters. According to this document, James and Adam had 130 hogsheads of tobacco on the snow[133] *Fanny*, docked at Metomkin Inlet, now Accomac. "The whole of the said Tobacco was brought by water from

---

[128] John Glassell (1736-1806) was the son of Robert Glassell and Mary Kellon of Dumfries, Scotland. In 1738 John and his brother Andrew arrived in Virginia where both employed themselves as tobacco merchants. John resided at The Chimneys on Caroline Street in Fredericksburg. A loyalist, at the outset of the Revolution, John deeded all his property in America to Andrew. Around 1780 he returned to Scotland and, well advanced in years, married Helen Buchan.

[129] James married Elizabeth Beverley (1725-1795), the daughter of Col. William Beverley of Blandfield. Mills and Hunter were business partners.

[130] George Frazier's will was recorded in Spotsylvania County. James Hunter was named executor (Spotsylvania Will Book D, p. 191).

[131] On Oct. 9, 1770 James Hunter and Company, Merchants, shipped 313 hogsheads of tobacco "& a parcel of Pig Iron & Lumber, from South Potomac in Virginia" to Greenock, Scotland. This was shipped on the *America*, Robert Park, master, of Ayr, Scotland.

[132] William Smallwood (1732-1792) was a general during the Revolution but saw little active duty. He was elected governor of Maryland in 1785 and served three terms.

[133] A small sailing vessel resembling a brig, carrying a main and fore mast and a supplementary trysail mast close behind the mainmast. Snows were often employed as warships.

Fredericksburg to xxx [sic] Nancock and Punckatee, both on the Bayside of Eastern Shore in the said province of Virginia from thence carried in Waggons to Folly Landing, from which place it was transported in Boats and Shipped on board the said Snow" (VCRP, High Court of Admiralty). Although bound for Bordeaux, France, customs agents seized the tobacco and it never left port.

Around 1771 James Hunter, Jr. (1746-1788), his brother William (1748-c.1773), their uncle Adam, and long-time friend John Taliaferro[134] (1745-1789) formed a partnership called Hunters and Taliaferro. It is unknown if James Hunter, Sr. provided any financial support for this venture, but he was not a member of the partnership. In September 1772 Adam traveled to England as a commercial agent and contracted with John Backhouse of Liverpool to do some shipping for the firm. He also made arrangements with the firm of Tate, Alexander, and Wilson, also of Liverpool, to send their goods to Virginia consigned to Hunters and Taliaferro. The latter shipped tobacco and other products back to Liverpool. Unfortunately, Hunters and Taliaferro was short lived, the company lacking adequate financial resources to support its activities. Creditors were soon pressing and much of the responsibility for settling the company's affairs fell to Adam. By 1773 Adam was in London attempting to interest merchants there to consign goods with Hunters and Taliaferro. Bankruptcies resulting from this type of consignment scheme made the London merchants reluctant to involve themselves with Hunters and Taliaferro. Unable to generate cash with which to pay their creditors, Adam surrendered himself for confinement to the King's Bench in November 1773.

Many years later, Rebecca Backhouse, widow and executrix of John Backhouse of Liverpool, brought suit claiming that as the surviving partner of Hunters and Taliaferro Adam owed her husband's estate $10,000. The court found in favor of Rebecca for $11,779.50, though there is no evidence that Adam or his executor ever paid the judgment (U. S. District Court Records, Backhouse admx).

At home in Virginia, William died and James, Jr., suffering from his own financial woes, left for Europe. John Taliaferro was left to settle the company's affairs here. James, Sr. provided no funds to save his family members from financial ruin and debtors prison. Writing to Adam, James, Jr. suggested that James, Sr. might be willing to invest in some type of partnership, thereby providing enough cash to pay off debts and enable the two to return to Virginia. Adam, frustrated with his brother's behavior and lack of support, answered young James in a letter dated Jan. 6, 1774:

> [I] can only promise, when I am so happy as to revisit Virga. I will speak to my Brother of the Things you mention, but will not take upon me to use any Arguments urging his Compliance with your Request if he does not voluntarily agree thereto, being determin'd in the future to have no altercation with him, nor further contact than can be avoided. You will know, I lived with him as a Menial, & after ten years puzzling myself in his intricate matters, what Emolument have I realis'd? (Coakely 15)

By this time, Adam had made it to Duns, Scotland where he was residing with his brother, John. He finally returned to Virginia and seems to have forgiven his grudge against James; for the remainder of their lives the two brothers were associated in various business ventures. Following his return however, Adam may have resided, at least for a time, in Fredericksburg as correspondence with the firm was addressed to that town.

Deteriorating relations between England and the colonies forced people to declare their loyalty to one side or the other. Many of the Scots merchants remained loyal to the mother country and returned home prior to the actual outbreak of the war. The Hunters, however, cast their lot with the colonial government. While James seems never to have been questioned with regards to his stand, Adam was suspected of disaffection, probably as a result of his earlier appointment as Stamp Inspector. No harm seems to have come to Adam, the authorities likely deciding that he was no threat to the patriotic cause.

Oddly, the newspapers contain few business notices from any of the Hunter partnerships. Two were placed in the *Virginia Gazette or American Advertiser*, published in Richmond. On August 31, 1782 James and Adam advertised:

> We have for Sale, about 2000 bushels of Salt, at Leeds and Hobb's Hole, for Cash or James River Tobacco: Also a pair of French burr Mill-Stones, four feet in diameter. For terms apply to us at Fredericksburg.

---

[134] John Taliaferro resided at Hayes in King George County.

James and Adam Hunter

The following Oct. 12 the brothers ran the following:

> Just imported and to be sold by the Subscribers, at Fredericksburg, by the package or piece, for specie, Rappahannock or James River tobacco—Oznabrigs, Checks, Cloths, Russia Duck, cotton Hose, Nails, Carpenters Tools, Glass Ware, Gin and French Brandy in cases, Patilla Linens, Green and Bohea Tea; also a large assortment of printed Linens and Handkerchiefs.
>
> James and Adam Hunter

On Aug. 16, 1783 James and Adam offered at their store in Fredericksburg "for sale by the brigantine *Friends Adventure*, from Madeira, a choice cargo of wine" (*Virginia Gazette*, Aug. 16, 1783)."

Recorded in the reports is one of the few known personal accounts of Adam Hunter. According to the report, "Adam Hunter died about three years ago but as he had long been afflicted with a palsy which deprived him of his limbs and speech, he had very little agency in the management of his brother's estate. After his death the whole business devolved upon Patrick Home" ("British Mercantile Claims," vol. 28, 222-224). On June 7, 1787 James Hunter, Jr. wrote to his uncle Archibald[135] regarding the estate. "Every part of the estate goes wrong for want of an active head...[Adam] can not be adequate to the task." Adam's affliction was the result of a series of strokes that left him incapacitated for some 14 years. His will, signed on June 6, 1787 and recorded on May 16, 1798, is on file in Fredericksburg. On Nov. 19, 1792 Thomas Jefferson wrote to Adam regarding work to be done on Philip Mazzei's musket. Adam was unable to answer the letter and it fell to Abner Vernon to reply to Jefferson's query. Abner wrote of Adam, "in His present Indisposition he has requested me to reply thereto as to Mr. Madzie's Gun" (Boyd, vol. 24, p. 637; vol. 24, p. 672; vol. 24, pp. 730-731). Adam's condition remained unchanged until his death nearly six years later.

First and foremost, James Hunter was a merchant and he continued shipping large quantities of goods between Virginia and Scotland. While none of his business ledgers are known to survive, from the few extant records it is learned that he shipped great quantities of tobacco, wood, and turpentine to England, Scotland, and other parts of western Europe. Problems with two such shipments were recorded in the papers of the Virginia Colonial Records Project. One letter dated Mar. 7, 1766 pertained to 235 hogsheads of tobacco sent to Ayr, the ship's manifest stating the product's weight as 259,507 pounds. When the tobacco was weighed in Scotland, it was discovered that it actually weighed only 200,000 pounds. Customs officials refused to allow the tobacco to be off-loaded until the merchants rectified the paperwork or produced the additional tobacco.

The second letter, dated Mar. 4, 1766, also reported a problem with a manifest and involved the importation into Scotland of "38 Loads, 9 feet & 11 Inches of Wood." This letter stated that those receiving the wood in Ayr refused to pay the bounty at the time of import because the ship's master didn't have a certificate from Virginia verifying the growth of the wood in that colony (VCRP, In-Letter Book).

Water Power at Rappahannock Forge

Not a nail, a pound of flour, or a single camp kettle could have been made at Rappahannock without waterpower. Most of the forge shops depended upon this essential and well-utilized power source. Around 1757/58 Hunter commenced digging a canal along the north bank of the Rappahannock River. Hunter found it expedient to import his own slaves, the task of digging the canal and building the first shops requiring a huge and expensive labor force (Mulkearn 206). By the time it was completed, his canal was some three-quarters of a mile in length, included floodgates, three tailraces, and incorporated the basic principles of hydraulics that maximized water flow to his machinery.

---

[135] Archibald Hunter (1738-1796) was also a merchant. He was born in Duns, Berwick, Scotland and died in Falmouth, Virginia. His first wife was Eleanor Dunn who died in Virginia. He married secondly Lainy Thornton.

By 1760 Hunter's forge was processing pig iron purchased from Principio's furnaces in Maryland and Tayloe's Occoquan and Neabsco furnaces in Prince William County, Virginia. A letter in the Alderman Library, one of the few surviving documents written in James Hunter's own hand, states that in 1761 he shipped iron and tobacco to Liverpool via his business partner, John Backhouse (died 1781) (Hunter Family Papers, letter, Nov. 6, 1761). James chose not to limit his business to iron production and quickly diversified his efforts at Rappahannock to include all manner of manufacturing.

To divert water into his canal the Rappahannock River was impounded by means of a crib dam, a deceptively simple but sturdy structure of logs and rocks built not straight across the river, but at an angle. The entrance of the canal was cut into a stone outcropping and fitted with heavy wooden gate3s that were closed during floods. The ability to close the canal prevented raging flood waters from cascading down the canal and damaging or destroying the machinery. The dam didn't span the entire width of the river; instead, it extended about two-thirds the way across, diverting a substantial amount of the river's water through the canal entrance. The stone wall at the canal's entrance prevented damaging erosion as the water made the slight turn from the river and began its trip down to the forge.

Nearly the entire length of the canal survives. The entrance is approximately 2,700 feet above the present Embrey dam. Some 520 feet of the canal were lost as a result of construction of the Embrey Dam and later flood damage immediately below the dam. The canal is currently 50-75 feet from the river bank, though the shoreline has undoubtedly changed over the years. The lower 1,580 feet of the canal includes the three tailraces where water exited the canal after powering waterwheels on the various shops. Today the canal is about four feet deep for much of its length and, during its years of operation, was probably only slightly deeper. The canal not only powered Hunter's mills but was also used by batteau traffic for hauling goods up and down the Rappahannock.

John Strode, manager and millwright at the forge, stated that there was a fall of 19 feet from the head of the canal to the mills at the end" (Thornton and Dunbar vs Stephen Winchester). Some basic principles of hydraulics were utilized in the planning and construction of the canal. The water entered from the river through an opening approximately 24 feet wide. The canal then broadened to nearly 35 feet. Somewhere in the vicinity of Embrey Dam the canal narrowed to about 16 feet, this width continuing for the remaining length of the canal. The reduction of width was deliberate. The wider part of the canal acted as a mill pond, allowing the gathering of a large volume of water. By narrowing the canal just above the mills, the force of the water was increased, thereby increasing the horsepower generated by the undershot waterwheels. The narrowing of the canal coupled with the 19 feet of fall created nearly unlimited power for the forge shops, so long as there was sufficient water in the river. On Sept. 19, 1800 John Strode made a deposition in which he noted that during extremely dry periods, there was insufficient water flow to power all the forge mills. Strode stated, "There [were] many times that we had not more water than sufficient for our grain Mill sometimes for two only the rest of the works all standing for weeks together for want of water" (Thornton and Dunbar vs Stephen Winchester).

A canal and millrace configuration similar to that seen at Rappahannock Forge was used at Col. Samuel Beall's[136] (1713-1777) Frederick Forge on Antietam Creek in Maryland. At this site a canal carried water from the dam on the creek and as many as four undershot spillways extended at right angles from the canal. These powered the gristmill, sawmill, and furnace bellows. Of the various buildings and shops at Rappahannock Forge, at least a dozen of them depended upon waterpower. The forge, wire mill, iron furnace, plating mill, rolling and slitting mill, armory, steel furnace, saw mill, fulling, grist and merchant mills were all powered by water.[137] Some of these undoubtedly had their own water wheels. Other shops or buildings may have shared wheels. A stream impounded on the hillside above the river powered the furnace and the slitting mill.

Water wheels were a necessary part of furnaces and forges, powering the huge bellows that created the high temperature required to separate the iron from its rock matrix and lifting the ponderous hammers

---

[136] Samuel Beall was born in Prince George's County, Maryland, the eldest son of John Beall (1688-1742) and Verlinda Magruder (born 1690). In 1734 he married Eleanor Brooke (1718-1785). Samuel served as sheriff of Frederick County 1753-1756, was a justice there 1763-1775, and represented Frederick as a delegate in 1776. By 1766 he owned a quarter interest in Frederick Forge (later known as Antietam Forge) on Antietam Creek.

[137] A survey of 1918 shows that one of these building housed Hunter's still (Stafford County loose surveys, Office of the Commissioner of the Revenue).

that converted pig iron to bars. Many 18th and 19th century mills utilized overshot wheels to which water was directed by means of elevated flumes. The overshot wheel and its several variations were usually considered more efficient than undershot wheels that turned no faster than the water that flowed beneath them. Based upon the configuration of the canal and its associated millraces, Hunter's primary power seems to have come from undershot wheels. The design of the canal created a tremendous flow that would have been sufficient to power all but the heaviest industrial operations. Those shops requiring more power were clustered together at the eastern end of the tract, about 1,900 feet below Embrey Dam, where a branch flowing down the hillside shows evidence of having been impounded well above the tops of the forge buildings. Water from the resulting millpond was probably channeled through elevated flumes to large overshot wheels. The forge[138] and hammers mentioned in John Mercer's letter as well as the rolling and plating mills required the additional horsepower of overshot wheels. The branch, which forms the eastern boundary of Hunter's property, appears to be the one referred to by Swank (Swank, 48).

The engineer behind most of this complex was John Strode. A brilliant man, he unquestionably made the most of the available resources and technology. Still, there are so many tantalizing and unanswered questions regarding the means by which he powered the mills and shops. Did he drive different machinery from shared wheels? How were his flumes designed to reach the multiple water wheels no doubt in use at the forge? How many shops were clustered along the canal and races? Only a detailed archeological study of this site will answer these questions, but the engineering of the water power system at Rappahannock Forge must have been as spectacular as the range of industries it supported.

Based upon the stone still littering the site, much of the building at Rappahannock was done with locally quarried hard stone as opposed to the sandstone so common in other parts of Stafford County. There was no shortage of stone in the immediate vicinity of the forge and stone for these buildings probably came from Hunter's own property. The area along and above his canal has long been quarried for stone, the pit from a modern quarry clearly visible on the south side of the river and on the west side of I-95. A deed between John Dixon and James Hunter included a stipulation that Dixon be allowed to take his batteaus up and down the forge canal, stone no doubt constituting some of these shipments (King George County Deed Book 5, page 646). Millstones were also quarried from Greenbank, just upstream from the forge:

> On Ten days notice, I will furnish Mill stones of any size that may be wanted (under 4 feet ½ diameter); I have a pair of Stones now in operation taken from my Quarry and find them to be of the very best quality for grinding Corn—the Rock is of the strongest granite, and sharpest grit, and continues sharp longer than any stones I have ever tried. The prices will be as follows $48 per Pair, for Stones of 4 feet diameter, other sizes in the same proportion. Any communication on the subject through the Fredg. Post Office, will be attended to.
>
> Geo. Banks[139]
>
> (*Fredericksburg Political Arena*, Feb. 23, 1830)

Stone for some of the forge buildings undoubtedly came from the granite quarries along the Rappahannock and just upstream from the iron works. Many of the stone foundations still visible are of granite rather than the sandstone that is so common in other parts of the county.

Transportation by water and land was vital to the economic success of the forge. Of major concern at Rappahannock were the transportation of supplies into the forge and the movement of finished work out to purchasers and waiting ships. On file in King George County is a deed of July 1766 in which "John

---

[138] A forge, also known as a fire, consisted of a hearth about four to five feet square the base of which was usually about 2 ½ feet high. The structure was often made of brick and had a bellows to the side or rear to blow the fire. Nearby stood a trough of water to quench hot iron and to cool the smith's tongs. Mineral coal, rather than charcoal, was usually used in a forge. The blacksmith had to be able to judge when stock was hot enough to work. This was accomplished by eye. Blood-red heat was used when only the surface required smoothing and the piece of iron was not to be re-shaped. A flame or white heat was necessary when the stock was to be hammered into a different shape, drawn down, or upset. Sparkling heat or welding heat was used only for the delicate and highly skilled process of welding.

[139] George Banks (1779-1737) owned Greenbank.

Dixon[140] of Gloucester Clerk and James Hunter of county of King George Merchant" struck a deal in which James paid £60 a seven-acre parcel, "also a wagon road or highway leading from the Town of Falmouth to the Iron Works of the said Hunter." By the time of this purchase, Hunter had built enough of a business to warrant needing a "highway" to Falmouth. A portion of this road is still clearly visible on the eastern end of the site. It is also included on the 1820 road map of Stafford (see Fig. 2-2).

On June 10, 1768 John Dixon sold Hunter nine acres of land and about ten acres of river rocks, the land to extend back from the river "a sufficient distance to make a road of a proper width for two waggons to pass each other thence keeping the same distance, from the river down a Parrallel course with the river to the said [Dixon's] Mill run, so that the distance between the river and the back line shall in no place exceed the distance that is below the Overseers House and the river." Dixon also sold Hunter "certain rocks, and Islands in the river aforesaid to the extent of fifty yards from the river side into the river to begin from the upper side of Falmouth Ferry and to go as far up the river as to the smooth water, between Vicaris's Island to the main land in King George County but so as not to extend into Vicaris's Island, for the purpose of rendering the said river navigable for Battoes, by a canal to his, the said James Hunter's Forges, reserving to the said John Dixon...the free use of the said Canal for the service of his Plantation, and liberty to have free access to the river for the benefit of the water." A memorandum added to this document stated, "It is the true intent and meaning of the parties that the said James Hunter, and his heirs shall have the priveledge of extending the waggon Road from the bottom of the hill opposite with the lower end of Vicaris's Island as far as his forge, and that exclusive of the within consideration, the said John Dixon, which the said James Hunter's Mill is kept up, to grind toll free when required, all the wheat and corn used by the said John Dixon's family either while he resides in Fredericksburg or King George County, and also to grind at the said James Hunter's Mill toll free, for the use of his plantation in King George County adjoining Falmouth" (King George Deed Book 5, page 646). These two agreements gave Hunter direct access from the forge to Falmouth and to the Warrenton Road.

Sources of Iron Ore, Pig and Bar Iron

While the presence of the two furnace sites (and the attendant waste material) proves that iron was smelted at Rappahannock, numerous historical records indicate that Hunter processed very large quantities of pig iron purchased from other furnaces. So far as his own furnaces are concerned, it isn't known precisely where Hunter obtained his iron ore. Occoquan Furnace in Prince William County shipped their ore from company-owned mines in Maryland. This was found to be more cost effective than mining it locally and paying Lord Fairfax, Proprietor of the Northern Neck, his 1/3 interest in ores mined on lands patented through the Northern Neck Proprietorship. Various contemporary documents record Hunter's purchase of pig and bar iron from the Principio and Baltimore[141] furnaces in Maryland and from Occoquan and Neabsco furnaces in Prince William County, Virginia,[142] and at least one shipment from Buckingham

---

[140] The Rev. John Dixon, was the son of John Dixon of Bristol, England and his wife Lucy Reade (1701-1731).

[141] On Mar. 26, 1772 Charles Carroll, Sr. (1702-1782) wrote to his son, Charles, Jr. (1737-1832), "Clem Brooke who is Here tells me you Have only ten Ton of Pigg Iron, may be Mr. Carroll's Clerk may sell to Hunter as you do, if so let Him give an order to Brooke & specify the Quantity tht Hunters Vessell may have a back freight" (*Maryland Historical Magazine*, "Extracts" vol. 14, p. 141).

[142] Though often owned by different partners, the various Chesapeake iron furnaces were inter-related in that they shared common workers and investors. These early works were not necessarily rivals and company ledgers regularly show purchases and sales between the different works. For many years, John Tayloe's Neabsco Furnace was the leading supplier of iron products in the Prince William area. Tayloe, of Mt. Airy in Richmond County, established this furnace in 1737. Tayloe's father, John Tayloe (1687-1747) was agent and attorney for a consortium of English businessmen who, in the 1720s, had built the Bristol Iron Works near Leedstown. Falkner, Neale, Triplett, and Ewall had all been involved with Bristol as well and, in 1746, contracted with the Baltimore Iron Company to build Lancashire Furnace on the Patapsco River in Maryland.

In 1749 Capt. Charles Ewell (1713-after 1749), backed by Ralph Falkner (died 1787) (who had previously worked for Principio), Edward Neale (died 1769), and John Triplett, opened another furnace in

Furnace[143] near Richmond. John Mercer's letter mentioned that James owed £3,000 to "Spotswood's' estate," indicating the possibility of his utilizing the old Tubal mines. These were well upstream from Hunter's works, approximately ten miles west of Tubal, and would have required transporting the ore overland by wagon for a considerable distance. However, Spotswood's mines were outside the bounds of the Northern Neck Proprietorship and, therefore, were exempt from Lord Fairfax's royalty. According to Thomas Jefferson "a forge of Mr. Hunter's, at Fredericksburgh, makes about 300 tons a year of bar iron, from pigs imported from Maryland" (Jefferson 28).

By the mid-1770s Hunter was desperate for iron and British gunboats were wrecking havoc on American ships. While Rappahannock had been importing pigs from Maryland, this was becoming more difficult. It may be, too, that he didn't want his own ships destroyed in the effort. Consequently, he began buying pigs from Occoquan and Neabsco, though they were facing the same shipping dangers as everyone else. A court case styled Lawson vs Tayloe, the records of which are on file in the Fredericksburg Circuit Court, reveals something of the relationship between James Hunter and Neabsco and Occoquan furnaces. According to one of the court reports filed in this case, "No ore could be procured in the neighborhood for [Occoquan] and that with which she was worked was brought always, by water from Patapsco River near Baltimore in Vessels, worked by slaves belonging to the Furnace, the hazard and risque attending the same during the war in all probability prevented the owners and managers from attempting it. It must be well remembered that during this time our Rivers and bays were often infested by the British Vessels and that

---

Prince William County on the Occoquan River just north of Neabsco. This was built on 1,520 acres on Hooe's Run on the Occoquan River. This venture did not experience long-term success owing largely to poor management.

In 1755 John Ballendine (died 1782) of Prince William, Charles Ewell's cousin, conceived of enlarging the Occoquan works and convinced his father-in-law, Richard Blackburn (1705-1757), and John Tayloe (1721-1779) to support the new venture. Blackburn put up collateral for Ballendine's share of the business and the latter became partners with Tayloe. This partnership operated successfully until 1760. At that time, Thomas Lawson (died 1793), manager of Neabsco, advised John Tayloe that Ballendine had overextended himself financially and was in danger economic collapse. John Semple (1727-1773), a merchant from Port Tobacco, Maryland took over the Occoquan works and under his poor management, the venture failed completely. Ballendine moved for a time to Amelia County, but soon returned and settled in Fairfax County where he died.

According to the contemporary account of Archdeacon Andrew Burnaby in 1759, ore for the Occoquan Furnace was shipped from Maryland mines. He wrote, "The ore wrought here is from Maryland: not that there is any doubt of there being plenty enough in the adjacent hills, but the inhabitants are discouraged from trying it by the proprietor's having reserved to himself a third of all ore that may be discovered in the Northern Neck." As relations between England and America deteriorated, Principio phased out most of their operations in Maryland. By 1777 Occoquan was forced to close, the result of British raids on American ships and an aging manager unable to maintain the facility. Occoquan's last blast yielded only 104 tons, not nearly enough to even pay the expense of running the furnace.

Some years ago, the present owners of the Rappahannock Forge site unearthed seven pigs, five of which were stamped with the name Occoquan and two with Neabsco. Since the latter venture ceased operations in 1765, it is safe to say that these pigs arrived at Rappahannock sometime prior to that year. The pigs were not found at the forge site, but on top of the ridge overlooking Embrey's Dam. They appeared to have been used in the bottom of a fireplace as part of a brick floor and foundation that were unearthed when the owners were putting in a driveway. This building may have been a kitchen, the pigs being used in the floor of the fireplace to retain heat.

[143] On June 14, 1776 the Fifth Virginia Convention appointed trustees to manage funds for a blast furnace "for making pigg Iron in the County of Buckingham." John Ballendine, part owner of the Occoquan Furnace, and John Reveley (1741-1806) were named managers of this new facility.

On July 10, 1778 the owners of the Buckingham Furnace sought "to inform the public that the pig iron from said furnace is not only fine for bars, but of an excellent quality for castings, which may be seen at the foundry below Westham [Westham was six miles above Richmond on the north side of the James River]. To prove its good quality in bar iron, &c. we sent a ton of pig to Mr. Hunter's forge, which has been wrought into bars, and yielded well. The bars have likewise been manufactured into gun barrels, horse shoes, axes, chains, nails, coarse and fine wire" (*Virginia Gazette*, July 10, 1778).

our Slaves were escaping to them by every opportunity." On Sept. 24, 1775 Thomas Lawson wrote of his concern about the impact of English raids on American ships, "I am every day in dread of hearing that Cutters are in the Bay and Rivers and will seize our vessels. If this should ever be the case, farewell all Virginia Iron Works." Eventually, bringing iron ore from Maryland became too risky an undertaking. Another note from the suit states that in 1782 pig iron shipments from Neabsco were half of what they had been in 1780, "the navigation of the bay and rivers afterwards came to be shut up, all the vessels taken and lost, and a total stop put to the furnace."

On Sept. 24, 1775 Occoquan and Neabsco manager Thomas Lawson (died 1793) wrote to owner John Tayloe[144] (1721-1779), "I have had many letters from Messrs. Hunter and Strode requesting me to put one or both furnaces in Blast so early this fall as to supply them with a sufficiency of Pig Iron before the Winter sets in...and was it not that I am in dread of not having a Sufficiency of Ore to carry her on through the course of the Winter brought home this fall, I would really begin to blow in a very few days." Lawson noted, "Mr. Hunter has given up the thoughts of getting Pigs from Maryland, on my telling him that I had done as much as I would do in order to obtain them." Of James Hunter he added, "I begin to think that he is one of the many to be found in the world who wants everything to be done for them without doing anything themselves, that he might not be under the necessity of buying Bar Iron for the use of the Smiths ship as he said he should if I did not get him Piggs" (Fredericksburg Circuit Court Records, Lawson vs Tayloe).

Exactly how much iron Rappahannock received from the Prince William furnaces and how long they did business with each other is unknown. The earliest mention of James Hunter in the fragments of ledgers included in the Lawson vs Tayloe suit is dated Dec. 31, 1772 and lists Hunter's balance for iron at £1,326.15. 3 ½. Hunter had been purchasing pigs prior to this time, however, as in October 1771 Thomas Lawson mentioned those purchases in a letter to John Tayloe. An entry dated Dec. 31, 1782 in the Occoquan ledger states that James then owed the furnace £2067.1.0 with £563.4.3 in interest from July 19, 1777. That amounted to £2630.5.3 to which was added a currency exchange fee of £876.15.1. The grand total was £3507.0.4 and, according to the records of the court case, as of 1786 the balance remained unpaid.

As relations with England deteriorated, however, obtaining ore and pig iron from Maryland became too costly and unpredictable. Shipping prices skyrocketed; the presence of English gunboats on the Potomac and the imposition of shipping embargoes made navigation hazardous. Because it was an English company, Principio was forced to close their operations at the outset of the war, much of their property being confiscated.

Hunter seems to have begun operations at Rappahannock with a furnace, forge, and merchant mill, the latter designed to grind large quantities of flour and meal for commercial export. He quickly expanded to include other areas of manufacturing. Perhaps inspired by others who had made profits by making spirits, Hunter built a still. According to John Mercer, "Fra. Thornton, I suppose encouraged by Woodford's[145] success & my disappointment has commenced brewer this year at the falls tho I imagine he has no copper, or only a very small one, as Mr. James Hunter informed me that for want of a good one, he had bought two or three of Mr. Dick's crack'd potash kettles,[146] which he purchased at the price of old iron,

---

[144] John Tayloe was educated in England. He built his home, Mt. Airy, in Richmond County, was a signer of the Treaty of Lancaster in 1744, a member of the Ohio Company, and a trustee of the town of Dumfries. In 1747 he married Rebecca Plater of Maryland and Middlesex County, Virginia. In 1757 John was appointed to the colonial council under Lord Governor John Dunmore and, following the reorganization of the state government was elected to the first Council of state under Governor Patrick Henry. He resigned on Oct. 9, 1776 and died in April 1779.

[145] Probably William Woodford, Jr. (1734-1780) of Windsor, Caroline County. In the summer of 1757 he became an ensign in the Virginia Regiment and eventually attained rank of lieutenant. During the Revolution, he served first as a colonel, then brigadier general in the Continental Army. He was captured in Charleston, South Carolina in 1780 and died as a prisoner of war in New York. In 1762 he married Mary Thornton (c.1744-1792).

[146] There was a potash factory at the foot of Amelia Street near Dick's home. Prior to the Revolution, this establishment shipped large quantities of soap to England. During the war, soap was made there and, later, fertilizer. On Oct. 3, 1775 the Committee of Safety paid Charles Dick £4.18.2 ½ for 22 ¾ pounds saltpeter and 346 pounds of sulfur (Virginia Committee of Safety, 1775-1776).

as it seems that after the smallest crack they will not do for potash, but will boil anything else" (Mulkearn 194).

Hunter's initial efforts were geared to satisfying domestic needs. He quickly expanded his operations to include rope[147] and textile manufacture, specifically wool and cotton and, perhaps, linen. He built a wire mill, the products of which were used for a number of purposes including chain and for making the wire teeth in textile cards. Sometime during or prior to 1776 Hunter employed as manager Alexander Hanewinkel who was skilled in textile and related manufactures. Technological advances in these fields were opening new export opportunities upon which Hunter hoped to capitalize. An undated petition of Alexander Hanewinkel to the House of Delegates[148] reveals something of this little known individual. He stated that he was "regularly bred by the Trustees for improving Manufactures in Great Britain, in the making of Heckles, Wool Combs, Tin Wool & Cotton Cards, Wire making, &c and [is] perfectly well acquainted with the dressing of Flax and Hemp" (Chandler 207-208).[149] In his petition he complained that he had "attempted to prosecute these branches, but by reason of his want of stock, and the excessive high wages of Workmen in erecting the Machinery & utencils &c [he] will be obliged to lay aside his designs or at least carry it on in such a small way, as will be but of very little use to the publick in general and entirely frustrate his first intention of rendering himself a useful member at a time when so much wanted." He continued by saying that he had formed a partnership with John Atkinson of Fredericksburg and had "erected a Hemp Mill, and Hecklery, which will undoubtedly reduce the exorbitant prices of linen in this part of the Country. But such a work will oblige your Petitioner and Partner to employ a number of spinners, Weavers and Ropemakers to expend such of the material as are unsaleable." Alexander, probably in hopes of gaining more notice of the members of the House, "begs leave to mention that he is the person who has the conducting of the Slitting Mill and the sole direction of the Steel furnace of Mr. Hunter's works."

Exactly how much flax and hemp were being processed at Hunter's Works is unknown. Both were extremely important and salable products during the 18th century. Flax was grown not only for its fibers that were woven into linen fabric, but for the linseed oil squeezed from its seeds. This valuable product was used in paints, as a wood preservative, for lamp oil, in the formulation of printing inks, and as an ingredient in certain medicines. Flaxseed contains 30-40% oil that can be forced from the seeds under very high pressure. Special water-powered mills were constructed for this purpose.

Because the flax was harvested after the seeds had matured, they were easily removed from the stems. Obtaining oil from the seeds required a multi-step process. Prior to crushing, the seed was cleaned to rid it of dirt, twigs, stones, or other foreign material. The seeds were then dumped into a heavy wooden trough or mortar and crushed in a water-powered stamping mill. Heavy vertical timbers were lifted by camshafts and let drop onto the seeds in the trough. Crushing the seeds into an oily meal greatly aided in the release of the oil during the subsequent pressing. Once the meal attained the correct consistency, it was removed from the mortar and transferred to a roasting pan. By gently heating the meal, oil-rich cells within the seeds were ruptured. After roasting, the meal was placed in woolen bags. These were wrapped in leather-backed horsehair mats and placed between heated iron plates in a water-powered wedge press. Massive wooden wedges were driven down between the plates using stamping-like vertical timbers lifted and dropped by camshafts. Oil from the seeds was collected from beneath the plates. After stamping, the woolen bags were peeled off the brittle presscakes. These were ground in the crushing mill and re-processed to recover an additional quantity of oil. Cakes from the second pressing were used for cattle feed or fertilizer. Flaxseed processing was carried out beginning in October or November after the flax was harvested and continued through the winter and spring. On average two men could process ten bushels of flaxseed per day producing 14-17 gallons of oil (Litchfield 50).

---

[147] In 1776, Hunter contracted to supply the Navy with naval stores, which included rope made with hemp. A single sailing ship could be fitted with 20 miles or more of rope.

[148] This petition probably dates from early 1776.

[149] The noted American engineer and inventor Oliver Evans was instrumental in improving the manufacturing of iron wire for making these cards. In 1777/78 he perfected his equipment and proposed to build a factory in Wilmington, Delaware and operate it under state patronage. He was not successful in this venture, but other inventors and businessmen carried forward with his ideas and by 1797 there were three manufacturers of iron wire cards in Philadelphia (Bining, Pennsylvania Iron Manufacture 104).

Hemp was an important raw material, being used not only for rope but for making a tough, durable fabric known as "hempen cloth." The hemp fibers were present only in the outer layer of the plant's stalk. Separating these from the unusable part of the stalk and processing them into cloth was a six-step process. Freshly harvested stalks were first "retted," or laid on the ground and weathered for three to six weeks until the non-fibrous part of the stalk began to rot. They were then thoroughly dried by heating over a fire. The dried stalks were crushed using a hemp brake, which resembled a flax brake but was larger. This part of the process shattered the partly rotted and dried portion of the stalk that fell away during the braking process, leaving the usable fibers. "Scutching" or scraping the fibers with a wooden blade removed the remaining woody bits. The fibers, now separated from the stalk, were softened by repeated pounding or rolling in order to make them pliable enough to be spun or woven. Finally, the fibers were pulled through a series of heckles, large combs with iron teeth that aligned the fibers and separated out the shorter lengths. The hemp fibers were then ready for spinning into coarse yarn.

Hemp seeds were also useful, being crushed and processed like flaxseed to produce oil. This was sometimes substituted for linseed oil in paints. Hemp was usually processed from late fall through early winter (Litchfield 51).

Prior to spinning into yarn, all fibers, wool, cotton, hemp, or flax, required combing. This was accomplished with combs or cards, little more than wooden paddles with metal teeth that were used in pairs to comb and straighten the fibers. Each type of fiber required a slightly different style of card, those being used for flax having larger teeth and known as heckles. During the early 1770s, automated machinery replaced hand labor in cutting and bending the metal teeth used in textile cards. The automated manufacturing of textile cards dramatically decreased their cost and made them a major American export. Hunter hoped to capitalize on this new industry by employing Hanewinkel who was familiar with the manufacture of these cards.

On Sept. 6, 1776, shortly after leaving Hunter's employ, Alexander placed an advertisement in the *Virginia Gazette*. Atkinson and Hanewinkel offered for sale "at their hemp and flax manufactory in Fredericksburg, dressed Flax and Hemp of all qualities...Wrought hemp from the brake[150] is taken in to mill and dress, at ten pounds in the hundred weight, the toll to be taken after milling." They also took the opportunity to offer employment to "regular bred Flax Dressers, Weavers, Spinners, and Rope makers. Gentlemen may have their negroes &c. instructed in Flax Dressing, for six months of their labour." Alexander added that he "finds himself under the necessity of specifying the prices of the heckles, as a great many people are disappointed who commission them, thinking that they are of the common kind, and of course send less money than they can be sold for; they being equal to, if not better than any imported from Britain" (*Virginia Gazette*, Sept. 6, 1776). The only other reference to Alexander yet located by the author, is contained in the records of Caroline County. On June 10, 1779 Alexander was named as a defendant in a suit brought by Robert Gilchrist (c.1721-1790) (Caroline County Order Book, 1778-1781, pg. 178). The records contain no details of the case. Beyond this, it is unknown what became of Alexander Hanewinkel and his heckles.

The Revolution

By 1776, activities at Rappahannock Forge changed dramatically. As relations with Britain deteriorated, Hunter and Strode shifted their emphasis from domestic goods to the production of military supplies and weapons. Henry Banks wrote, "After the war had commenced with Great Britain, [Strode] obtained the leave of his employer to establish a manufactory of small arms and other utensils necessary for the portending war" (Banks 81). Strode then "made an excursion to Philadelphia and other parts of Pennsylvania, and brought with him many mechanics, of different avocations. These were soon fully employed, and most rapid progress in making muskets, pistols, carbines, horsemen's caps, camp kettles, spades, shovels, &c. was made" (Banks 81). It was John Strode, then, who was largely responsible for the growth of Rappahannock Forge in 1776-77.

The military was in desperate need of all manner of supplies the colonies having no organized military infrastructure capable of supplying, coordinating, or maintaining a war. At the outset of the

---

[150] A flax brake was a heavy wooden contraption that broke the stems of the flax plant, freeing the long fibers within.

Revolution it was obvious that there were insufficient weapons available with which to fight a war, no accepted standards for weapons production, and few facilities capable of repairing damaged guns. While there were many firearms in the colonies, most were owned by individuals rather than by the military. Further, most of these weapons were obsolete, survivors from the French and Indian Wars or old English weapons imported prior to the Mother Country's embargo of English arms. Further, there were relatively few gunmakers residing in the colonies. Of the gunsmiths who did live in America, the most highly skilled were the Quaker craftsmen[151] of Pennsylvania and made the popular "Kentuckies," rifled long guns noted for their accuracy. These were not suitable in battle situations, however. Because they were rifled, the balls were tight fitting, making the gun slow to load by comparison to the English smoothbore guns that could be loaded and ready to fire every 15 seconds. Additionally, the Kentuckies lacked a bayonet, leaving the rifleman defenseless in the inevitable charge. Finally, each of these weapons was hand made by a single craftsman, making production painfully slow, and standardization of breakable parts non-existent. What was lacking, then, was a ready supply of firearms that were built to similar standards and suitable for the battle tactics of the day. Also missing was a network of gun manufacturers and repairmen who could build new guns and quickly and efficiently repair those that were damaged. Desperate for guns, Virginia contracted with small companies and private individuals for weapons. These were made in haste and under the very lax and diverse requirements of the various militia units that contracted for them. In essence, then, there was no standardization and no system for efficiently producing and repairing the quantity or quality of arms needed to fight a war. By rights, the colonies were so deficient of weapons and the organized political and military structure to defend themselves that what began as an armed uprising in Boston should never have escalated to a years-long war. The critical factor that allowed the colonies to wage successful armed opposition against the world's most powerful nation was the well developed iron industry based primarily in the Chesapeake region. Colonial authorities had always been deliberately vague about the subject of iron production when making official reports to the Crown. Consequently, England probably had little concept of the production capabilities of American ironmakers. It was a relatively simple matter for iron masters to shift from the manufacture of domestic items to military supplies.

---

[151] Many of Hunter's skilled craftsmen were Quakers, their vocations ranging from gunsmiths to millwrights. They established their own settlement and meeting house on Poplar Road (State Route 616) centered around Poplar Grove farm. There are few surviving written records pertaining to the Quaker enclave and most of the residents moved west in the early 19th century after activities at Rappahannock Forge slowed. The Friends held a monthly meeting in Stafford from c.1769-c.1807. The Stafford Meeting was also known as the Potomac Creek Meeting and probably was held at or near Poplar Grove (Worrall 539).

As a point of interest, the Quakers seem to have preferred building with hard stone as opposed to using the local sandstone. The old Poplar Grove house (pulled down c.1900) was built of hard stone as was the handsome spring house which is still standing. These structures, all of which are attributed to Quaker builders, were unique in Stafford because nearly all other stone buildings or foundations utilized the easily worked local sandstone. Most of the foundations remaining on the forge site appear to be of granite or other igneous rock that was hauled down from a quarry upstream on the Rappahannock, probably at Greenbank. The limited quantity of sandstone at the forge site suggests that this material was used primarily as quoins or other decorative accents on a few of the buildings. About one mile southeast of Poplar Grove is the Milton Christie farm on which still stands a very old stone house, possibly of Quaker construction.

At Hopewell on Aug. 7, 1769 Friends living near Potomac Run in Stafford requested permission to hold a meeting for worship. The next mention of the Stafford Quakers was on Oct. 4, 1780 when Hopewell decided that Stafford, Bear Garden, Culpeper, and Mt. Pleasant "be further indulged as heretofore."

Little is known about the Stafford Quaker meeting house. A meeting was held there in 1807 and in June of 1817 Abraham Branson (1754-1827) and William Jolliffe of the Hopewell Meeting (Frederick County, Virginia) were appointed trustees to sell the meeting houses in Stafford, Southland, and Smith Creek, reserving the cemeteries at each place. Money from the sale was to be applied to enclosing the graveyards and paying the expenses of the sales. In September 1822 Robert Painter conveyed title to the Friends of Stafford but no deed was executed at that time.

Abraham was the son of William and Elizabeth Branson of Stafford. He married first Catherine Reese (born 1760) of Frederick County and, secondly, Sarah White (1755-1832), also of Frederick.

To cope with the weapons crisis, the Second Continental Congress ordered the organization of Committees of Safety. Virginia's Committee was chosen at its third convention in August 1775. The committee essentially became a working government, replacing the former colonial organization. In addition to its function as an executive council, the committee was also responsible for obtaining weapons for the local militias and setting minimum standards for the production of new arms. In December 1775 the Convention of Delegates ordered the Committee of Safety "to contract, upon the best terms they can, with such gunsmiths, or others, as they may approve, for manufacturing or supplying such quantity of arms as they shall judge proper for the defence of this colony" (Virginia Committee of Safety Ledger, 1775-1776). Established facilities such as Rappahannock Forge were supported by the creation of new armories designed to meet military needs.

In July 1775 the Third Virginia Convention established a gun factory known as the Fredericksburg Manufactory (Hening, vol. 9, 71-72). This complex was constructed on the old Dixon's Mill tract on Hazel Run, utilizing several existing buildings being "within 200 yards of Fredericksburg and convenient to the landing where sea vessels may quickly load."[152] This facility consisted of the main factory, a powder magazine managed by Charles Washington (1738-1799), a cartridge works, repair shops, and Dixon's Mill. The first public gunnery established during the Revolutionary period, this establishment not only made and repaired arms, but also prepared cartridges and other ammunition. Fielding Lewis (1725-1781), Charles Dick[153] (1715-1783), Mann Page, Jr. (1719-1781), William Fitzhugh[154] (1721-1798), and Samuel Selden[155] (1725-1790) were appointed to oversee the operation (Hening, vol. 9, 71-72). Lewis and Dick were from Fredericksburg, Page from Spotsylvania, and the latter two from Stafford. Although there were five commissioners appointed to establish the manufactory, the day-to-day operation of this facility was under the direct control of Fielding Lewis and Charles Dick.

Fielding Lewis resided at Kenmore (now within the confines of the city of Fredericksburg) and was well acquainted with James Hunter. Charles Dick lived at what is now 1107 Princess Anne Street, between Amelia and Lewis Streets. He was a Scots merchant and operated a potash factory below his house near what is now the Central Rappahannock Regional Library. His house is considered the oldest surviving home in Fredericksburg. Scottish by birth, Charles was a major landowner and patriot who did much to promote the Revolutionary cause. He was also well acquainted with his fellow Scotsman, James Hunter and all three had had dealings with Accokeek Furnace. During the Revolution, these three men had a close working relationship.

On June 6, 1776 James Hunter delivered several well-made muskets to the Virginia Convention of Delegates. The members were pleased with the quality of the weapons and their minutes record:

> Ordered, That the Committee of Safety be directed to engage with Mr. James Hunter, of Fredericksburg, for as many good muskets, with bayonets, sheaths, and steel ramrods as he can manufacture within twelve months from this time, at the rate of six pounds for each stand; and that they allow the same price to any other person who shall manufacture arms

[152] Recent archeological excavations around the gunnery revealed its presence on the hill overlooking the old Walker-Grant Middle School.

[153] In 1750 Charles Dick married Mary Roy, the daughter of Dr. Mungo Roy of Caroline County.

[154] William Fitzhugh built and resided at Chatham. He owned thousands of acres in Stafford, Westmoreland, King George, Richmond, Spotsylvania, Fauquier, Prince William, and Fairfax counties. In 1763 he married Ann Randolph (1747-1805), the daughter of Col. Peter Randolph (1713-1767) and Lucy Bolling of Chatsworth, Henrico County. William was an intimate friend of George Washington and spent his later years on his Fairfax County plantation, Ravensworth, where he frequently entertained Washington.

[155] Samuel Selden resided at Salvington on Potomac Creek. He was the son of Joseph Selden (died c.1726) and the grandson of Samuel Selden the immigrant who settled in Elizabeth City County. This early Selden patented Salvington in 1699. Young Samuel seated the property and married in 1751 Mary Thomson Mason (1731-1758), the daughter of Col. George Mason (1690-1735) of Stafford. He married secondly his first wife's cousin, Sarah Ann Mason Mercer (1738-1806), the daughter of John Mercer (1704-1768) of Marlborough. The 1782 land tax records list this tract as then consisting of some 1,700 acres. Col. Selden died at Salvington and the plantation was inherited to his son, Wilson Cary Selden (1772-1822). The fine brick home was destroyed during the War Between the States.

within this colony, of equal goodness with the sample now produced by Hunter (Palmer, vol. 1, p. 440).

The Board of War approved this resolution and increased payment to £8 for each weapon "providing they shall be as well filled [sic] and finished as those formerly purchased by this Board of the said James Hunter" (Palmer, vol. 1, 440).

It is the author's belief that during 1776 and 1777, Virginia made a substantial financial investment in Rappahannock Forge. This money was used for building and expanding the already considerable milling and production facilities Hunter had previously established. The author is of the opinion that, while in 1775 the Third Virginia Convention ordered the building of the Fredericksburg Manufactory, the members also saw the advantage of helping Hunter expand what was already an established industrial complex. It would take a year or more to set up a factory, hire gunsmiths, and make productive this new facility in Fredericksburg, time that simply could not be spared. According to Paula S. Felder, "As early as August 1776, it was clear that Hunter's forge could outstrip the gun manufactory in output" (Felder, Fielding Lewis 233). In his report to the Council Charles Dick acknowledged that Fredericksburg's first year of production was 1776 and that the manufacture of new arms was very limited (Executive Communications, Jan. 23, 1781). Initially, Fredericksburg was primarily concerned with repairing damaged weapons and Rappahannock with the production of new ones. There are several documents that support the theory that Virginia made financial investments in Rappahannock, as does some logical thinking.

On Feb. 4, 1776 Fielding Lewis[156] wrote to his brother-in-law, George Washington, regarding the Fredericksburg Manufactory. He stated, "Our little manufactory improves daily. I expect by the last of March we shall be able to make ten muskets completed per day; we have been mostly employed in repairing old guns since we began, and have only one gunn lock maker, who has instructed many others who begin to be very expert, we make now 35 per week, and increasing. Most of the locks which Lord Dunmore stole away from the muskets in the magazine [at Williamsburg] are now replaced by our workmen" (Abbot, Revolutionary War Series, vol. 3, 244-246).

At the outset of the Revolution, Hunter already had skilled gunsmiths on hand and, prior to that time, may have been making a few guns to meet local demand. He was certainly manufacturing high quality muskets by the spring of 1776. He also had the proven capability of producing a wide variety of iron tools, hardware, and many other products necessary to the war effort. By July he had contracted with the state to produce and obtain naval stores, requiring further expansion of his facility. The state's investment in Rappahannock Forge would result in a faster output of much needed weapons. While it has long been believed that every building, every employee, and every gun, sword, and kettle produced at Rappahannock were personally financed by James Hunter, this cannot have been so. James was, without doubt, a wealthy man. He seems to have arrived in Virginia with money and, using his keen sense of business, increased his fortune as a merchant, shipper, and slave dealer. Unquestionably, he used his own wealth to build the initial part of his forge and mill complex on the Rappahannock. However, during 1776 and 1777 there was an explosion of building and production at Rappahannock Forge. Almost overnight, Rappahannock Forge was transformed from an independently owned business into one of the largest industrial complexes in America. The financial outlay for this complex was far beyond what a single individual could have managed, regardless of his business savvy or wealth.

Hunter's shift in production from domestic goods to war materiel is reflected in a newspaper notice of June 1776:

---

[156] Fielding Lewis and James Hunter were well acquainted and considered themselves friends, not competitors. Their relationship continued beyond the realm of arms manufacture as evidenced by a letter dated Apr. 11, 1774 from George Washington to Hugh Mercer (1725-1777). Hugh, who already owned Pine Grove just across the Rappahannock River from Fredericksburg, wished to purchase part of Washington's Ferry Farm tract. The conveyance was dependent upon James Hunter who had leased and was farming the property. Washington wrote, "When possession can be given, I am not altogether clear in, as I believe Mr. [William] Fitzhugh and Mr. Hunter look upon the tillable and Pasture Land as engaged to them till the Fall; but Colo. Lewis can give you the best information on that head, as it was with him the Agreement was made. I apprehend their time was up last Fall, but on my way from Wmsburg. in Decr. last, Mr. Hunter told me, that by his agreement with Colo. Lewis, he was to have notice of my wanting the place by seed time" (Abbot, Colonial Series, vol. 10, pp. 27-28).

Wanted, at Hunter's Iron Works, at the Falls of Rappahannock River, in the Manufactory of small Arms, a Number of Hands who understand the File, also Anchor Smiths, Blacksmiths, and Nailers, to whom suitable Encouragement will be given; where may be had, for ready Money only, Bar Iron, Fullers Shears, Files, and a Variety of Articles made in the Iron Branch, and where Tradesmen of every Kind, who incline to settle, may have Half Acre Lots of Land at a small Ground Rent for ever, and may be assisted in building, Provisions, and Materials, and the Produce of their Labour taken in Payment for the same. Apply to John Strode, Manager of said Works (*Virginia Gazette*, June 22, 1776).

This notice was run again on Nov. 15, 1776 at which time Hunter offered to purchase "Negro Tradesmen in any of the above Branches, also Coopers" (*Virginia Gazette*, Nov. 15, 1776).

On Feb. 19, 1777 John Strode, forge manager, answered an earlier letter from Gov. Patrick Henry who had inquired how the facilities at Rappahannock might be enlarged. Strode's response to Henry may reveal the source of some of the funding for Rappahannock. He wrote,

I observe His Excellencies disposition *to encourage these manufactures in a more extensive plan of operation.* Happy should I be on the Occasion to receive and carry your Commands for that Laudable purpose into execution. Provided it can be done on Terms whereby you will not be looser. Hands I believe under proper regulations might be obtained, especially as most of those at present employed here are master workmen and could each take in a prentice or unexperienced hand under him if we had shoproom. Tools and the different Machinery enlarged. An New shop for the Gunsmith 300 feet in length will be absolutely necessary, where a Clerk or two to assist the Head Workman could receive and deliver work and minute down each mans performance and loss of Time. But the expence of Building and procuring (in these times) the Materials for the different apparatus would in the whole amount to a Sum too large I apprehend for you *or any private person in America* to advance without Government [assistance] (Executive Communications, Feb. 19, 1777).

Strode also suggested that some "provision by Law for price and payment" be made that would be "a just equivalent for your expence and trouble at least." John expressed further concern regarding the procurement of metal:

For I am well convinced that you are no gainer by it even now, when your stock of Pig Iron, Steel, Brass, Copper, Spelter(?) and most all the Materials hath been laid in in the very best terms, long since purchased when plenty & cheap paid for in specie and Bills. Its true muskets can be made without Brass counting. But without Pig Iron you cannot make even the Scalps for your Factory or any other State not to mention the Vast consumption in Ship building. And anchors for the Navy, and variety of Smith work you are already (and in a great expence) prepared to do and so deeply engaged in. And tis plain from Mr. [Thomas] Lawson's Letters you can have but small dependance in Colo. Tayloes Furnaces for any supply. If you are to furnish your works from Maryland with Pig metal it must come excessive high. Should the Iron Masters there agree to sell it at all, which I must doubt as each Furnace which makes good Pig for Bars are connected with Forges belonging to same owners who will perhaps have it in their power soon to oblige their extensive State to purchase from them Manufactured Iron only. The encouragement which that province has at all times given to works of the king enables the proprietors to carry them on with such facility, dispatch and advantage as you nor no Iron Master here can accomplish untill countenanced and enabled by the Legislative power, but were you sure of purchasing Pig Iron from Maryland on any terms you are not always certain of conveyance by Water and that conveyance extravagant as freight is raised is not always safe, at best difficult and never certain (Executive Communications, Feb. 19, 1777).

Strode also asked Henry to guarantee Rappahannock a ten-year contract with the state and recommended that further improvements might take the following form, "consider of some expedient for working the Iron mines within the state[157] [and] Protecting the Artificers, Labourers, Teams and other property of the Adventurers employed therein and the Works, Factorys, &c. depending thereon."[158] The Virginia authorities immediately addressed all the issues raised by Strode. We know from the adjutant general's report of Jan. 13, 1809 that the huge armory had been built (though that report states that the building was 350 feet long) (Cromwell 1). In his journal entry of June 30, 1777 Ebenezer Hazard mentioned the newly finished steel furnace, slitting and plating mills (Shelley 417). Such expansion simply could not have occurred without a substantial infusion of money from a source greater than James Hunter's pocketbook.

In his letter Strode also warned the governor that Hunter had made little or no profit on the previous run of arms. He closed his letter by commenting on some of the frustrations he faced in carrying on his day-to-day operations. He wrote, "You cannot then promise much to yourself or country while every recruiting officer has it in his power to envigle and enlist the hands from the works and at any time press the waggons out of your Service, which has happened frequently and in the Very throng of my business. Once last Summer they carried off Six Waggons. Just as I was endeavoring to remedy that loss by replacing those with Three more, all I could get, they in like manner pressed them; and divers other times they have taken waggons some on Journeys for grain and other Supplies for the Works."

Strode summed up his plan for expansion with an itemized list:

1. That no recruiting officer be allowed to enlist any of your workmen that has not obtained your (or yr managers) discharge, or been absent from the place Three months.
2. That no officer whatever without your consent be allowed to take a Team employed at your works whether your own property or hired.
3. That if at the instance of the Legislative powers you extend the different machinery for gunns and other Arms they agree to take the produce at certain reasonable rates for some fixed time, or if it be found expedient to discontinue the Manufactory for any reason, cause or event, which may arrise before the expiration of yr Term, in that case you to have an adequate allowce [sic] made for the loss and disappointment which you must otherwise suffer.
4. The liberty to raise ore on any place convenient and carry it off, paying the same as before mentioned and
5. To erect Furnaces for Smelting said ore on the Next and most convenient place for wood and water.
6. Some small encouragement or priviledge to Artificers, particularly for ingenuity and constant faithful application to the branch they profess—particularly gunlock's.
7. That the most effectual Measures be taken for prosecuting and enlarging the different branches you have already engaged in viz 1st Bar Iron for Army & Navy, 2d Arms, 3d Slitting and plating mills, 4th Wire mill, 5th Steel Furnace (Tyler 84).

---

157 On Nov. 1, 1777 a jury met at Garrard's Ordinary near Stafford Courthouse to begin escheat proceedings on the Accokeek tract. Although the jury was well aware that the land in question belonged to the Washington family, Hunter was granted 200 acres including the mines and old furnace site. Hunter's move to escheat the Accokeek land created a controversy that long outlived him and many of the Washington family who fought unsuccessfully for the return of their land.

158 On May 1, 1777 the General Assembly passed an "Act for the encouragement of iron works" which provided for Strode's suggestions. This act allowed Hunter to claim 200 acres of the old Accokeek tract, vested in him "half an acre of ground for a landing, situate at some convenient place on Aquia or Potowmack creek...so long as the said James Hunter his heirs and assigns, shall continue to keep up and carry on his furnace and works aforesaid." The act also prohibited the impressing of horses, wagons, or drivers employed at any lead, copper, or iron works.

Shortly thereafter, James Hunter submitted a now lost memorial or petition to the Governor and Council of Virginia reiterating Strode's plan of expansion. Patrick Henry summarized the memorial in a letter in which he wrote to the House of Delegates:

> The Memorial of James Hunter humbly sheweth that your Memorialist actuated by the warmest zeal for the good of his Country, has with very great Labour & expence erected a Variety of Works, such as Forges, steel Furnaces &c & begun others such as slitting, plating & wire Mills, & established Factories for fabricating small Arms, entrenching Tools, Anchors & other things necessary in the Army & Navy, Works evidently essential not only to the welfare, but to the very existence of this State.
>
> That he has been encouraged to overlook in the Prosecution of these works Difficulties which seemed to private Abilities insurmountable by assurances from your Excellency of Publick Countenance & Support.
>
> That he has been heretofore supplied with Pig Iron, the Basis of all his Manufactures, from Maryland; but that this mode of supply, at best are improper as well as an unbecoming dependence for a great State, is now become exceedingly expensive & precarious.
>
> That Nature has made ample Provision amongst ourselves for these our necessities; but that no advantage can be derived from this circumstance either to the Publick or your Memorialist, these necessary Materials being the property of Persons who either have not the Power or the Inclination to work them.
>
> That he has more than once suffered by the enlisting of his workmen & the pressing of his waggons, & that he dreads the greatest Detriment to the Publick as well as to himself if he be not secured from such injurious Proceedings in future (Executive Communications, Correspondence, May 31, 1777).

Henry took Strode's plan to heart and his response to the plan provides further evidence of Virginia's financial involvement at Rappahannock. On May 31, 1777 Henry wrote, "As there was no Manufactory of Iron in the State which was carried on to such an Extent, and to Purpose of such vast Importance as Mr. Hunter's near Fredericksburg, I took the Liberty of promising him the *Assistance of the Public* in the Prosecution of his works on a more enlarged Plan" (Executive Communications, Correspondence, May 31, 1777). Henry continued by saying of Strode's letter, "The subject of these Papers was of so much consequence to this State, and Mr. Hunter himself *so deserving of the attention of the Publick* that I thought it my duty to lay them before the General Assembly, *who alone can enable him to carry on these extensive & valuable Works*. We must strongly recommend Mr. Hunter is, that he asks for no pecuniary assistance, but merely for Materials to work. He requires only what the good of the State most evidently points out, which is to, open Mines within the same, and not to depend on our Neighbours for so necessary an article as Iron" (Executive Communications, Correspondence, May 31, 1777). While strained, Virginia's economy did not collapse under the stress of war until 1779.

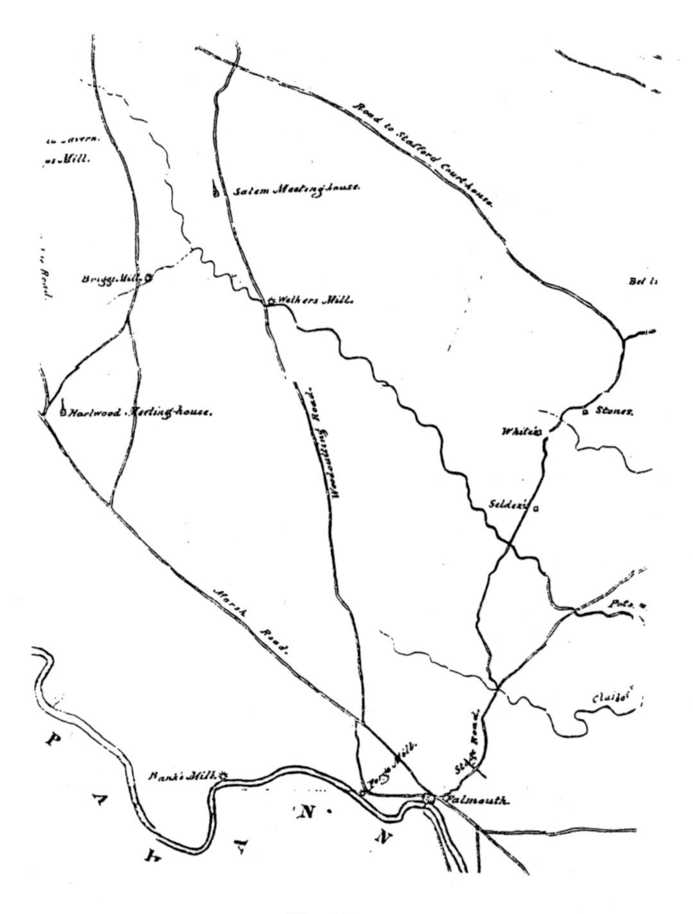

Fig. 2-2
The "Woodcutting Road," now State Route 627, made a nearly direct line between the Accokeek Furnace property and Rappahannock Forge ("Stafford County" c.1820. BPW 711(33), from sheet 3. Reproduced with permission of the Library of Virginia).

Further support of Rappahannock's receiving substantial state funding is found in a reference to a letter to the governor dated Aug. 27, 1791. Written by Abner Vernon and Jesse Hargrave[159] the letter was in reference "to the imputed indebtedness of James Hunter, deceased, as superintendent of the *Public Foundry*" (Palmer, vol. 5, 360). That letter is now lost, but the reference to the public foundry is certainly significant in that it indicates that Rappahannock was the recipient of public funds.

Ebenezer Hazard[160] (1744-1817) was a resident of several northern states and was an ardent patriot. During the Revolutionary period, he made two trips south and kept wonderfully descriptive

[159] Jesse Hargrave (1752-after 1819) was a Quaker and the son of Samuel and Martha Hargrave. In 1776 he married Mary Pleasants, the daughter of John and Agnes Pleasants of the Henrico Monthly Meeting. In 1819 Jesse served as deputy sheriff of Sussex County.

[160] Ebenezer was a New York bookseller who began receiving and forwarding mail in May 1775 when the royal postal service between New York and New England was discontinued. On Sept. 21, 1775 Continental postmaster, Benjamin Franklin, appointed Ebenezer postmaster of New York City. When the Continental army abandoned that city in Sept. 1776, Ebenezer moved his operations to George Washington's

journals of both. When he visited Falmouth and made his journal entry of June 30, 1777, Rappahannock Forge was a major enterprise. He wrote,

> Crossed Rappahannock in Company with Mr. Smith & rode with him to see some Works erected by James & Adam Hunter: on our Way we passed through a small Village called Falmouth, situated on the North Side of Rappahannock, about 1 ½ or 2 Miles higher up than Fredericksburg. The Falls of the River are so trifling as not to be worth mentioning. A little above Falmouth are Mr. Hunter's Works which, with the Dwelling Houses for the Workmen, form a small Village. He is now erecting a Mill for slitting & plating Iron, & is about building a Furnace for melting Iron Ore. At present he makes (from Pig Iron) Bar Iron, Anchors, all Kinds of common Blacksmith's Work, Small Arms, Pistols, Swords, Files, Fuller's Shears, and Nails. He has a Grist Mill & Saw Mill, a Cooper's Shop, a Saddler's shop, a Shoemaker's Shop, a Brass Founder's Shop, & a Wheel-Wright's Shop. All these, except the Grist Mill are constantly employed in his own Business, & not to supply the Wants of other People. Besides all these, Mr. Hunter has erected Works for making Steel, (this Business he is just beginning upon) & raises large Quantities of Wheat, Corn, Oats, Hay, &c: in short he is a great Farmer. He informs me that his different Works, & the Negroes he employes cost him £40,000 Virginia Currency. He has cut a Canal, ¾ of a Mile in Length by which the Water is conveyed (Part of the Way through Rocks) from the River to his Mills &c, & whereever Water can be used he takes the Advantage of it, which saves a vast deal of manual Labor. There is always Plenty of Water, & he has Flood-gates, fixed in the Rocks, which are used in the Case of Freshes, to carry off the superfluous Water. Mr. Hunter has hitherto imported the Pig Iron he used from Maryland, but as the Navigation is now obstructed, he proposes to collect Ore in Virginia and making Pigs at his own Works...The Steel Manufactory[161] is situate on a high Hill which commands a beautiful & extensive Prospect (Shelley pp. 400-423).

At this time, there were very few slitting or plating mills in colonial America. Though regulations imposed by Parliament varied greatly during the course of the 18th century, the construction of slitting and plating mills was generally discouraged or outright forbidden. The Iron Act of 1750 clearly stated that no new slitting or plating mills or steel furnaces were to be erected in the colonies, though those built prior to the act could continue in operation. This act was intended to restrict three important areas of iron

---

headquarters. In January 1777 he became a surveyor of postal roads for the Continental post office and, in January 1782 Congress appointed him Postmaster General of the United States.

[161] Hazard stated that the steel mill was built upon a hill suggesting that it may have been an air furnace similar to that built by Alexander Spotswood at Massaponax. Air furnaces were built with tall chimneys and created their own draft without the aid of bellows but required a steady flow of air over the chimney. The placement of the furnace on top of a hill facilitated a continuous draft. William Byrd described Alexander Spotswood's air furnace at Massaponax:

> There is an opening about a foot square for the fresh air to pass through from without. This leads up to an iron grate that holds about half a bushel of charcoal and is about six feet higher than the opening. When the fire is kindled, it rarefies the air in such a manner as to make a very strong draft from without. About two foot above the grate is a hole [that] leads into a kind of oven, the floor of which is laid shelving toward the mouth. In the middle of this oven, on one side, is another hole that leads into the funnel of a chimney, about forty feet high. The smoke mounts up this way, drawing the flame after it with so much force that in less than an hour it melts the sows of iron that are thrust toward the upper end of the oven. As the metal melts, it urns toward the mouth into a hollow place, out of which the potter lades it in iron ladles, in order to pour it into the several molds just by...The inside of the oven is lined with soft bricks made of Sturbridge or Windsor clay, because no other will endure the intense heat of the fire {Wright 370).

manufacturing: nail making, sheet metal or tin plate iron production, and the making of blister steel. American ironmasters freely ignored the Iron Act producing whatever items they could sell for a profit.

The very narrow strip of level land that parallels the north side of the river severely limited the placement of buildings at Rappahannock. Most of Hunter's mills were placed along the canal though special arrangements were made for a few shops. Slitting and plating mills were decidedly "heavy industry" and normally required the power of an overshot water wheel. From evidence on the forge site, as well as from contemporary documents, it appears that all shops requiring overshot wheels were clustered on the east end of the tract and were powered by water from a branch that was impounded within the steep walls of a ravine. This branch also formed the boundary between the lands of James Hunter and John Richards (1734-1785). By the mid-1770s there were at least three buildings standing on the west side of this branch. There was no room for an additional building on that side and Hunter had little choice but to put the slitting mill on John Richards' side of the branch. The 1777 act for the encouragement of iron works allowed him to purchase a half acre from Richards, noting, "whereas it is represented that the said James Hunter cannot erect his dam and slitting mill at his aforesaid works without a small quantity of land adjoining thereto, the property of John Richards, gentleman, be set apart for that purpose" (Hening, vol. 9, 303-306). Foundation B (see site map, Fig. 2-9) is on the east side of the branch and is probably the remains of the slitting mill. The reference here to a dam further supports the impoundment of the branch for powering these works.

Hazard stated in his journal that there was a brass founder's shop at Rappahannock. Brass was a highly versatile metal that was utilized in many everyday utensils as well as for gun parts. As with any industrial process, brass making was a dangerous endeavor. The 1797 edition of Encyclopedia Britannica, includes the following description of the process:

> Brass is frequently made by cementing plates of copper with calamine, whereby the copper imbibed one-fourth or one-fifth its weight of the zinc which rifes from the calamine.[162] The process consists of mixing three parts of the calamine and two of copper with charcoal dust in a crucible, which is exposed to a red heat for some hours, and then brought to fusion. The vapours of the calamine penetrate the heated plates of copper, and add thereby to its fusibility. It is of great consequence for the success of this process to have the copper cut into small pieces, and intimately blended with the calamine.
>
> In most foreign founderies the copper is broken small by mechanical means with a great deal of labour; but at Bristol the workman employ an easier method. A pit is dug in the ground of the manufacture about four feet deep, the sides of which are lined with wood. The bottom is made of copper or brass, and is moveable by means of a chain. The top is made also of brass with a space near the center, perforated with small holes, which are luted with clay; through them the melted copper is poured, which runs in a number of streams into the water, and this is perpetually renewed by a fresh stream that passes through the pit. As the copper falls down it forms itself into grains, which collect at the bottom. But great precaution is required to hinder the dangerous explosions which melted copper produces when thrown into cold water; which end is obtained by pouring small quantities of the metal at once. The granulated copper is completely mixed with powdered calamine, and fused afterwards. The process lasts eight to ten hours, and even some days, according to the quality of the calamine.
>
> It is a wonderful thing, says Cramer, that zinc itself, being simply melted with copper, robs it of all its malleability; but if it be applied in form of vapour from the calamine, the sublimates, or the flowers, it does not cause the metal to become brittle.
>
> The method mentioned by Cramer to make brass from copper, by the volatile emanations of zinc, seems to be preferable to any other process, as the metal is then preserved from the heterogeneous parts contained in the zinc itself, or in its ore. It consisted in mixing the calamine and charcoal with moistened clay, and ramming the mixture to the bottom of the melting pot, on which the copper, mixed also with the charcoal, is to be placed above the rammed matter. When the proper degree of heat is applied, the metallic vapour of the zinc contained in the calamine will transpire through the clay, and attach itself to the copper, leaving the iron and the lead which were in the

[162] Calamine is an oxide of zinc mixed with a small amount of ferric oxide.

calamine retained in the clay, without mixing with the upper metal. Dr. Watson says, that a very good metallurgist of Bristol, named John Champion, has obtained a patent for making brass by combining zinc in the vaporous form with heated copper plates; and that the brass from this manufacture is reported to be of the finest kind.

Brass is sometimes made in another way, by mixing the two metals directly; but the heat requisite to melt the copper makes the zinc burn and flame out, by which the copper is defrauded of the due proportion of zinc. If the copper be melted separately, and the melted zinc poured into it, a considerable and dangerous explosion ensues; but if the zinc is only heated and plunged into the copper, it is quickly imbibed and retained. The union, however, of these two metals succeeds better if the flux composed of inflammable substances be first fused in the crucible, and the copper and zinc poured into it. As soon as they appear thoroughly melted, they are to be well stirred, and expeditiously poured out, or else the zinc will be inflamed and leave only the red copper behind (Encyclopedia, vol. XI, p. 466).

It is unknown which system of brass making was utilized at Rappahannock Forge, nor is the location of the brass foundry known.

Ebenezer also mentioned in his journal that Hunter had been using Maryland pigs and, due to the difficulty of obtaining them, was finishing a furnace where he could smelt his own. Without constant repairs, these structures deteriorated. According to John Mercer's letter to his son, Hunter had operated a furnace prior to early 1768 (Mulkearn 206). It is unknown if the furnace mentioned in Hazard's journal was a replacement for the original or was a second furnace. Though purely conjectural, it is possible that the old furnace suffered from structural defects, lack of water, or some other technological difficulty. The present forge site contains what appear to be the remnants of two iron furnaces.

Throughout the Revolution, there was a working relationship between Rappahannock Forge and the Fredericksburg Manufactory. In his February 1777 letter to Patrick Henry, John Strode stated that during the Fredericksburg Manufactory's early years, that operation was somewhat dependent upon Rappahannock. Strode wrote, "with your positive orders to supply the publick gunnery [Fredericksburg] from the coal here to continue coaling all this winter and now in the severest frost & snow (to your great loss) weather neither fit for Man nor Beast to go out in, or otherwise suffer all your works to stand idle, on the whole, as laying in a Stock of Coale and every other part of the business here performed by Waggon stands on so precarious a footing, nothing now, nor indeed ever will without absolute security in this service be carried on with advantage to yourself or the State." Rappahannock, then, was supplying all the charcoal used by the Fredericksburg Manufactory, requiring Strode to keep his men charcoaling all winter in order to meet the needs of both facilities. Hunter provided other necessary items to the gunnery as well. On June 15, 1776 the Committee of Safety ordered "that Mr. James Hunter be requested to furnish the commissioners of the Fredericksburg Gun Manufactory with Spades & Shovels, which will be paid for by this Board upon their Certificate of having received them" (McIlwaine, vol. 1, p. 24).

The official records of the various state agencies charged with carrying out the Revolutionary effort indicate that Virginia's iron works and gunneries operated cooperatively and communicated frequently with each other. From at least 1776-1779 Hunter maintained a working relationship with Westham Foundry near Richmond. In May and June of 1776 Rappahannock purchased from Westham such things as bar iron, forge hammers, wire, and tallow. These were paid for with bar iron, spade and shovel molds, and finished hand tools (Westham 1779-1781). While the relationship between the foundries was normally cooperative in nature, the relationship occasionally became competitive. An interesting notation appeared in the minutes of the Board of Trade in which the Board of War ordered "that Mr [David] Ross' proposals of delivering the Iron contracted for at the Foundry at his charge and risque at one third of the price of Bar Iron at Mr. James Hunters works per Ton be accepted. Warrant delivered the State Agent for Twelve thousand pounds for part payment of Iron purchased of Mr. David Ross" (Virginia Board of Trade Minute Book). Hunter's old friend at Westham, John Ballendine, underbid him on bar iron and won a substantial contract with the state. Ballendine and Westham soon went bankrupt, however, and it is unknown how much of this iron Hunter's competitor actually delivered at the bargain price.

1780-81 marked a turning point in the history of the forge. For all intent and purpose, Virginia was bankrupt. By late 1780 the act exempting iron workers from the draft had been rescinded. Many of Hunter's craftsmen were either in the military or had moved over to the Fredericksburg Manufactory where they remained exempt. The loss of these workers forced James to close his small arms factory. In January 1781 Jefferson warned Hunter that Tarleton appeared to be headed towards Fredericksburg, probably with the intention of destroying both gunneries. Hunter dismantled his shops and moved the equipment to a safer location. Tarleton chose not to raid the forge, though whether Hunter ever returned all of his equipment to Rappahannock is unknown. The state still owed him at least £180,000 and there is no indication that this debt was ever paid. Pressed for cash and short of skilled craftsmen, James was forced to severely limit his operations at Rappahannock. The business never fully recovered, though Hunter continued to produce military supplies. A financial recession followed the Revolution. The fledgling United States was deeply in debt and the Virginia economy was devastated. James Hunter died in 1784 and although some of the forge shops continued in operation, production never again equaled that of the 1770s.

The Virginia records abound with references to James Hunter and his forge, but there are also several vague though intriguing possible references to him in the Maryland records. Based upon a letter from Adam Hunter to Gov. William Smallwood (1732-1792) Maryland officials attempted to purchase arms from Rappahannock Forge. Smallwood's letter arrived at the forge well after the removal of equipment occasioned by Tarleton's threatened advance. On July 31, 1781 Adam wrote,

> In my brothers absence who has resided at his plantation in Culpepper ever since our late alarm of the enemy coming this way, I have your favour of 29th currt by express, to which am sorry I cannot return a satisfactory answer. We have long discontinued the manufactory of small arms of which none remain on hand, and the workmen employed in that branch, are so dispersed that we believe it would be difficult matter to engage them again. I shall tomorrow transmit your letter to my brother, to whom I beg leave to refer you for satisfaction on this subject, but am apprehensive he will not be able to furnish the accoutrements mentioned. I am with the greatest respect—
>
> Your Excellenceys Most Obed Svt
> Adam Hunter
>
> (Pleasants, vol. 47, 376)

The Maryland records also contain several references to payments made to a James Hunter, but provide no specific information proving that these payments were made to James Hunter of Rappahannock Forge.

Though exact quantities cannot be ascertained from the surviving records, Rappahannock Forge produced firearms, swords, camp supplies, a wide variety of iron, brass, and steel military accouterments and naval stores utilized by the American army and navy during the Revolution. Despite the lack of definitive production figures, research indicates that Rappahannock was a major contributor to the Revolutionary effort. Yet the forge is essentially unknown to historians or to the general public. Rappahannock's historical obscurity is due to a loss of records rather than a lack of effort by owner James Hunter and manager John Strode. These men provided a struggling fledgling military with the supplies necessary to defeat the most powerful nation on earth, and they did so while facing daunting economic and logistical hardships.

The Naval Contract

On July 26, 1776 the Virginia Navy Board:

> Resolved that Mr. James Hunter be appointed Agent or Contractor for purchasing provisions Ship Materials and Naval Stores and all other necessaries which he may be directed to purchase by the Board for the Navy in the District of Potowmack River; and that he be allowed for his Trouble and expenses therein and for the Storage of all Goods by him purchased and committed to his care by the Board and Commissioners of two and one half percent upon all Sums of Money by him disbursed or paid for the use of the

Navy who entered into Bond with Benjamin Harrison Esquire[163] his Security in the Penalty of Ten thousand Pounds conditioned for the faithful discharge of the several Duties of his office and for Complying with all such Instructions as may be from Time to time given him by the said Board (Virginia Navy Board, Journals).

The awarding of this contract was a major factor in the explosive growth of the forge in 1776/77. Hunter was responsible for supplying shipbuilders within his district with such materials as pitch, tar, turpentine, sail cloth, rope, anchors, iron and brass hardware, weapons, spars, masts, bowsprits, etc. The records of the Virginia Navy Board abound with orders placed with Hunter (see Appendix A), some of these materials being made at Rappahannock Forge and some being purchased from other sources.

Hunter traveled to Williamsburg to sign the naval contract, then returned to Falmouth (Shelley 416). Early on in the contract there was a problem at Hunter's end either with production and acquisition or with shipping ordered items. George Mason of Gunston Hall wrote to Thomas Whiting[164] stating that Hunter was not supplying naval stores as he had contracted to do. Whiting responded to Mason, "We are surprised to hear that Mr. Hunter has made no Provisions for the [use] of the Navy and cou'd not conceive he wou'd doubt of his Authority to mak[e] such Provisions after having contracted and enter'd into Bond for that Purpose. We shall write him immediately on the Subject which we presum[e] will prevent all further difficulties in the business" (Virginia Navy Board, Letterbook). Whiting's letter to Hunter was dated Sept. 7, 1776 and read:

> Sir—We have received a Letter from Colonel George Mason respecting the supply of the Navy in the Potowmack District. He informs us you have not laid in any Stock of Provisions or any kind of Stores for want of a particular order from us for that purpose. We presumed that immediately after your appointment and return home you would have made yourself acquainted what quantity of Stores and Provisions was on hand in your district and would have laid in such further supply as you judged was necessary and did not conceive you would wait for a particular instruction from us directing the kinds or quantities you were to provide. We therefore now desire you will immediately procure on the best Terms you can such quantity of Clothing, Beding and Provisions as you may find on inquiry will be necessary for the present supply of your department and your draught on the Board will be duly paid (Virginia Navy Board, Letterbook).

The sheer volume of naval materials being manufactured and purchased for this contract far exceeded Hunter's warehouse facilities at the forge. At about the same time Hunter was negotiating his contract with the Navy Board, he began work to take over Cave's Warehouse near Belle Plains on the south side of Potomac Creek. Proximity to navigable water may also have been a consideration in choosing this particular location. The Journals of the Virginia Navy Board record, "Resolved that the following places be appointed as Magazines or Repositories for the reception of provisions and Naval Stores for the use of the Navy in the Several Rivers, to wit, for York River Fraziers upon Mattapony, for James River Low Point upon Chepokes Creek, for Rappahannock River—Bathursts Landing upon Piscataway Creek, and for Potowmack River Potowmack Creek in Stafford County...James Hunter upon Potomac River, entitled to a

---

[163] Benjamin Harrison (1726-1791) was the son of Benjamin Harrison (died 1745) and Anne Carter of Berkeley, Charles City County. He attended the College of William and Mary, but did not graduate. In 1749 Benjamin was elected to the House of Burgesses, a position he held until that body was dissolved by Lord Dunmore in 1775. During his years with the House, he was frequently chosen speaker. He was part of a committee that, in 1764, drew up a vigorous protest against the Stamp Act. In March 1776 Benjamin was placed on the marine committee and in June joined the newly established board of war and ordnance. After the creation of a new government in 1776 Benjamin was elected to the House of Delegates and served there until 1781. He was one of the signers of the Declaration of Independence and served twice as governor of Virginia. Benjamin married Elizabeth Bassett.

[164] Thomas Whiting (1712-1781) was the son of Maj. Henry Whiting (died 1728) and Ann Beverley. Thomas was a burgess from Gloucester County from 1755-1776 and a member of various state conventions and the Virginia Navy Board. He married first Elizabeth Beverley (died 1749), secondly Elizabeth Thruston (1740-1766), and thirdly Elizabeth Sewell.

commission of 2.5% on all sums disbursed on authorized purchases." An entry in these same records dated July 30, 1776 directed "that Mr. James Hunter receive into his care whatever Quantity of Cannon Ball may be sent him from time to time by Isaac Zane Esqr.[165] and forward the same in equal Proportions to the several Districts of Potowmack Rappahannock, York and James Rivers as soon as possible" (Virginia Navy Board Minute Book).

In October 1776 the House of Delegates passed "An act to establish publick storehouses, at the head of Potowmack creek for the reception of naval stores." This act authorized the "use of the said land for a publick warehouse...for the reception and safe keeping of the naval stores and materials for ship building." According to a legislative petition dated Nov. 5, 1776, "James Hunter states that the Commissioners of the Navy have contracted with him to take care of the naval stores and other materials in the upper part of the country. He has pitched on Cave's Warehouses[166] on the head of Potomac Creek as a place not accessible by the enemy. He asks use of the publick warehouses there" (Virginia Legislative Petitions, 198-P). Cave's was, at that point, still an official tobacco warehouse, but shipments there had drastically declined as the population had moved away from Potomac Creek. There were at least two existing warehouses on the site and Hunter proposed the construction of several additional buildings.

The tobacco warehouses stood within the bounds of a plantation owned by Andrew Edwards[167] (died c.1788) who later petitioned the Assembly to close them as little tobacco was being received there.[168] In October 1776 the Assembly passed an act stating in part, "Whereas it is found necessary that publick storehouses, for the reception of naval stores, be established at Cave's warehouse, near the head of

---

[165] Isaac Zane (1745-1795) was the son of Isaac Zane, Sr. (1711-1794), a prominent Philadelphia Quaker. Isaac, Jr. was not a model Friend, being far too outspoken, too militant, and too careless in his religious obligations. As a further insult to the Friends' sensibilities, he kept a mistress, Miss Betsy McFarland (died 1833), with whom he lived for some twenty years and who bore him a son, also named Isaac Zane. Isaac owned and operated Marlboro Iron Works in Frederick County, Virginia located on Cedar Creek about 12 miles from Winchester. They were near the junction of modern State Routes 622 and 628. In 1773 Isaac represented that county in the House of Burgesses where he spoke out against British tyranny. At the outbreak of the Revolution he converted his iron works to the production of four and six pound ordnance, shot, kettles, salt pans, camp stoves, and cannons. In 1794 Isaac was made brigadier general of the Frederick County militia.

In November 1776 Isaac contracted with the state to cast carbooses (ships' stoves) and pots for the navy. He was ordered to send these to James Hunter for distribution (Virginia Navy Board, Ledger, 1776-1779, Miscellaneous Reel #302). On Mar. 3, 1777 Thomas Whiting wrote to Zane asking if he had cast any cannons and, if so, of what sizes. Isaac was directed to immediately send any available cannons, carbooses, and pots to James Hunter for distribution (Virginia Navy Board, Ledger, 1776-1779, Miscellaneous Reel #302).

Jefferson commented on the remarkable toughness of Zane's castings. (Jefferson 28). By 1828 Zane's works had been abandoned.

[166] Cave's Warehouses were built on the old Spilman tract that is still shown on modern maps as Spilman's Landing. In an indenture dated Aug. 14, 1759, Peter Daniel (1706-1777) and his wife Sarah (Travers) (171_-1788) leased to William Spilman, planter, and his son Thomas (born 1751) "all the houses, orchards, woods, underwoods, profits and advantages (all mines and Quarries excepted)." The Spilmans agreed to pay an annual rent of 800 pounds of tobacco. William agreed to plant an orchard of 100 apple and 400 peach treed and to keep the property well fenced "from harm of Cattle or other Creatures and to keep the same well trimed [sic] from Superfluous branches and succors" and "to build Good & Sufficient Dwelling Houses and Tobacco Houses and keep the same in good repair."

In 1751 a William Spilman married Elizabeth Reeds. In 1755 Jeremiah Spilman married Bridget Edwards. The somewhat hazy relationship between William and Jeremiah Spilman and the Edwards family may account for Andrew Edwards residing on the property at the time James Hunter petitioned to take over the warehouses.

[167] Andrew Edwards married Elizabeth Withers.

[168] On Nov. 1, 1779 Andrew petitioned the Assembly claiming that only a small quantity of tobacco had been brought to Cave's Warehouses for the last four years "not exceeding 100 hogsheads [and] has resulted in the buildings being neglected. He asks that the inspection be ended there. The public lot is in the midst of his dwelling plantation" (Virginia Legislative Petitions, 197-P).

Potowmack creek, in the county of Stafford, to be under the care and direction of such person as shall be appointed by the commissioners of the navy" (Hening, vol. 9, pp. 235-236). A second petition by Edwards, also dated Nov. 5, stated, "the commissioners for providing for the Navy on Potowmack have thought fit to appoint Cave's as the most convenient place for storing the different commodities for the supplies of the Potowmack part of the Navy. This will cause him many inconveniences and impositions. Cave's is part of the plantation where he and his family lives" (Virginia Legislative Petitions, 197-P). Edwards claimed that that his pastures and orchard "would be under the mercy of lawless and ungovernable people, the men belonging to the different ships."

Although the authorities rejected both of these petitions, Hunter was granted permission to use the site. The act establishing those warehouses ordered "any two justices of peace in the county of Stafford...are hereby required, to issue their precept to the sheriff of the said county, commanding him to summon a jury...to meet the said justices at the warehouses called Cave's...[and] view and examine one acre of land whereon the warehouses aforesaid stand, and value the same exclusive of the said warehouses." That being accomplished, the court was to "give to the proprietor of the land a certificate of the valuation...and thereafter the said acre of land shall be vested in the governour of this commonwealth...for the use of the publick." The commissioners of the navy were authorized to "either cause proper houses to be built thereon, at the publick expense, for the reception and safe keeping of the naval stores and materials for ship building, and appoint a proper person to take care of the houses and stores, or may let the said acre of land to any person or persons for the purposes of building and keeping such storehouses thereon, and contract with them for the receipt, safe keeping, and delivery of such stores and materials, as they shall judge most for the publick good."

On June 9, 1777 the Board of War:

> Ordered that James Hunter Keeper of Naval Stores and Materials for ship building at Caves Warehouse on Potomac River, be permitted to Land & Store on the ground appropriated to the public use Pigg Iron & all such articles as may be transported thither by Land or Waters, for the Navy, Publick Factorys, his own Iron works & other uses: provided the same do not interfere with the necessary welfare of persons carrying Tobacco, to the Warehouses at Caves, nor hinder the free egress or regress to the same, for inspecting, delivering & carrying away Tobacco (Virginia Board of War, Journal, 1777-1780).

Fig. 2-3

Forging an anchor. Here the anchor claw is joined to the arm. Both pieces were kept as close to red heat as possible while several forgemen pounded them together with blows from sledge hammers (Diderot, Vol. VII, Marine, Forge des Ancres, Plate X).

On Sept. 24, 1775 Thomas Lawson, manager of Occoquan and Neabsco furnaces, wrote to furnace owner John Tayloe. From this letter it is evident that Hunter was using Cave's Warehouses prior to his official petition requesting such use. Lawson stated that he had "sent [Hunter] lately to Caves Warehouse upwards of ten tons of Piggs and scraps collected at the Occoquan and Neabsco furnaces with [illegible] which he must be satisfied till I make Iron" (Fredericksburg Circuit Court, Lawson vs Tayloe).

On Aug. 22, 1776 Thomas Whiting informed Fielding Lewis that Capt. Willis Cooper had been ordered to deliver 230 barrels for pitch and turpentine "to your care, half of which is to be detained for the use of your River and the remainder forwarded to Mr. James Hunter for the Patowmack department" (Virginia Board of War, Journal, 1777-1780).

There are few references to Hunter supplying cannons to the Navy. Those he did provide were obtained from outside sources, the casting of cannon being a highly specialized task not conducted at Rappahannock. Rappahannock Forge did manufacture muskets for the Navy. The Council of Virginia directed Col. George Weedon[169] (1734-1793) of the 3rd Battalion to apply to Hunter for these weapons. The "Guns are desired to be forwarded to this place without delay for the use of the marine Companies in York and James river" (Virginia Navy Board, Minute Books).

The naval contract authorized Hunter to produce supplies as well as act as a transfer agent for naval stores being routed through his district. On July 18, 1777, the Navy Board wrote to James:

> The board of Commissioners for directing Naval affairs desire to know whether you can supply them with Iron of different sizes and Anchors which you'll please to let them know by the first opportunity, and also your price. They would also be glad to know whether you have sent the Iron ordered by the Committee of Safety to the care of Mr. Richard Adams at Richmond. If you have not already sent it out, you'll please to send it as soon as possible as the Galleys wait for it (Virginia Board of War, Journal, 1777-1780).

Hunter supplied naval stores for ships built on the Eastern Shore, at Mattaponi, Smithfield, and Fredericksburg (Morgan 142, 188, 1120, 1312). The best known of these was the *Dragon*, built at Fredericksburg and the inspiration of Fielding Lewis. On Feb. 4, 1776 Lewis wrote to George Washington outlining plans to build warships at Fredericksburg. He wrote, "we are also preparing a Naval force Two Row Gallys one to carry one 18 pounder and the other a 12 pounder. Mann Page Esqr. and myself are to build immediately at this place."[170] In December 1776 Capt. Eleazer Callender, formerly commander of the sloop[171] *Defiance*, was assigned to supervise the building of the galley. He was appointed captain of the *Dragon* on Oct. 9, 1777. The original crew included as first lieutenant—John Hamilton; midshipmen—Francis Webb, Alvin Wilson, Joshua McWilliams, and Benjamin Rust; sailing-master—Wolling Smith;

---

169 George Weedon was born in Westmoreland County. He was appointed an ensign in the French and Indian War, rose to the rank of lieutenant in 1757, and to captain lieutenant in 1762. After the war he moved to Fredericksburg and married Catherine Gordon (died 1797). George operated a tavern in that town until January 1776 when he was appointed a lieutenant colonel in the 3rd Virginia Regiment. In June of that year he was made colonel. In August he and his troops joined the Continental army in New York and, on Feb. 21, 1777 was appointed brigadier general. He retired due to a dispute over his rank and for health reasons but was recalled to active duty in 1780 to organize and command Virginia troops. During this tour he served at the siege of Yorktown. After the war, he returned to tavern keeping in Fredericksburg. George also served on the Fredericksburg council and as mayor.

170 Lewis and Page were actually overseeing the building of two row galleys at Page's home, Mannsfield, near the present Fredericksburg Country Club. Additional ship building was also being done almost directly across the river below Snowden. A narrow gut on the north side of the river made an ideal boat ramp upon which the boats were built, then pushed into the river.

171 Also called shallops. These were large, heavy boats fitted with one or more masts fore and aft for lug sails. They were sometimes furnished with guns.

gunner—Iveson Nuttall; gunner's mate—John Nuttall. Later midshipmen were Samuel Eskridge,[172] Edwi
Eskridge,[173] and James Tutt.[174] Theophilus Field[175] was a later lieutenant on board (Stewart, Robert 57).

The keel of the *Dragon* was laid in 1776 and she was launched in October 1777. Ebenezer Hazar saw this ship during his visit to Falmouth. His journal entry of July 1, 1777 read, "Went to see a Galle built by the State of Virginia; she is called the Dragon, has three Masts, is to be rigged in the Manner of Schooner, & to mount two 18 Pounders forward, two 12 Pounders aft, & 16 double fortified 4 Pounder amid-ships. I think her too narrow for her Length, & her Masts are too taunt" (Shelley 419).

Dragon's first year of service was spent on the Rappahannock during which time an interestin event occurred. As the *Dragon* cautiously sailed along the Rappahannock, the crew awaited confrontatio with the enemy at each bend of the river. With little warning, a violent storm erupted, the force of the win being sufficient to snap and topple one of the *Dragon's* three masts. Facing imminent capture by the Britis if discovered in such a vulnerable situation, the ship's carpenter, Daniel George, went to work. The rive was lined with dense forest. Daniel rowed ashore, felled a tree, roughly fashioned it into a mast, and withi the space of a few hours, the ship was fitted with a new mast and ready for action, much to the amazemen of his shipmates. While the *Dragon* spent a good deal of time patrolling the Rappahannock, she also sav duty along the Eastern Shore (Stewart, Robert 57-60).

Some idea of the extent of Rappahannock's manufacturing capabilities is provided in a receip from the papers of the Continental Congress. This document was included in a volume of unpaid bill compiled in October 1785 and includes quantities of naval stores ordered and their prices (Continenta Congress). Considering the fact that a few of the items were not priced, the total of this one bill fa exceeded £11,000.

50 dozen broad axes	£600
4 dozen and 2 grindstones	50
2 ½ dozen crosscut saws	30
25 dozen adzes	300
25 dozen mauls	300
6 dozen and 8 whip saws[176]	80
10 Compass whip saws	80
25 dozen hand saws	300
171 dozen hand sawfiles[177]	2,052
60 dozen whip sawfiles	720
15 dozen whip sawfiles	780
10 dozen augers from 1 ¼ In. to ½ In.	120
4 dozen claw hammers	48
4 dozen sledges	48
4 dozen crow bars	48
40 dozen Spoke gimblets[178] of different sizes	480
15 dozen small gimblets of do. do.	180
66 dozen and 8 Rules	800
66 dozen and 8 compasses	800
140 dozen chalk lines	1,680
Oakum[179] three tons	60

[172] Samuel F. Eskridge was killed or died on the ship *Protector Gally* on Nov. 3, 1780. His son, George Eskridge received a land grant from Samuel as "heir-at-law" for 2666 acres granted to Samuel for his service as a lieutenant in the Virginia Navy. Samuel married Mary Foushee Lewis.

[173] In 1786 Edwin Eskridge was granted land allowed to a midshipman of the Virginia Navy for three years of service.

[174] James Tutt (c.1755-1821).

[175] In 1787 Theophilus Field married Susan Thweatt in Prince George County.

[176] Similar to a buck saw.

[177] Used to sharpen saw teeth.

[178] A type of plane or file with a curved blade used for shaping wagon spokes, ladder rungs, etc.

[179] Loosely twisted hemp or jute fiber impregnated with tar and used for caulking seams in wooden ships.

5 dozen Mallotts	60
30 gin falls & Blocks[180]	80
20 pr. hand saws	80
8 dozen hand saw setts[181]	180
5 dozen and 10 whip saw do.	70
4 dozen and 2 iron squares	50
12 ½ dozen saw bars for whip saws	150
33 dozen and 4 chisels from 2 In. to 1 In.	400
8 dozen and 4 gouges from 2 In. to ¾ In.	100
1000 pounds of Chalk	200
16# Red lead for moulding	16
1 set of tools to make pumps	
50 dozen falling axes	60
13855 ½ pounds spikes	
41366 ½ pounds Bolts Chains & Ruder Irons	
4000 Bushells of Coats to be made of woolen	
20 Barrells Tar	
20 Barrels Pitch	
20 Barrells Turpentine	
30 scrapers	
6 dozen Tan__ (illegible)	
1000 pounds Brimstone[182]	

The tantalizing references to Hunter's production rates and ship building efforts are scattered among various collections of old papers. Piecing together the history of the forge is possible only by diligent searching through vast quantities of records, the rewards of this labor being the discovery of a mere line or two. Contained in the papers of the Continental Congress and dated Aug. 20, 1778 is an order to issue a warrant on the treasurer "for 3000 dollars, in favour of the Committee of Commerce, to be by them transmitted to Messrs. James and Adam Hunter, of Fredericksburg, in Virginia, to pay the wages due the seamen on board the brigantine Morris, now under their care, laded with tobacco on public account, and other charges attending that vessel; the said committee to be accountable" (American Memory, Journal, Aug. 20, 1778).

Another similar reference is also found in the papers of the Continental Congress and dated Nov. 13, 1778. This was an order to "pay a bill drawn on [the Committee of Commerce] by Messrs. James and Adam Hunter, of Virginia, in favour of Mr. Amos Strettle [1720-1791], dated 26 Oct., 1778, for seven thousand seven hundred and thirty-five and 45/90 dollars; the said Committee to be accountable" (American Memory, Journal, Nov. 13, 1778). While it is unclear what role Amos played in the Revolution, he and his father, Robert Strettle (1693-1761) were also mentioned in the Accokeek ledger. On Jan. 20, 1779 Congress authorized payment of another bill from James and Adam to Amos for $460 (American Memory, Journal, Dec. 21, 1778).

Based upon the surviving official records, it is impossible to determine exactly how much the state owed James Hunter, much less how much remuneration he actually received. Hunter supplied material for the Committee of Safety, the Board of War, and the Navy Board, each of which maintained separate accounts with the forge. It is difficult to impossible to determine the difference between warrants and actual payments and little if any correlation exists between orders and payments/warrants. The Commercial Agents may have fully paid their debt to Rappahannock as on June 27, 1786 they paid £15,000 for bar iron and nail rod previously omitted (Virginia Commercial Agent, Day Book, July 18, 1782 - June 27, 1786). Tradition has long held that James Hunter was not fully compensated for purchases made by the state, but the extent of that official debt is unknown. Although Virginia's economy collapsed in 1779, warrants and at least some payments continued being directed to Hunter throughout the war.

---

[180] Block and tackle sets.

[181] Saw sets are used to adjust the angle of the saw teeth.

[182] Sulfur. May have been used for fumigating or gunpowder.

John Strode

It is important to realize that James Hunter was a merchant and a businessman, not an industrialist or manufacturer. He was well aware of his limitations and sought highly qualified people to conduct the various activities at the forge. Three major contributors to the success of Rappahannock were Frederick Klette, a German gun lock maker from Philadelphia and acknowledged master of his trade, Joseph Purkin,[183] a master gunsmith, and John Strode (c.1735-c.1820), a Pennsylvania Quaker who was a brilliant millwright and engineer. The latter was a Quaker from Chester County, Pennsylvania whom James hired as a millwright but whose energy and abilities soon carried him to the position of general manager of the entire operation. The success of Rappahannock Forge and the explosion of building that occurred there in 1776 and 1777 were direct results of Strode's efforts.

Little is known of John's early life prior to his arrival in Virginia. According to research done by Quaker historian, Jay Worrall, John Strode's great grandfather was a London barrister who pleaded for George Fox before the King's Bench in 1674. In 1682 his grandparents, George and Margaret Strode settled on 500 acres in Concord Township, now Delaware County, Pennsylvania. His parents, John and Magdelen, belonged to the Birmingham Meeting of Quakers in Pennsylvania. In their "Historic American Buildings Survey" the Library of Congress has a report on Strode's Grist Mill built in 1721 by George Strode (Historic American Buildings Survey). This mill is still standing with much of its original interior. It is located on the Lenape and Birmingham Roads in Sconnelltown, Chester County, Pennsylvania. According to this report, George Strode bought 236 acres in East Bradford Township in 1715 upon which this mill was built. The mill descended from George to his son John, then passed successively to Richard Strode, Joseph Strode, and finally to Caleb Strode who in 1853 sold it out of the family.

John Strode had a wife named Anne, though neither the Quaker records in Pennsylvania or Virginia nor records from the counties of Stafford or Culpeper reveal her maiden name. He also had one known son, Capt. Thomas Strode (died 1829) who was active in Culpeper and eventually shared his father's financial ruin. Thomas married Harriett Somerville Richards,[184] the daughter of Capt. William Richards (1755-1817)[185] of Culpeper and his wife Eliza Miller (died c.1815).[186] The announcement of their wedding appeared in the *Virginia Herald* of Aug. 16, 1808. On Aug. 5, 1809 a marriage contract was recorded between John's daughter, Mary, and Hugh Wallace Wormeley (born c.1786) of Frederick County ("The Wormeley Family" 901). Hugh was the son of John Wormeley (1761-1809) of Cool Spring, Frederick County, Virginia. The Strodes also had another daughter named Elizabeth who married Mordecai Barbour[187] (1763-1846). Perhaps encouraged by his father-in-law, Mordecai also became involved in

[183] Joseph Purkin was born in England. In 1779 he worked as an artificer in the Pennsylvania line before coming to Rappahannock Forge. By 1784 he was back in Pennsylvania working with Samuel Coutty repairing arms belonging to the city of Philadelphia. From 1793-1796 he was master armorer at Harper's Ferry (Hartzler, The Southern Arsenal, 75).

[184] There were two other Harriet Somerville Richards. The immigrant William Richards, who came to Virginia with Alexander Spotswood, married Harriet Somervile, a native of Wales. The name was passed down in the family. William Richards (1755-1817) named his daughter Harriet Somerville as did William's son, Maj. James Richards (1774-1844). This last Harriet (1820-1908) married in 1846 Philip Smith Hale (1814-1875).

[185] Capt. William Richards was the son of William Byrd Richards (born c.1697) and owned a good deal of real estate in Fredericksburg, Culpeper, and Stafford, including a grist mill and ferry on Deep Run. He lived at Mount Pleasant at the junction of the Rappahannock and Rapidan rivers in Culpeper County. A copy of his will was recorded in Fredericksburg.

[186] Eliza was the daughter of Simon Miller (171_-1799) and Isabella Miller, said to be the daughter of Simon's uncle John Miller (16_-1743) of Essex County.

[187] Mordecai Barbour was the son of Col. James Barbour (1775-1842) and Lucy Johnson. Mordecai joined the Culpeper County militia in May or June 1781, serving as a private, then lieutenant under Gen. Mechlenburg. His first assignment was to protect Rappahannock Forge. He also served under John Nicholas, John Woodford, and John Stewart. He was with Lafayette at the Battle of Jamestown, then moved on to Williamsburg and Richmond. He was also present when Cornwallis was captured.

milling and manufacturing. There are several deeds in Culpeper that document Mordecai's involvement in and ownership of water-powered mills. John Strode purchased at least two of his mills from him. Mordecai's manufacturing interests extended from cotton gins to making nails. On May 29, 1805 the *Virginia Herald* carried an advertisement for his cut nail manufactory in Fredericksburg "where they will sell, Cut and wrought Nails, Brads, Springs, Sadler's Tacks of all sizes." The Barbour family gave their name to Barboursville, Virginia.

In his research Mr. Worrall found that several of John and Magdelen Strode's children settled in Virginia. In addition to son John who worked at Rappahannock, his aunt, Margaret (Strode) Bryan (died 1748) lived in Frederick County, Virginia as the wife of Morgan Bryan. They were married in Chester County, Pennsylvania. Their granddaughter, Rebecca Bryan (1738-1813), married Daniel Boone (1734-1820). Edward Strode, Sr. (died c.1749), also of Frederick County, was probably John's uncle and Edward's sons, Edward, Jr. and James, were John's cousins. During the French and Indian War, Indians burned James Strode's home near Martinsburg. He was one of the eleven men appointed to organize Berkeley County (now West Virginia) when it separated from Frederick County.

John's first cousin was Jeremiah Strode. According to Strode genealogist Peggy Chapman, Jeremiah was a surveyor in the western country. He was mentioned in Col. William F. Gray's diary in which the latter recorded his conversation with Jeremiah:

> He is now a surveyor in the Mexican service, and living in Texas...He thinks there is a negotiation on foot, if it has not been concluded, for the sale of that country to the United States for $14,000,000, and if the United States does not buy it, that it will soon be independent, for that it is fast settling with people from the United States, who know their rights, and are determined to defend them; and they will never submit to Santa Anna's project of a central government...He ran the line between Texas and the United States, and says the maps are all wrong—that the line strikes the Red River about the *middle of the Raft*, instead of at *Pecan Point* (Gray, William 22).

On Sept. 19, 1800 John Strode was called as an expert witness in a court case involving two mills on the south side of the Rappahannock River. This deposition provides clues as to his birth date as well as the year of his arrival at Rappahannock. When asked if he was "vers'd in the science of Hydraulicks in general and of erecting water grain Mills in particular," he said that he had been doing that work "for about forty odd years." Assuming that he began millwrighting at about the age of 20, he was probably born in the mid-1730s. Strode was also asked if he had superintended the "erecting and managing the water works at

---

In 1833 or 1834 Mordecai applied for a Revolutionary War pension (Pension Application S8043). The authorities seemed dubious about his rank and he wrote a letter explaining that his lieutenancy was a camp appointment. He also said that he served the entire tour at the siege of Yorktown, then escorted British prisoners to the barracks beyond Winchester. He was granted his annual pension though the records state two different sums of $53.33 and $73.33.

Mordecai resided in Culpeper from his birth until 1806 when he removed to Fredericksburg. He remained there until 1808 and then moved to Dinwiddie County before moving into the town of Richmond. A deposition in the pension records states that he had four children, one of whom predeceased his father. Two others were dead by Aug. 17, 1847, the date of the deposition, leaving only a son, John Strode Barbour (1790-1855) whose name appears frequently in 19th century Virginia records. For unexplained reasons, when a very old man, Mordecai moved from Richmond to Greene County, Alabama where he died. Elizabeth Strode predeceased her husband and he married secondly Mrs. Sally (Haskell) Byrne (Revolutionary War Pension Applications).

Mordecai was heavily involved in milling and manufacturing, though whether this involvement was a result of his father-in-law's influence or whether he was already so inclined. On Feb. 26, 1808 Mordecai executed a deed of trust in Culpeper County (Deed Book CC, page 287). He owed John Strode $5,078.03 and secured the debt by conveying to John 7,000 acres in Kentucky, 200 acres on the Hazel River in Culpeper on which there was a merchant mill known as Chilton's Mill, 50 acres on Grindstone Mountain, 70 acres in Culpeper, a 2/9 interest in slaves belonging to his father's estate, 6 other slaves, 1 clock, 4 beds and furniture, 1 sideboard, 1 desk, and 2 bureaus.

the Rappahannock Forge oposite [sic] to, and adjoining on Doctr. Mortimers[188] Island, and how long.' Strode replied, "I had for the most part the superintendency of [those] works while they were erecting and continued to reside on them as the chief manager for upwards of fifteen years." Since he left the forge in 1779, he must have begun work there c.1764 (Fredericksburg Circuit Court Records, Hord vs Richards).

As manager of Rappahannock Forge, John had contact with many important government officials including Patrick Henry, Thomas Jefferson, James Madison, and Thomas Ludwell Lee.[189] During the Revolution, John and Thomas Jefferson developed a friendship and the Library of Congress has numerous letters between the two. One, dated Monday evening, Apr. 18, 1808, was written in response to then-President Jefferson who had asked John's advice regarding re-building the mill dam at Shadwell. This letter revealed that Strode built a dam at Rappahannock in 1772, most likely a replacement of the original dam built c.1758. He carefully detailed how he had constructed what is, essentially, a crib dam:

> At the Rappahannock Iron works the River has a narrow passage with amazing rapidity between two Steep Rocky hills. At that place in the year 1772 I erected a Dam precisely the height which you have described yours, which dam (except the top log which has been some renewed) is without any other expense now standing firm and intire. Altho the expense thereon was not at first proportionally near as much as yours must at once building have cost. That is according with your description of it; nearly as high up stream as below. There is one error that has been the defect and cause of its destruction. My dams are not raised upstream any more than I can possibly help. On such a bottom as the Bed of the River on which your dam rests, Long large hew'd logs should be laid from one side of the River to the other longitudinally, another Row in like manner 15 feet distant from them further up the River and in a parallel line with the first Row or line of Logs. From one of those line of logs to the other cross pairs of Timber which we call Ties at the distance of Eight feet from One another should be well dovetailed into them lines of logs forming at right angles, sort of T__ as you have been pleased to call them. So far I presume your ___ and I agree. Then I lay on another Row of Logs on the down stream line and then put in as before another set of Ties, Dove tailing them to the line of upstream logs at midway between the first ties, and so right across to the second line of logs down stream which for distinction I call the front of the dam, thus then the Dam will be two logs high in front and one log high up stream. The up stream line of logs having two sets of Ties Dove tailed in it. Then as before laying another line of logs in front and also another line of logs upstream, but now take notice that the second line of upstream logs is not to rest on the first but to lay down on the bed of the River by the side of them immediately upstream which will receive your third set of Ties which must be the

---

[188] Dr. Charles Mortimer (c.1727-1801) was a noted physician in Fredericksburg. Originally from Mount Sligo, Ireland, he graduated from the University of Edinburgh and came to Fredericksburg in 1749. He lived at what is now 213 Caroline Street and his house survives. Charles was one of the signers of the Westmoreland Resolutions, opposing the Stamp Act. He was the first mayor of Fredericksburg after its incorporation in 1782. Charles married Sarah Griffen (died 1804).

During the Revolution, Dr. Mortimer established and conducted at his own expense a hospital for sick and wounded soldiers. On June 18, 1781 he wrote to Gen. Weedon, "I have acted since last August merely through humanity for soldiers and prisoners, and have sent off near one hundred from this place without a death—found them all my own medicines, often things from my house and rendered them every service in my power, and make no doubt but you will assist me when opportunity offers to get some compensation for medicines, &c. I have made frequent applications North and South for medicines to no purpose—not even an answer to my letters" (Balch 157). Dr. Mortimer was one of Mary Ball Washington's physicians and attended her in her last illness. Additionally, he was engaged in extensive trade and owned several ships. Charles was buried in what is now Hurkamp Park.

[189] The Virginia Historical Society in Richmond has a letter dated Nov. 23, 1777 from Thomas Ludwell Lee (1730-1778) to James Steptoe (1750-1826) of Bedford County in which the former writes, "our friend Strode has stood by me, and that I shall certainly challenge his spanish barrels, when we meet at Belview this Christmas, at partridge, duck, wild goose, or swan." Belview was on the south side of Potomac Creek just west of Belle Plains (Lee, Thomas Ludwell, letter).

thickness of a log longer than the first or second set. Your dam will then be three logs high in front and but one log high up stream. Then proceed with your fourth log in front and its Ties as with your second. There ends the direction of the timber work of your Dam, that is the front logs being previous got a size useful to raise your Dam as high as it is needful. R.B.—The two last set of Ties will perhaps be 20 inch or two feet longer than the two first. The ties should be of good white oak and nearly as large in size as the length way logs, or they will not admit of a good Dovetail tennon at each end. Much depends on the care and workmanship of this part. The ties are not only to be let into the logs below them, but they are also carefully to be let into the line of logs above them, that is in front. That is half the thickness of the tie goes into the log below and half thereof into the log above, and so with every one of them in front.

The Stone Work should commence as soon as the row of ties are put in and be performed by Masons or some handy men who understand building a stone wall and laid in layer by later with great care and going over each layer of the wall and progressing with small stones and very course gravell carefully filling up every chink, joint and vacancy between the stones untill the wall be completely filled and perfectly solid, and so on layer by later, untill the wall be very nearly level, that is nearly as high up stream as below. Reserving however some of largest and best Blocks of Stone for the last layer, then instead of the monstrous quantity of earth you mention on the upper side, let small stone be put in smaller and smaller untill it arrives at the size of very course gravell. I then put in a lyer of straw or tender pushes neatly thatched and put on a coat of mud if it can be had or thick soil, not very M__, thou any sort of gravelly or muddy stuff will do. I use no plank for spoiling as it is called or any other purpose about the dam, it is too buoyant, may serve to swim or float off down stream but do no good whatever! (American Memory, Jefferson Papers, Apr. 18, 1808)

Strode and Jefferson shared much in common, Jefferson's inventive mind seeking the opinion of his technically skilled Quaker friend. Both men were also keenly interested in agriculture and carried out experiments on their own plantations. On Mar. 11, 1805 Jefferson wrote to Strode asking his opinion of a new type of plow he and his son-in-law had designed. Along with the letter, Jefferson sent two small models, his working diagram, and an encyclopedia containing an article that he had used as a reference (Boyd, vol. 16, p. 371). John's opinion of the new plow design is unknown. From time to time, Jefferson also lodged at Fleetwood. Enroute from Philadelphia to James Madison's home in Virginia, he wrote, "According to the stages I have mapped out I shall lodge at Strode's on Friday the 20th and come the next morning" (Catanzariti, vol. 24, pp. 151-152).

Exactly when John assumed management of Rappahannock Forge is unknown, but by June 21, 1776 Alexander Hanewinkel had returned to his flax business in Fredericksburg and a newspaper notice named John as manager (*Virginia Gazette*, June 21, 1776). Henry Banks[190] (1761-1836), clerk of the iron works, wrote the only known biographical sketch of John Strode. Henry was the son of Gerard Banks (c.1725-1787) and his wife, Frances Bruce (1735-1818). He grew up at Greenbank, the ancestral Banks home located just a few miles northwest of the forge between the Warrenton Road and the Rappahannock.[191] By his own account, at the age of 16 Henry "left a literary establishment at Fredericksburg, and placed myself, with my father's consent, under [Strode's] direction, as a clerk, so that I knew Mr. Strode well." Henry continued, "This extraordinary man was a Pennsylvanian by birth, a quaker by profession, and was bred to industry and usefulness from his infancy. Whether we should consider him more of the farmer, mechanic, or skillful and accurate accountant, I cannot say; because I knew him well, and also knew him to excel in all these things, but in which the most, I could not distinguish; nor can I

---

[190] After leaving Rappahannock, Banks removed to Richmond where he became a wealthy businessman and lawyer. After the Revolution, he engaged in extensive land speculation and acquired large holdings in Kentucky and 77,000 acres in western Virginia. He removed to Frankfurt, Kentucky where he was a political essayist. Copies of his two books of essays are contained in the rare book collection at the Virginia State Library. In 1796 Henry married Martha Koyall Read (1775-1804) and died without issue.

[191] The Geico Insurance Company building now stands on part of this plantation.

prefer, of the several mechanical callings, in which he was the greatest adept, because it was his singular fortune to be, or soon to become, the master of whatever he undertook" (Banks 81).

Banks continued, "After the war had commenced with Great Britain, [Strode] obtained the leave of his employer to establish a manufactory of small arms and other utensils necessary for the portending war. Thus authorized, and without any other aid, he made an excursion to Philadelphia and other parts of Pennsylvania and brought with him many mechanics of different avocations. These were soon fully employed and most rapid progress in making muskets, pistols, carbines, horsemen's caps, camp kettles, spades, shovels, &c. was made.[192] Thus he supplied all articles necessary for a camp, on very short notice, of the best quality." Banks noted, "such talents, such an opportunity, and such a devotion to public utility, enabled [Strode] to render to the United States, in arms and accoutrements, more essential service, than any other individual of Virginia" (Banks 81).

In his biography Henry made practically no mention of James Hunter's role in the development of Rappahannock Forge. Henry was not necessarily a neutral biographer where the Hunters were concerned. He had a strong personal dislike for James Hunter's nephew, styled James, Jr. (1746-1788) of Fredericksburg. This younger James had had business dealings with Henry's older brother, John Banks (1756-1784) that had left John bankrupt. The venture had nearly bankrupted James, as well, who heaped as much blame upon John as did John upon James.[193] Henry's series of essays titled "The Vindication of John Banks" is a lengthy, spurious account of James Hunter, Jr.'s character and life. One must, then, give latitude to Henry when it comes to the Hunter family as he cannot be considered unbiased in his evaluations.

According to Henry Banks, John Strode's health declined and he "retired from his usefulness to his own private property, about thirty miles from Fredericksburg, in the county of Culpeper, on the road from Washington City to the residences of Presidents Jefferson and Madison, was often known and honored by many other leading men of Virginia. He had established some manufacturing mills near his residence, and by the misfortunes and losses incident to the nonintercourse and embargoes, and to extensive speculations in flour, and in endorsing for his friends, he received a shock, which terminated with his ruin, and being far advanced in life, he never restored his affairs. He went afterwards to North Carolina, and employed as an improver of some of the waters of that country, where he died, and now lies among those who knew not his early worth and value, nor how much the United States owed to the exertions and successes of this truly great and useful man, and most devoted patriot" (Banks 82). Henry lamented that no monument had ever been raised to honor John Strode and his contributions to America's freedom.

All of what Henry Banks says of Strode may be documented. John's name first appeared in the Culpeper County records on Oct. 18, 1769 when he paid George Hume[194] £190 for 466 acres of land.[195] During this time, John was actively involved with and residing at Rappahannock Forge and he leased the 466 acres to Zephaniah Nooe[196] of Culpeper. In 1771 James Hunter "of King George" and John Strode "of the same" entered into an agreement for 200 acres of Hume's land, which Hunter had purchased, then

---

[192] It is unknown if any of these caps, kettles, or tools survive.

[193] This partnership was styled Hunter & Banks. On Mar. 29, 1776 the Committee of Safety authorized James Hunter, Jr. to operate a public store in Fredericksburg for the purpose of supplying the army and navy. In June of that year the Virginia Convention authorized the issuance of treasury notes of £100,000 for this store (Palmer, vol. 8, p. 144). The firm of Hunter & Banks appears frequently in commissary records; one order alone amounted to £295,099 and included such items as 1,972 yards of cloth, 3,065 yards of canvas, and 1 ton of lead (Virginia Commissary of Stores).

[194] Interestingly, George Hume and Patrick Home (died 1803), "devisee and executor of James Hunter," were cousins. The Hume/Home family was from Widderburn Castle in Duns, Scotland and had had for at least two generations extensive interaction with the Hunter family both here and in Scotland. Patrick was James Hunter's nephew, the son of his sister, Jane (Hunter) Hume and her husband, Patrick Hume, Sr. According to family genealogist, Edgar E. Hume, the spellings "Hume" and "Home" were used interchangeably, often within the same document, but was always pronounced "Hume."

[195] Culpeper Deed Book F, p. 59, Oct. 18, 1769.

[196] Zephaniah Nooe was originally from Charles County, Maryland. On Jan. 6, 1791 he placed a notice in the *Virginia Herald and Fredericksburg Advertiser* stating that he had opened a "House of Entertainment" in the upper end of Fredericksburg. Essentially, this was a tavern licensed to sell liquor and conduct gambling as well. In 1803 Zephaniah married Sarah Kirtley in Culpeper.

conveyed to Strode. Between 1769 and 1803 John purchased over 4,775 acres in various parcels, most of it lying along Mountain Run and on the north branch of the Rappahannock River. Approximately 2,300 acres of this became Fleetwood[197] and was located off Route 685 in southeastern Culpeper.

Henry Banks stated that John Strode became ill while working at the forge and was forced to retire. John seems to have settled permanently in Culpeper sometime in 1779 at which time Abner Vernon (died 1792), the company bookkeeper, assumed management of the forge. A Culpeper County deed of Oct. 20, 1778 names Strode as "of Stafford." Almost exactly one year later, on Oct. 23, 1779, another deed recorded that he was a resident of Culpeper. Although no longer actively involved in manufacturing supplies for the Continental army, John continued to support them. Between September 1780 and November 1781 John, then residing in Culpeper, submitted several claims for a substantial quantity of goods utilized by the Continental Army. These included 70 bushels of oats, 11 ½ bushels corn, 3 ½ gallons whiskey, 60 pounds of hay, 1,200 pounds of flour, and 250 pounds of beef.

Unable to deny his love for mills and industry, Strode was actively involved in those businesses in Culpeper. Just as he had been responsible for the explosion of building and development at Rappahannock in 1776-77, John quickly built a milling and manufacturing empire in Culpeper. On Aug. 22, 1787 Edward Voss and wife Jane, then of Richmond, sold John 53 acres on the south side of Mountain Run containing a large water grist mill and saw mill along with one acre on the opposite shore for the dam. This was described as adjoining the lands of Joseph Sanford and John Strode's other land. The following August he executed a 99-year lease with Philip Rootes Thompson[198] (1766-1837) for a small strip of land five feet wide on the upper side of Rogues Road "together with the water which shall pass down Flat run after the said Thompson has Emptied as much thereof as he may think fit to his own proper use." This strip was probably part of a tail race for one of his mills.

A few years later John built a mill on the north branch of the Rappahannock that he called his "New Mill" and locals came to call "Strode's Mill." In order to abut his dam on the opposite shore, land that was owned by Landon Carter of Fauquier, it was necessary to file a writ of ad quod damnum with the Fauquier County court (Fauquier County Mill Records). This new mill was located about 1 ½ miles above Norman's Ford and the road in Culpeper leading to it retains the name, Strode's Mill Road. For some unexplained reason, perhaps due to his advancing years, Strode hired John Shelton[199] of Stafford, a millwright, to work on his new mill. By 1811, John was so deeply in debt that he had no cash with which to pay Shelton for his work and he executed a deed of trust with Shelton, conveying to him 500 acres near the mill to secure payment for the latter's work (Culpeper Deed Book CC, page 44).

John's interests extended beyond industrial matters and he was a member of the Culpeper Agricultural and Manufacturing Society that had been founded in 1792. He spoke at a June meeting of the Society and reported on "loosing experiments" using ground stone lime, sea salt, and river mud to improve the quality of heavy soils. He stated that by using manure, gypsum, and 40 wagonloads of "loosing material," he had increased his wheat yield four to five times. He added that he had previously tried these experiments along with resting the land and irrigation as early as 1762 in Chester and Berks counties in Pennsylvania and, later, in Stafford (Scheel 98).

---

[197] Fleetwood was located on what is now State Route 754, northeast of Brandy Station. Between 1810 and 1812 John apparently transferred ownership of this property to his son, Thomas, who was listed in Culpeper documents as owner in 1812. In later years it belonged to the Barbour family, into which John's daughter married. On June 9, 1863 war erupted on Fleetwood and much of the Battle of Brandy Station was fought on John Strode's old fields. In 2002 the Brandy Station Foundation purchased Fleetwood hill and 18 acres.

[198] Philip was the son of the Rev. John Thompson (died c.1772), rector of St. Mark's Parish, and his second wife Elizabeth Rootes. Philip graduated from William and Mary and was admitted to the bar in Fairfax County. He represented Culpeper County in the Virginia House of Delegates from 1793-1797 and again from 1808-1809. He was a congressman from 1801-1807. Philip married first Susanna Davenport (1768-1798), the daughter of Burkett Davenport (1738-1817) and secondly Sarah Slaughter, the daughter of Robert Slaughter. When Philip decided to leave Culpeper, he sold his home, Holly Farm, to John Strode Barbour (1790-1855). He died in Kanawha County, West Virginia.

[199] There were two John Sheltons residing in Stafford County during this period. John, Sr. (1740-1805) was probably too old to have undertaken to build a mill. It is more likely that this was John Shelton, Jr. (1774-1818), son of the elder John and his wife, Susan Hord (1742-1799). John, Jr. married Lethe Conyers (1780-c.1867), the daughter of John Conyers (c.1754-1819) and Mary Davis.

In 1795 John purchased "the Limestone Quarry" from Robert Brooke Voss (died 1811) of Culpeper (Culpeper Deed Book S, p. 321). Limestone was essential to agriculture. Once quarried, the limestone was ground into powder, and then baked in a kiln before being sold to farmers. This baking process, or calcining, drove off unwanted organic compounds present in the limestone. The kiln was located at or near the quarry. That same year John paid Armistead Green of Culpeper £430 for ten slaves that he probably put to work in his quarry and mills.[200] By then he owned several custom, merchant, saw, and oil mills and was a partner in a blacksmithing business with Baylor Banks[201] (1755-1815). In an agreement dated Feb. 1, 1795, John and Baylor agreed to expand their operations so long as John could maintain the use of a road through the mill property. The agreement read:

> Whereas John Strode and Baylor Banks hath been some time engaged in prosecuting the Blacksmith business and manufacture of Nails in Partnership and which partnership they are desirous to continue under the firm of Baylor Banks and company, and also enlarging and extending the same to the containing of a Mill for the manufacture of Linseed Oil and perhaps to some other branches of Business and Trade, and as the Ground whereon the present Shop for Smith's work and making of Nails &c. are erected and which is also the Ground from its situation best calculated for erecting an Oil Mill is a Lott or part of the said land on Mountain Run near his Mills...to be and remain for the use and benefit of the Partnership...John Strode keeping and saving...the use and exercise of the Road...where it now passes through the said Lott (Culpeper Deed Book S, p. 183).

John's blacksmith shop appears to have been a "full-service" facility. Edward Voss's account with the shop records the sale of horse collars, horseshoes, oak planks, shoe leather, flour, and chain and repairs made on plows, coulters, harnesses, and various types of tools (Account ledger, Edward Voss).

By the mid-1790s John's financial situation was becoming troublesome. While he was unquestionably a brilliant millwright and engineer, he seems not to have had a good business sense. As his debts mounted he sought to meet them by taking on more mills and land, perhaps believing that the increased revenue from such ventures would exceed his debt and enable him to pay his ever-growing number of creditors. While it is unlikely that this approach would have succeeded in an ideal economic setting, Jefferson's shipping embargo effectively sealed John's financial fate. His poor business decisions continued until he bankrupted. No doubt aware of his friend's predicament, Jefferson tried to convince John to assume management of the Harper's Ferry iron works. The letters pertaining to this offer were written between 1805 and early 1806. In one of his letters Jefferson wrote:

> The iron works belonging to the U. S. near Harper's Ferry are now unoccupied. Mr. Foxhall the last occupant & owner of a furnace here having sometime ago given them up, it is of importance to the public that they be worked, because they furnish a metal which ___ the same would ___ brass field pieces stand the same and greater proofs, consequently they are more valuable than a brass cannon insomuch as they will last so much longer. Having understood that you wished again to engage in the iron business, I have advised General Dearborne[202] to write you the ___ proposition. I can assure you you will find us disposed to be very accommodating with you, should you think the matter owing of your attention. You will of course come on, & I think you had better take them also on the way, as an examination of them will enable you to propose terms specifically suited to the state of things. Mr. Foxhall thinks that going to the works a stranger, there are persons there into whose hands you might fall, and from them you would receive information biased by their particular interests. No one is so capable of giving you good information as himself, having worked the establishment himself and he authorized me to

---

[200] In the 1783 Culpeper tax lists, James Hunter owned 45 slaves in that county and John Strode owned 31. In the 1810 Culpeper census John Strode was listed with 50 slaves and Thomas owned 73. Both men were among the five largest slave holders in Culpeper during that period.

[201] Baylor was the son of Tunstall Banks (c.1722-c.1785) and Sarah Baylor(?). Baylor married Anna Slaughter (1770-1818), the daughter of Lawrence Slaughter and Susanna Field.

[202] Henry Dearborn (1751-1829), Secretary of War.

say that if you would let me know the day on which you will be there, he will go on purpose to meet you and let you fully into every circumstance you may wish to know. Be so good as to let me hear from you as soon as you can make up your mind (American Memory, Jefferson Papers, Dec. 15, 1805).

By this time, Strode was deeply in debt and extremely busy with his own mills and related businesses. To undertake yet another project, especially one as involved as managing an iron works, was impossible. On Jan. 12, 1806 he wrote to Jefferson, "I continue to be of the mind, that I cannot possibly avail myself of the benefits held out to me respecting the Old Furnace & Iron Mine on the Potomac, especially for this Year." He suggested that Jefferson approach John Tayloe (1771-1828) of Mt. Airy "who has Slaves of his Own, complete artisans in some branches of that business, having broke up and discontinued his Smelting Furnace at Neabsco a little way from Colchester" (American Memory, Jefferson Papers, Jan. 20, 1806).

Jefferson continued to press the question. John, with little money of his own and well aware of the hazards and expense of doing business with the government, wrote to Jefferson again on Jan. 20, 1806 saying:

> Without aid I cannot attempt the business, which ought to be carried on to its fullest extent. Any thing less, will make it a losing & and disgraceful business. In an undertaking immediately in this part of the country, my credit might go considerable length, especially in procuring Labour, in that part, every article must be Cash, frequently in advance! There is not any private person at this time to loan me 10 or 12,000 Dollr. and Government not in the habit of making loans or advances to individuals—I know not how to propose it; but if that (on the present occasion) with propriety can be done, I can for the small sum wanted, give more than a Ten fold Security. Should I attempt the important business with that aid, loss and disgrace to my self and disappointment to Government would inevitably ensue! Pray Sir be pleased to pardon this plain but true Statement, and through the Secretary or as you think proper, honor me with your Answer (American Memory, Jefferson Papers, Jan. 20, 1806).

Jefferson seems finally to have accepted his friend's position with regards to the Harper's Ferry works and not other correspondence on the subject survives.

John's financial situation steadily worsened. In his brief biography of Strode Henry Banks mentioned that John had overspeculated in the flour industry and had been sorely affected by Jefferson's imposition of a shipping embargo during the Napoleonic Wars (Banks 82). Strode was not alone with regards to the flour exports. Many a merchant miller had invested huge sums of money (often borrowed) in flour mills, anxious to capitalize on the seemingly insatiable European demand. Jefferson's poorly conceived embargo on exports bankrupted many of these men and nearly crippled the nation's economy. By 1809 Robert Dunbar, apparently friend and business associate of John's, owned four of Falmouth's five flour mills. Both men suffered the same economic collapse.

Culpeper's court records document John's gradual slide towards bankruptcy. In an effort to generate cash and secure earlier debts, John executed a number of deeds of trust with various individuals. On July 3, 1789 he conveyed to Philadelphia merchants John Head, Jr. and William Sansome seven tracts totaling 911 acres in Culpeper agreeing to pay them by July 3, 1790 £900 money of Pennsylvania "equal to Two thousand four hundred spanish milled dollars" (Fredericksburg Deed Book A, page 16).

On October 16, 1795 John placed a notice in the *Virginia Herald* stating that he was "unexpectedly necessitated to raise a sum of Money, [and] Offers for sale, singly or in Families, fifteen to twenty slaves...also, Fifteen Hundred to Two Thousand Acres of Land laid off with, or without, Improvements." Whether he succeeded in selling any of the slaves or land at this time is unknown.

His earliest mortgage on file in Culpeper is dated Oct. 31, 1797. In this instrument John secured a debt of £771.16.6 due to Alexandria merchants Jonah Thompson and Richard Veitch (c.1770-1835) with two tracts of 388 and 400 acres each.

In order to generate some cash, John began assuming a variety of county offices. He served as a justice in Culpeper in 1783, 1794, and 1795 and probably during some of the intervening years as well. In

1797 he was a founding trustee of the Stevensburg Academy,[203] one of Culpeper's two state-chartere schools. The Academy opened its doors in 1802. On Sept. 17, 1798 John Strode, French Strother,[204] Phili Slaughter,[205] Robert Latham, Jr.,[206] and Mordecai Barbour put up a $30,000 bond for John's faithful servic as sheriff of Culpeper. He also served as a state inspector of firearms, making numerous visits to Georg Wheeler's gun manufactory in Culpeper (Palmer, vol. 9, p. 217, 283).[207] In 1810 and 1811 John Strode an Moses Green represented Culpeper County in the Virginia General Assembly. In 1811 he and Zephanial Turner were appointed to supervise Culpeper's collection of escheats, fines, confiscations, and penalties du to the state. While it is easy to praise John for fulfilling these important public duties, one must remembe that as a member of the Society of Friends he was forbidden from swearing any oaths, an act that wa requisite to his serving in several of these public offices. The choice between his faith and his ability t keep food on the table must have been a painful one.

John considered making his own arms. On Mar. 18, 1796 he wrote to the governor and stated, "I conversation which I had with the Honorable Colo. Burnley relative to the arms which I contemplate t make for the public, it appeared that specific proposals are expected from me; to make which I am unde difficulty, because the honorable Board has not furnished me with a pattern to operate as a rule, nor am informed of the length and quality which will be required." In his letter Strode suggested certai specifications as to length of barrel and bayonet as well as "neat Brass Mountings, Steel Ramrod, neatl Stocked of Black Walnut, a Cartouche Box suitable to contain 24 rounds, with neat Black Leather Belts mounted with Brass Buckles, complete; wiper for the pan, and picker for the touch-hole; as well finishe and as good in quality as those of the United States." He asked to be allowed $15 per stand. At the time h wrote the letter, Strode had not yet established his manufactory. He asked to be allowed four months "t prepare the machinery, procure hands, and fix for the business" and felt he could make 4,000 stand withi sixteen months (Palmer, vol. 8, pp. 356-357).

Thomas Jefferson expressed his concern about John in a letter to another long-time friend, Jame Madison. Dated July 13, 1813 the letter read:

> You know the present situation of our friend Strode, entirely penniless. How he comes to be left to subsist himself by his labours in subordinate emploiments, while his son is at ease, I am not informed, nor whether they have had any difference. Yet the fact is he is in indigence, and anxious to get his living by any services he can render. You know his qualifications. The public iron works, the Armory; the army or some of the sedentary offices at Washington may perhaps offer some employment analogous to his talents. His wish is to earn a livelihood; and altho in his letter to me he does not propose to sollicit anything, yet the expressions of his situation shew that some decent emploiment could not fail to be very acceptable (Smith1725-1726).

Another deed of trust, dated Oct. 9, 1806, is on file in the Fredericksburg Circuit Court records. I this instrument John conveyed to James Ross of Fredericksburg 350 acres of land in Culpeper (purchase from Ross) on the northwest branch of the Rappahannock River. John agreed to pay Ross four installment of $583.33 on Jan. 1 of 1807, 1808, 1809, and 1810 (Fredericksburg Deed Book E, page 513). H managed to pay for this tract and upon it John built what he called his "New Mill."

---

[203] Mordecai Barbour was also a trustee.

[204] French Strother (1733-1800) was the son of James Strother (died 1761) and Margaret French. Frenc was prominent in Culpeper County which he represented for more than 25 years in the General Assembly He was a member of the Constitutional Conventions of Virginia in 1776 and 1788, served as Count Lieutenant and as a justice of the peace. He married Lucy Coleman, the daughter of Robert Coleman an Sarah Anne Saunders of Culpeper County.

[205] Philip Slaughter (1758-1849) was the son of Col. James Slaughter and Susan Clayton. He marrie Margaret French Strother.

[206] Robert Latham, Jr. (born c.1760) was the son of John Latham (c.1730-c.1789) and Frances Foster (die 1789). He was styled "Junior" to distinguish him from his uncle.

[207] Wheeler had been a gunsmith at Rappahannock and may have served some administrative function ther as well. On Dec. 20, 1800 Wheeler submitted "proposals of self and Mr. John Brent of Maryland, t manufacture four thousand stand of arms for the State by June 1st, 1802." (Palmer, vol. 9, p. 192.)

As John's many creditors pressed him for payment he was forced to mortgage most of his property. The Culpeper records contain numerous trusts and mortgages, not only for land but later for household furniture as well. His son, likewise in debt, attempted for some years to sustain his father. On Feb. 17, 1808 John and Thomas Strode executed a mortgage to James Green, Jr. to secure a debt of $1,171 with interest (Culpeper Deed Book CC, p. 163). On July 20, 1809 John mortgaged to Mary Stuart, executor of John A. Stuart of King George, two tracts in Culpeper of 391 and 100 acres respectively to secure a debt of $1,512.30 (Culpeper Deed Book DD, p. 216). On Sept. 3, 1809 John executed a deed of trust to John W. Green[208] of Fredericksburg (Culpeper Deed Book DD, p. 158). Thomas had agreed to act as his father's security for several unspecified debts to James Calhoun,[209] John Randall, Armistead Gordon,[210] Heartshorn Large, William Major, Maj. ___ Yancy, James Green, Dr. John Willis, Thomas Woolfolk,[211] and Gooding Hutchings. To secure Thomas against any losses, John made another deed of trust on the Paoli Mills with 2,000 acres, the Pamunkey Mills containing 600 acres in Orange County, and Alum Springs Mills,[212] the latter two having been purchased from Mordecai Barbour. The debts and mortgages continued. On Sept. 23 of that same year, he again mortgaged his Paoli Mills and 2,300 acres to secure a debt of $9,536.53 owed to the Bank of Virginia (Culpeper Deed Book DD, p. 210). He also sold several tracts of land in an effort to pay off creditors.

John's financial demise is reflected in a series of letters to another friend, James Madison. When he wrote to Madison on June 20, 1789, he was considering the possibility of moving west to make a new start. He wrote, "Yet Sir [I] have some times sincerely Contemplated an advantage whc. I might in a new country derive to myself & Children from the experience and mechanical knowledge I possess, particularly in all the different branches of manufacturing iron, from the taking it out of the earth, through each State and process untill converted into every device in which Pot metal, Bar Iron & Steel can be made useful, from the largest to the minutest article, and the Construction of every water works and machinery necessary thereto. The unsettled Country to the westward, present a large field (as yet unexplored, or at least not properly Scrutinized) for the founding in some part or other thereof, a General Iron manufactory to public utilization" (Presidential Papers, June 20, 1789).[213] For whatever reasons, however, Strode didn't leave Culpeper at that time.

---

208 Judge John Williams Green (1781-1834) was the son of William Green and Lucy Williams. From 1822-1834 he served as chancellor of the Virginia Supreme Court of Appeals. He married first Mary Browne and secondly in 1817 Million Cooke (1785-1842), the daughter of Col. John Cooke (1755-1819) of Stafford. John resided at Greenwood in Culpeper County.

209 Probably the James Calhoun of Charlotte County, Virginia who married in 1788 Martha Claybrook and in 1792 Mary Lessly.

210 Armistead Gordon (born 1773) was the son of John Gordon and Lucy Churchill of Middlesex County. He resided in Spotsylvania County and married Elizabeth Clayton.

211 Possibly Judge Thomas Woolfolk (1755:73-1848) who was born in Orange County.

212 John Strode purchased this mill from Mordecai on Dec. 5, 1807, paying $5,000 for "a merchant grain mill" on 20 ¼ acres (Spotsylvania Deed Book R, p. 357). On July 5, 1809 John used Allum Springs Mill and Pamunkey Mills (Orange County) to secure several debts to Murray, Grinnan, and Mundell of Fredericksburg (Spotsylvania Deed Book S, p. 468). When John defaulted on these debts, the mills were sold. John Mundell purchased Allum Springs Mill on Mar. 30, 1819, paying $1,500 (Spotsylvania Deed Book W, p. 118).

213 On May 22, 1789 John purchased two tracts in Kentucky of 300 and 700 acres respectively. These were located on the Cumberland River adjoining the lands of his old clerk, Henry Banks, and William Roberts. John may have been considering moving to this land when he wrote the above letter, though he never did so. John accumulated a considerable amount of land in Kentucky, most of which is designated in Jillson's Kentucky Land Grants as "Military;" that is, it was part of a special military district established in Kentucky south of the Green River. As Strode had never served in the military, he didn't qualify for grants due for military service; thus, his land in Kentucky was acquired by purchase. Some of these purchases were recorded in Culpeper and some in Kentucky. On Mar. 13, 1790 William Green of Mercer County, Kentucky sold John 1,000 acres on the Cumberland River in Lincoln County. This was land that William had previously acquired from his brother, Gabriel Green. John seems to have sold some of his Kentucky lands and it is unclear how much if any land he owned there at the time of his death.

On Oct. 15, 1808 John again wrote to Madison, asking that his son Thomas and his militia company be stationed at Norfolk rather than in some distant place. Strode's desperate financial situation is evident as is his embarrassment over asking a favor of the Secretary of State. He wrote:

> With all due respect & regard I beg leave good & worthy Sir to approach your hand and implore that you will be pleased from matters of infinitely more importance to be good enough to spare one single minutes attention to my singular situation and kindly favour me with your advice. My son has nearly made up His Company and most probably will be ordered to some distant place, in which case in my imbarrassed situation Lord knows what I am to do! Never did poor man stand more in need of the aid & advice nor any one more destitute of that aid, except indeed you will condescendingly deign to favour me therewith. Do you believe Sir that if I was to petition the President and the commander in chief, that my Son might be stationed at Norfolk? Could that favour be obtained without any loss to the Service of our Country. He might beside performing the duties of his office find some leisure moments to assist me in things which I might at that place confide to his care & which might in some degree contribute perhaps to relieve me from the severe imbarrassment which the present unforeseen convulsions on the other side of the Atlantic has produced in general, but more especially to my self in particular. If that favour be too great for me to expect or any way improper for me to ask I shall be Silent and Submissive! At any rate pray Sir be good enough to honor me with a line I intreat you. Pray commend me to your Angel like lady (Madison, James, Papers, Oct. 15, 1808).

John had purchased the 1807 wheat crop from three of Madison's Orange County plantations anc had made arrangements with Madison, through the president's estate manager, Gideon Gooch,[214] to pay foı the wheat when he was able to do so. That debt lagged on for nearly three years, the shipping embargc eliminating most of Strode's cash flow. A newspaper article in the *Virginia Gazette* of May 25, 180ε claimed that Madison had "dunned him for the sum due." According to the article, Strode had said, "Gc back and tell Mr. Madison if he will take off the Embargo I will pay him for his wheat." Infuriated by the article, John wrote a firm letter to the editor of the paper denouncing the printing of such an article anc stating that Madison had "sent me word that he should during the present times be among the last whc would ask me for payment" (Madison, James, Papers, June 11, 1808). The incident further embarrassec Strode who did not wish his long-standing friendship with Madison undermined.

Strode's debt to Madison continued unpaid through early 1810. On Feb. 7 of that year, John wrot to his old friend informing him that he still did not have money to pay him. Instead, he offered to draw up a deed of trust for a piece of property in Fauquier County to secure the debt of £320.13.10, "equal to On Thousand & Sixty Eight Dollars & ninety seven cents with interest." That was done on July 6, 1810 conveying 162 acres to President Madison (Madison, James, Papers July 6, 1810).

Paoli was a merchant mill, meaning that wheat was purchased in bulk, ground into flour, and sold often on the European market. On July 3, 1807 John placed an advertisement in the *Virginia Heral* (Fredericksburg) seeking men to haul flour from Paoli. It read:

> Every person who will Waggon Flour from my Mills to Fredericksburg, shall be paid on the delivery thereof, Four Shillings and four pence per barrel. Provisions and Provender of every kind will be furnished on reasonable terms. The distance is 25 miles—three trips a week can easily be made.
>
> John Strode
>
> Culpeper, June 7, 1807

Between 1808 and 1809 John Strode and Robert Dunbar (c.1745-1831) of Falmouth jointl borrowed from the Bank of Virginia some $22,680 drawn on sixteen notes (Fredericksburg, "Bank o Virginia vs Dunbar and Strode"). The only known records of these loans are found in the Fredericksbur

---

[214] Gideon (1773-c.1837) was the son of Thomas Gooch (died c.1804) of Orange County. In 1794 h married in Louisa County Sarah/Sally Madison, the daughter of John Madison.

Circuit Court, as the borrowers never seem to have satisfied their creditors. The court records give no indication of where or how this money was used. While Strode may have been in business with Dunbar in the old Rappahannock Forge mills, which Dunbar purchased from Hunter's executors, it is perhaps more likely that Dunbar was involved with Strode's milling and manufacturing empire in Culpeper. Robert left several indentures in the Culpeper land records, none of which provide any clues as to the relationship between these two men. Nor is it known if this enormous debt was ever paid.

The Culpeper court records contain an interesting indenture in which John conveyed nearly everything he owned to his son, Thomas (Culpeper Deed Book DD, p. 127). In what appears to be a deed of trust, Thomas paid his father £1,000 and was deeded his father's mansion house and about 2,300 acres; 1,200 acres on Great Marsh Run; 40 slaves, 20 horses, 40 cattle, 40 sheep, 1 ox wagon, 2 horse wagons with harnesses, the tannery at the mansion along with its stock of hides and tanned leather; all the tools belonging to the smith's, carpenter's and wheelwright's shops; the copper stills; all farming implements on the Fleetwood tract and the "Marshfield Estate;" and all the beds and household furniture on both tracts. It is unclear whether this conveyance was made because of Thomas' marriage, which occurred at about this time, or if John feared he would lose everything to his creditors and sought to vest Thomas with as much of his estate as possible. Some months later, John sold Thomas his new mill on the north branch of the Rappahannock (Culpeper Deed Book DD, p. 162). Thomas paid his father £4,200 for the mill and 350 adjoining acres described as bounding on the lands of William Wheatley[215] and Robert Beverley, deceased (Culpeper Deed Book DD, p. 162).

Although Jefferson's shipping embargo was repealed in March 1809, the damage had been done and Strode was so far behind financially that he would never recover. On Mar. 14, 1810 he attempted to sell his real estate. In just over twenty years he had amassed a remarkable quantity of industrial properties and acreage. He offered to sell the 2,300-acre Paoli[216] Mill tract, including "several hundred acres of rich low Grounds...a large Quarry of the best quality lime stone and a well constructed lime kiln." On this same tract was "a very large, commodious, well constructed manufacturing mill with two pair of best French burr millstones and one pair of country stones...together with a most complete Saw-Mill, a well constructed Flaxseed Oil Mill, a Wool-Carding Machine, a large well built Bake House,[217] [and] a Kiln for drying Indian Corn to make meal for exportation." He had three dwelling houses "with good stone chimneys, and stone wall cellars...also, Smith's, Carpenter's and Cooper's Shops." In addition, he was selling 700 acres on the Pamunkey River in Orange County that included a manufacturing mill, a good dwelling house with kitchen, stable, out houses, and several shops. Within two miles of Fredericksburg he had "a newly rebuilt well finished manufacturing mill, with two pair French burr millstones, known as Alum Springs Mill."[218]

---

[215] In 1792 William Wheatley married Susanna Grigsby in Culpeper.

[216] On State Route 672 in Culpeper County is Stone's Mill. It was built on Mountain Run and a deed of 1812 reveals that Abner Vernon conveyed it to Thomas Strode sometime prior to that year (the earlier deed not having been recorded). The 1812 deed refers to the building as "Gaeli Mills," possibly a corruption of the word "Paoli."

[217] John ground flour and made his own bread and biscuits for commercial sale.

[218] On Jan. 15, 1805 Mordecai Barbour, John Strode's son-in-law, placed the following notice in the *Virginia Herald*:

> A MACHINE for separating COTTON from the seed, is just erected at the Allum S[prings] Mills, where seed cotton will be taken in & picked for the eighth part—Customers may calculate on dispatch, as the Machine works 37 Saws. Cash will also be given for good Seed Cotton delivered in this place.

In 1793 Eli Whitney developed a practical cotton gin that consisted of a roller with wire teeth that perfected through slits in an iron guard. Cotton seeds, which were wider than the slits, fell from the guard into a collection box beneath the gin. The cotton fibers were swept from the teeth by a roller brush that turned in the opposite direction. Whitney's one-horsepower machine could separate fiber from seeds faster than 50 men working by hand. Almost as soon as Whitney patented his cotton gin, other inventors began modifying it. In 1795 Hodgen Holmes replaced the wire teeth with small circular saws. This change in design freed inventors from Whitney's patent and thenceforth the gins were made by local carpenters and blacksmiths.

The last two properties advertised were a 500-acre tract in Culpeper about four miles above the mansion house of William Richards and 170 acres in Fauquier County (*Virginia Herald*, Mar. 14, 1810).

John's effort to sell his property was unsuccessful and on June 20, 1811 he conveyed everything he owned to Thomas Spilman, sheriff of Culpeper, and availed himself of the "benefit of the Act of Assembly entitled an act to reduce into one the several acts concerning executions and for the relief of insolvent debtors." His holdings included the Paoli Mills with its manufacturing mill, sawmill, flaxseed oil mill and 2,300 acres; 500 acres adjoining William Richards and Benjamin Barnett; a ½ moiety on 200 acres on a branch of the Hazel River with manufacturing and saw mills; and a ¼ interest in 2,500 acres that was part of the estate of Joseph Roberts, deceased (Culpeper Deed Book EE, page 323). As was customary, John's property was sold to pay his creditors. Thomas was the highest bidder and, on Aug. 13, 1811, he received a deed for the Paoli Mill (Culpeper Deed Book EE, page 321). For some time prior to the sale, he had been leasing the mill from his father for $1,200 per year.

John's financial demise took many years to reach closure. There were notices placed in Fredericksburg newspapers announcing two cases heard in the Superior Court of Chancery in Fredericksburg. Defendants named in the first case, held on Apr. 27, 1815, were John and Thomas Strode, Henry Field, Jr. (died 1785), and Isaac Hite Williams[219] (c.1770-1828). Plaintiffs in the case were George Murray (died 1819), Daniel Grinnan[220] (1771-1830), and John Mundell,[221] merchants of Falmouth and Fredericksburg who traded under the firm name of Murray, Grinnan & Mundell.[222] The defendants were given until Nov. 27, 1816 to pay three notes (due since 1808) totaling well over £754. The second case involving the Strodes, heard in January 1820, was brought by James Green against John and Thomas Strode and Charles Carter (Fredericksburg Circuit Court Records, "Green vs Strode"). No details of this case are known.

John's last appearance in the Culpeper records was in August 1811 when he executed a mortgage to John Shelton (Culpeper Deed Book FF, page 117). According to his old friend and biographer, Henry Banks, after his bankruptcy, John removed to North Carolina where he died c.1820, far from those who appreciated his contributions to the founding of the United States (Banks 83). To date the author has been unable to determine where he died in North Carolina.

---

[219] Isaac was the son of John Williams and Eleanor Hite and was a lawyer and clerk of the Superior Court of Chancery of the Fredericksburg District. He married Lucy Coleman Slaughter (born c.1760), the daughter of Capt. Philip Slaughter (1758-1849) and Margaret French Strother.

[220] Daniel was the son of Daniel Grinnan, Sr. (born 1739) of Accomack County and Helen B. Glassell, the daughter of Andrew Glassell of Torthowold, Madison County, Virginia. The elder Grinnan moved from Accomac to Culpeper County and lived on a handsome estate on Cedar Run. He served in the Revolution under Gen. Edward Stevens and was at the Battle of Guilford Courthouse in North Carolina. Around 1792 Daniel, Jr. moved to Fredericksburg where he became clerk for James Somerville, a Glasgow merchant who had established himself in the town some 40 or 50 years earlier. Daniel was named executor of Somerville's estate and succeeded the old gentleman in his business. As part of the settlement of British Mercantile Claims, he was an appointed agent for collecting debts due in Virginia to many English and Scottish merchants. Around 1800 Daniel formed a partnership with George Murray of Norfolk under the name of Murray, Grinnan and Mundell. This firm had counting houses and warehouses in Fredericksburg and Norfolk and acted as agents for the Argentine Confederacy in their war with Spain.

In 1804 Daniel married Eliza Richards Green (1787-1813), the daughter of Timothy Green (born 1763).

[221] In 1809 John Mundell served on the Fredericksburg City Council.

[222] This mercantile firm also acted as a bank, making small loans to their customers. Daniel Grinnan's papers are on file in the Virginia Historical Society and reveal numerous such loans. On July 18, 1805 John Strode sent a note to the firm requesting a loan of $70. He wrote, "I am under the extreme necessity of begging the favour of You to lend me Seventy Dollars, or my Credit must very much Suffer. I shall not be necessitated to trouble you any more on sight, untill I have Sufficient funds in Yr. hands, nor even then, except on very particular occasions where it can't be avoided." He added a postscript, "If you can conveniently spare me the Money please give it to Richd...and if convenient send me Ten Dolls. more in the whole Eighty" (Grinnan papers, letter). On Apr. 14, 1803 John also gave a bond to the firm for £200 (Grinnan papers, bond).

Thomas Strode[223] shared his father's financial crisis and the Culpeper court records contain numerous documents pertaining to him. Based upon these records, Thomas' financial problems became evident in 1809. As he, too, was heavily involved in milling and manufacturing, Jefferson's shipping embargo was financially devastating. But added to Thomas' burden was his effort to help his father, an effort recorded in the numerous deeds of trust and conveyances of property between the two. From November of that year until August 1813, Thomas and his wife executed five deeds of trust securing debts to as many creditors (Culpeper Deed Books CC, p. 333; EE, p. 247; EE, p. 534; FF, p. 327; GG, p. 13).

On Jan. 12, 1812 Thomas advertised his 2,600-acre home plantation in Culpeper "to satisfy his debts" (*Virginia Herald*, Jan. 12, 1812). Also up for sale was a tract of 350 acres on the Rappahannock "22 miles above Tidewater called the New Mill" which included a large merchant mill. He also offered his part of the Paoli Mill[224] tract of 2,500 acres and "my farm lying on Marsh Run containing about 1,200 acres" (*Virginia Herald*, Jan. 12, 1812).[225] No buyer was forthcoming for Paoli and the sale of the other property failed to satisfy his need for funds. The Oct. 9, 1813 issue of the *Virginia Herald* carried a heart-breaking notice of the sale of Paoli Mill[226] and all the personal property of Thomas Strode "consisting of his stocks of Horses, cattle, Hogs, Oxen, Waggons, Carpenter's, Cooper's, Blacksmith's and Plantation Tools—a quantity of Grain and Hay, and a large stock of Barrels and barrel timber partly dressed, and a great variety of other property." Perhaps in an effort to help his son-in-law William Richards paid Thomas $6,000 for

---

[223] Thomas Strode owned a considerable amount of property in Kentucky though, like his father, he acquired it by purchase rather than by military grant. On Jan. 31, 1810 Thomas paid his father-in-law, William Richards, $6,000 for 981 acres on Otter Creek in Hardin County, Kentucky along with another 4,155 acres, also in Hardin County. In 1816, just prior to his death, William bought a one-quarter interest in 50,000 acres in Kentucky owned by Thomas Carneal of Hardin County. In his will, William asked that the land be sold to pay his debts. In October 1817 Thomas Strode purchased William's interest in the tract. Jillson's Old Kentucky Entries and Deeds lists Thomas Strode as owning 13,481 acres in that state in 1817.

Thomas was also involved in the attempt to build a Rappahannock River Canal. On Oct. 30, 1793 a group of Culpeper citizens petitioned the General Assembly asking that a canal be opened above the falls of the river. On Dec. 11 the Assembly passed an act "for opening and clearing the navigation of Rappahannock river" from the falls to Norman's Ford, the crossing place of the Carolina Road. Notices were placed in newspapers offering stock in the company but, due to lack of widespread enthusiasm, little money was raised and the project never began. On Feb. 9, 1811 the Assembly passed a second act, again authorizing the opening of the canal. Thomas was one of six commissioners appointed from Culpeper and the act specified that the canal was to run from Fredericksburg to Strode's Mill, one mile upriver from Kelly's Ford. Due to concerns over the War of 1812, subscriptions were again low and insufficient money was raised to fund the project. By 1816 the canal project was still pending but Thomas' finances were in ruins. On July 1 of that year Richard Norris and John McNeale replaced him as commissioner (Sheele 145).

[224] Paoli was in Chester County, Pennsylvania and was the birthplace of Anthony Wayne (1745-1796). It was also the site of a bloody British massacre on the night of Sept. 20-21, 1771. American soldiers, led by Gen. Wayne, were attacked and put to the sword by British soldiers under the command of Maj. Gen. Charles Grey. 53 Americans died in the battle. For an unknown reason, the soldiers were not buried in the field on which they died but were moved a short distance to what is now known as the Paoli Memorial Grounds.

[225] This was James Hunter's old Marsh Tract in Fauquier that John and Thomas jointly purchased from Patrick Home.

[226] Judge John Williams Green (1781-1834) purchased Paoli Mill and mortgaged it through the Bank of Virginia. Judge Green died before repaying the loan and the property was sold by Commissioner William Green to satisfy the debt. The advertisement that appeared in the *Fredericksburg Political Arena* of Feb. 5, 1836, described the tract as "supposed to contain 2300 acres; but, on a survey of it, after a purchase by the late Judge Green, it was found to contain 1523 acres, of which 1157 acres were held by a title that was unquestionable, and accordingly accepted by him; the other 366 acres were held by a *doubtful* title, which he was not bound to accept, and did not; but if it be a good one, the purchaser at this sale, will be entitled to them, as well as the 1157. The Arable Land on this estate is very good and productive, and it contains a very fine Mill seat, now improved with a Saw and Grist Mill: besides a Dwelling House, Barn, and all other convenient buildings."

property that he had conveyed to him "in part of the Marriage Portion of his Daughter...and also all the lots in the Town of Fredericksburg and in the County of Spotsylvania which said Richards conveyed to said Strode" (Culpeper Deed Book FF, p. 119)."

Thomas knew his finances were ruined and on Aug. 14, 1813 he sold some land in Fauquier County to Daniel Grinnan and John Mundell so that Thomas could settle several slaves on Harriet for her own use and as an inheritance for his children (Culpeper Deed Book GG, p. 14). On Aug. 14, 1816, Thomas and Harriet deeded to her father, William Richards, all Harriet's interest in her mother's estate, which Eliza had inherited from her father, Simon Miller (Culpeper Deed Book HH, page 203). At the time of this conveyance Thomas and Harriet were residents of Henrico County, but it is unknown when they removed to Henrico or how long they lived there. They eventually returned to Culpeper and were residents there in 1819 when the Superior Court of Chancery in Fredericksburg heard a case against Thomas. On Apr. 18 of that year James Williams complained against Charles Carter, Thomas Strode, and George Pickett in plea of debt. The court ordered John Stannard Marshall, commissioner, to advertise the property in question for three weeks and then to sell it. On Nov. 15, 1819 Marshall sold "all those several tracts or parcels of land lying and being in the County of Culpeper contiguous to each other containing in the whole somewhere about [2200 or 2300] acres which John Strode conveyed to Thomas Strode in pursuance of a contract made between William Richards and said John Strode" (Culpeper Deed Book LL, p. 123). High bidder for the property was Thomas Goodwin who offered $1,896.65.

Thomas' financial empire finally collapsed and, on Feb. 28, 1822 Thomas Strode, "now in the County of Culpeper" and "now in the custody of the said Sheriff, by virtue of a writ of capias ad respondendum," signed over his property to Sheriff William Broadus (died c.1836) and declared bankruptcy (Culpeper Deed Book NN, p. 228). William M. Thomason had brought suit against Thomas and he "not being able to provide appearance bail as the law requires wishes to avail himself of the benefit of the insolvent debtors act for which purpose he confessed Judgment." The property conveyed to the sheriff consisted of one tract in Spotsylvania that was "one moiety of the land conveyed to the said Strode by William Richards deceased situated on the waters of the Rapid Ann River and the adjoining lands of Ludwell Lee; 7 lots with appurtenances in the town of Leeds on the Rappahannock formerly belonging to Clapham Richardson, a tract conveyed by William Richards to Daniel Grinnan on the Kanawha River and later purchased by Thomas Strode, and all Thomas' interest in the estate of William Richards.

Thomas died in 1829 leaving Harriet and several children.[227] Harriet's later life seems to have been one of difficulty and continuing debt. On Apr. 2, 1830 she executed a deed of trust with trustee Philip Alexander (1783-1817) of Fredericksburg (Culpeper Deed Book WW, p. 381). She owed merchant/banker Thomas Seddon (1779-1831) three debts totaling $1,247.05, the debt to be paid from the hire of the Negroes held in trust for her. What became of Harriet is unknown as she disappeared from the records after that date. That was her last appearance in the Culpeper records and nothing further is known of her. The Strode era was past and a new nation began to mature.

Manufacturing at Rappahannock Forge

But to return to Rappahannock. As mentioned earlier, a massive building program instituted during 1776-77 resulted in Rappahannock Forge becoming one of the largest industrial complexes in America. As early as 1770 the Commissioners of Customs described Rappahannock Forge as "the greatest iron works that is upon the continent" (VCRP, Treasury Papers—In Letters, 1770-1771). By 1777 Rappahannock was staggering in its magnitude. Hunter and Strode employed craftsmen of nearly every avocation and established well-furnished shops in which they worked. Based upon the known contemporary accounts of Rappahannock Forge included in letters, journals, and government documents, the following is a list of the known buildings at the forge and this list is, no doubt, incomplete:

[227] Thomas Strode's obituary appeared in the May 6, 1829 issue of the *Virginia Herald* and stated that he died in Philadelphia.

company store	grist mill[228]
slitting and rolling mill[229]	merchant mill[230]

[228] This mill was described in a Mutual Assurance Policy of 1810 as 30' by 18' and built completely of wood. Constructed c.1790, it was probably built as a replacement for an earlier grist mill. It seems to have disappeared by 1820.

[229] Rolling mills were an improvement over the traditional forging of bars by blacksmiths using hand-held or water-powered hammers. One of the problems of working iron by the hammer was that it was difficult to make long thin rods such as were needed by blacksmiths and nailmakers. The iron had to be worked red-hot; hammering it was slow and the metal cooled more quickly the longer and thinner it became. The problem was first addressed in England in 1590. A piece of iron was hammered out to a flat strip, then passed between a pair of flat rollers turned by water power, thus being flattened and extended. The strip was then passed between rotating disc cutters that cut, or slit, it into a number of narrow rods. The 1797 edition of Encyclopedia Britannica says of rolling mills, "Cast iron has of late been brought into the malleable state by passing it through rollers instead of forging it. Indeed this seems to be a real improvement in the process, as well in point of dispatch, as in its not requiring that skill and dexterity which forgemen only acquire by long practice. If the purposes of commerce should require more iron to be made, it will be easy to fabricate and erect rolling machines, though it might be impracticable to procure expert forgemen in a short time" (Encyclopedia, vol. IX, p. 351).

Frequently, rolling and slitting mills were constructed side by side as the bars produced by the rollers could be easily processed into nails by the slitting mill. According to James Swank, "Near the close of the sixteenth century there was introduced into England an invention for slitting flattened bars of iron into strips called nail rods. This invention was the slitting mill [which] greatly benefited the nail trade of England" (Swank 48).

Schoepf described the rolling and slitting mills at Rappahannock Forge. He wrote, "The rolling-mill is adapted for drawing iron-plate, that is to say, the machine is such that between the two smooth, steel cylinders the plate is drqwn with more rapidity, more easily, and with greater uniformity than it is possible to do with hammers. The slitting-mill is another ingenious mechanism for splitting broad iron bars at a stroke into many narrower bars, which is a much slower process by the customary method under the hammer" (Schoepf, vol. 2, pp. 42-43).

One of the most useful items produced by the rolling mill was sheet metal. This was easily transformed into tin plate by the application of a thin layer of tin over the sheet of iron. Tin plate was used for pots, kettles, canteens, saucepans, milk churns, cash boxes, and tea urns. The Germans originated the idea of tin plating and made it by forge welding the two layers together under the blows of a hammer. The bonding of the tin and iron was easily accomplished in the roller mill and the finished product had a smoother surface that the old hammered work.

In a slitting mill cold bars, which had been cut with heavy shears into pieces 13-20 inches long, were heated in a furnace to a white heat, then run between smooth rollers which drew each piece to about three times its former length and to a thickness suitable for nails. Immediately after leaving the rollers, the flattened bar was run between grooved rollers which slit the strip into nail-sized rods. These were sold commercially or converted in a nailery into headed nails. Flattened bars that had not been slit could be shaped into shovels or other tools.

When Thomas Jefferson returned to Monticello after many years' absence, he faced severely limited resources and depleted soil no longer suited to agriculture. He turned to the making of nails, a business he could conduct with little capital. About a dozen slaves made nails from three to four tons of nail rods that Jefferson imported from Philadelphia, thus providing local residents with a source of much sought after nails.

Hunter's rolling and slitting mill was probably building B on the site map (Figure 2-9). On May 1, 1777 the General Assembly authorized "An Act for the Encouragement of Iron Works" in which Hunter was authorized to take ½ acre from John Richards for which he was to pay a price as determined by a jury.

[230] This was distinguished from the custom mill by the extent of the grinding and the manner in which grain was purchased for the mill. Hunter's custom mill would have ground locally produced wheat and corn, providing flour and meal for his workers. The merchant mill was a much larger facility that had multiple stones. Wheat and corn were purchased by the ton from outside sources and ground into meal and flour that

cooper's shop	warehouses (an unknown number)
forge[231]	wheelwright's shop
furnace	coal house
nailery	blacksmith's shop
plating mill[232]	tannery[233]
saw mill	tan yard
carpenter's shop	still
wool mill and house[234]	stables[235]
armory	counting house
steel furnace[236]	houses for managers and workmen

---

was then sold, much of it being shipped to Europe. The merchant mill at the forge was described in a Mutual Assurance Policy of 1822 as 68' by 36', of stone construction with a wooden roof.

[231] A forge consisted of at least one chimney (separate from the furnace stack), two hearths, and one large hammer. Larger forges contained more chimneys and hammers. Water powered the bellows and hammers. Forge workers were called forgemen, hammermen, or finers and were highly skilled individuals who processed pigs into bars at a refinery forge or finery. These skilled workers were assisted by several laborers and apprentices. In this process, the pigs were brought from the furnace and heated by the finer on a charcoal hearth, also called the refinery. The finer worked the pig with a finer's bar until it became a plastic bloom. The bloom was then hammered, driving out impurities. The process of hammering, re-heating, and hammering was repeated several times until the iron was converted into an ancony, an elongated bar with a knob on each end. Much of this hammering was done by powerful water-driven trip hammers. Iron was occasionally sold in this form but more often, was taken through several additional heatings and poundings at a second or chafery hearth where the anconies were finally converted to bars. These usually weighed about 30 pounds each and could either be sold or directly manufactured into finished goods.

On Oct. 15, 1767 John Strode placed an advertisement in the *Maryland Gazette* seeking "Workmen who understand the FINERY and CHAFERY Business. Masterly hands only need apply, to whom suitable Encouragement will be given."

[232] At plating mills, also known as plating forges, iron bars were heated and hammered under huge water-driven hammers into thin sheets. These sheets were often tinned and the resulting tin plate made into a variety of utensils such as fish kettles, stewing pans, coffee pots, canteens, and camp kettles.

[233] Animal hides were used to make a wide variety of items such as shoes, boots, saddles, bridles, harnesses, straps, containers, etc. Processing raw hides into usable leather was a seven-step process. First, the hides were washed and trimmed. They were then soaked in a lime solution for several weeks. After being removed from the lime, they were "beamed" or scraped to remove the hair and any traces of remaining flesh. The fourth step involved soaking the hides for several weeks in a "bating" solution, a foul-smelling combination of water, salt, and manure that neutralized any remaining lime and made the hides more flexible. The actual tanning took place in a vat of tanbark in which the hides were left for 6 to 12 months. Tanbark was obtained from oak and hemlock trees, though tanners also used oak galls. The bark and galls were dried and ground so that they mixed well with water. Soaking in tanbark increased the durability of the leather. After removal from the tanbark solution, the leather was fulled in a mill or pounded in order to make it pliable. Finally, the hides were carefully dried at which point they were ready to be made into finished products.

[234] Wool mills, also known as fulling mills, were used to finish the rough wool cloth woven on the giant looms. The raw cloth was soaked in warm water and beaten by heavy water-powered wooden hammers that caused the wool fibers to mesh and tighten.

[235] In 1783 James Hunter paid taxes on 94 horses, most of which were probably left out in open pastures unless actively working at the forge. These animals would have been used primarily for hauling the endless wagon loads of firearms and supplies being manufactured at the forge.

[236] Limited amounts of steel were produced during the colonial period. This was often referred to as "cementation" or "blister" steel because the process used to make it left the metal's surface rough and covered with blisters.

According to the 1797 edition of Encyclopedia Britannica, "The best steel is usually made by cementation from the best forged iron, with matters chiefly of the inflammable kind. Two parts of pounded

wire mill[237]	anchor shop
kitchen	iron house

---

charcoal and one of wood ashes is esteemed a good cement. The charcoal dust may be made of bones, horns, leather and hairs of animals, or of any of these ingredients after they are burned in a close vessel till they are black: these being pulverized, and mixed with wood ashes, must be well mixed together...A deep crucible, two or three inches higher than the bars, is to receive part of the cement, well pressed at the bottom, the height of 1 ½ inch; and the bars are to be placed perpendicularly, about one inch distant from the sides of the vessel and from each other. All the interstices are to be filled with the same cement, and the whole covered to the top with it; then a tile is applied to cover the vessel, stopping the joints with thin lute [clay]" (Encyclopedia, vol. IX, p. 351). The containers of bars were then placed in a furnace or "air oven" and kept red hot for 6-10 hours. If the process worked according to plan, the bars were transformed into steel through the absorption of carbon from the crushed charcoal. Steel produced in this manner was suitable for the manufacture of swords and knives as well as for musket ramrods.

[237] On Dec. 4, 1779 Abner Vernon, company accountant placed an advertisement in the *Virginia Gazette* seeking "a masterly hand in a wire mill, also nailers, card makers, and other artists in iron industries, where may be had, bar iron, nail rods, rolled hoops, plates and cast iron, anchors, and a variety of other articles in the iron branch, with pots, kettles, and other castings." This ad was placed shortly after John Strode retired to Culpeper, likely explaining why the company bookkeeper was placing a notice in a newspaper. At the time this notice was printed, John Strode had retired to Culpeper and Abner Vernon had assumed the position of manager.

Wire was used for such things as chain, bolts, washers, nuts, and the teeth in textile cards, a major Revolutionary-era export item. The process of making wire began with rods formed at the rolling and slitting mill. These were produced in diameters of ½, 3/8, or ¼ inch. For heavy wire products, such as chain, the diameter might be larger. The rod was then drawn into a wire-drawing machine, which in the 18th century was powered by an overshot water wheel. Iron wire was drawn cold, that is, it entered the wire-drawing die cold. The metal was drawn through a succession of wire-drawing dies, each one progressively smaller than the preceding one, until it was reduced in diameter to the desired finished size. To aid in passage through the dies, the metal was lubricated. Iron was "dry drawn," that is, the wire was passed through a lubricant box that was placed immediately ahead of the die and contained powdered soap, tallow, or grease mixed with lime. This lubricant was essentially dry, but moist enough to adhere to the wire in its passage through the box.

As might be imagined, drawing cold strips of iron through the dies could not be accomplished by hand. Overshot wheels were required to pull the rods through the ever-smaller dies that reduced the diameter of the metal and increased its length.

Fig. 2-4
A combination rolling and slitting mill. Bar iron heated to a plastic stage was flattened by passing through rollers on the right. Flattened strips were then fed through the slitter and emerged as rods (notice bundles of rods stacked against the wall on the far left (Diderot, Vol. IV, Forges, 5th Section, Plate III).

Fig. 2-5
Forging iron with a trip hammer. Water poured on the hot iron during hammering made a shrieking hiss as it exploded into steam (Diderot, Vol. IV, Forges, 4th Section, Plate VII).

Not counting housing for the workers and an unknown number of warehouses, this list constitutes a minimum of 28 industrial buildings on one site. Warranting special attention is the armory. Strode mentioned in his letter to Patrick Henry of February 1777 that he needed a large building for his gunnery. This was in place shortly thereafter and was included in the report of the adjutant general dated Jan. 13, 1809 (Cromwell 1). According to the report, "The armory of the Rappahannoc [sic] was 350 feet long, Four Stories high; in the Lower floor was all the water machinery and the workmen employ'd to carry on the different branches for which they were constituted: the Second floor without any partitions from end to end was occupied by all the workmen each class form'd a separate division by themselves but without any wall or partition between, about the middle of this large space on an elevated station sat the Clerk (who was also an engraver) and noted every man's conduct the exact time he labored and all he performed etc--The Third floor was laid off in Lodging rooms & store rooms for part finished work, and the fourth Story was the repository for finished work, of every description of arms on hand & Sea Service Camp kettles, intrenching tools &c Saddles, Bridles, caps, boots and spurs for horsemen etc." (Banks 82).

This huge building was the gunsmith's shop where the boring of barrels and assembly and finishing of weapons was carried out. The first and sometimes second floors of multi-storied mills were often constructed of stone, while the upper levels were usually of wood. The stone base was better able than wood to withstand the violent vibration generated by the water-powered machinery. Unfortunately, the adjutant general's report does not provide the width of the building or the types of materials used in its construction. There are also no known surveys showing the location of the armory in relation to the other buildings at Rappahannock Forge. Considering the short time it took to build the armory and the fact that at least 32 other stone foundations remain on the site (none of which are obviously part of the armory), it is possible that the entire structure was built of wood. Time, termites, and/or Union vandalism could have effectively removed it from the landscape.

By any standards, the armory was a mammoth structure. Upon the fourth floor were stored finished products ready for shipment. Some of these finished products were firearms that began the process of assembly and finishing on the lower level and progressed upwards. One might be left to wonder why finished items would be hauled to the upper story, only to have to be carried or lowered back down again where they were loaded onto wagons and hauled away. It is critical to remember that Hunter's people were nothing if not practical and efficient. No extra steps would have been taken to move heavy crates of kettles, swords, or guns. No extra laborers would have been employed to carry out these unnecessary steps. The answer to this puzzle has to be found in the orientation of the building with the hill.

No discernible single foundation for the armory is visible on the site. A cursory study of the present forge site suggests that the armory may have included what appear to be two separate foundations labeled C and D on the site map (Fig. 2-9). Based upon the topography and the placement of the canal and road relative to the river, it would appear that the armory ran parallel to the river and stood to the east of the last tail race between the river and the old road. The branch forming the eastern boundary of the forge tract is lined on both sides with carefully laid stonework and appears to have been a wheelpit and tailrace. If this was indeed the tailrace for the armory water wheel, then that stonework may have extended upward to form the east wall of the armory. Measuring west 350 feet from this point brings one to within about 30 feet of the coal house (Foundation E). The coal house measures 40 feet by 84 feet and its floor is covered with a thick layer of coal dust and chunks of charcoal. Coal for the iron furnace would have been stored on the hillside above the furnace. This lower coal house probably supplied the forges that, logically, would have been located very near the armory. This suggested location for the forge armory is purely speculation and is based upon various physical features of the area. Only a detailed archeological survey of the site will prove if this speculation is accurate.

If the armory did stand in this approximate location, a bridge from the upper floor of the armory would have led straight across the road to the level ridge on the north side of the building. Loaded crates could have been rolled through a large door, across the bridge, and loaded onto waiting wagons parked on the ridge.[238] Coincidentally, at this point the ridge is approximately 41 feet high, the height of a four-story building, and a network of old roads can still be seen along the ridge's top and south side. The visible

[238] Gordon Barlow, Revolutionary War firearms historian, believes that most of Hunter's weapons were shipped from Rappahannock Forge to Point of Fork Arsenal from which they were distributed to the military. This facility was further inland and much easier to defend than Rappahannock.

remnant of an old road still leads from the edge of this ridge (overlooking foundations C, D, and E) out to Route 17 and this road is marked on several early maps and surveys. A quantity of stones also remain along the edge of this ridge, possibly from another building foundation or from the piers of the bridge that led from the top floor of the armory to the ridge and road.

James Hunter Junior and Senior also operated a ferry that crossed the Rappahannock at the foot of present-day Wolfe Street. Two other ferries operated during this period, one from Francis Thornton's land across to Falmouth and Dixon's ferry from lower Fredericksburg to Ferry Farm. The issue of ferries became a topic of heated debate for Fredericksburg residents who found it difficult to cross the river. In various petitions submitted to the Virginia legislature in 1779, residents claimed that travelers were often "detained at Mr. Hunter's ferry by the number of waggons which frequently pass that way" (Felder, Fielding Lewis 254). There were many animated debates about moving the ferry to a more central location. Not surprisingly, James Hunter Junior and Senior opposed changing the location of the ferry or opening another one. They countered the argument saying that "[A new ferry] could serve no purpose than depriving them of a fund useful in repairing streets and wharfs." They added that roads in the vicinity of the ferry had been built "at their own expense and a new skow built to transport large numbers of troops" (Virginia Legislative Petitions, 937-P).

Production

There are no known surviving ledgers from Rappahannock Forge that would provide a definitive picture of production at that facility. Lacking that, an understanding of the forge's production *capability* must be gleaned from a combination of sources. In 1770 British customs officer, John Williams, toured the customs stations along the James, Rappahannock, and Potomac rivers and wrote reports on each that he submitted to his superiors in England. Of Fredericksburg and Falmouth he wrote:

> ...30 Miles above [Port Royal] is situated on the same Side Fredericksburg, where the greatest Trade upon this River is carried on. Vessells of 40 & 70 Tons may load & unload their Cargoes. Two Miles above this is Falmouth near to which is reckoned the greatest Iron Works that is upon the Continent, which manufactures one & half Tons of Pig Iron into Barrs per Day (VCRP, Treasury Papers—In Letters, 1770-1771).

John Williams' statement is especially impressive considering the fact that in 1770 production had not yet peaked at Rappahannock. He then listed exports from the Rappahannock ports for the previous year, which included 113 tons of bar iron and 364 tons of pigs. How much of this iron originated at Rappahannock is unknown.

Over its 40-odd years of operation, Rappahannock Forge fabricated a wide range of products that seems to excluded little beyond glass, linseed oil, and cannon.[239] The following list, while extensive, is probably far from complete. The items contained on this list have been gleaned from primary source documents pertaining to the works and from the present property owner's personal collection of excavated artifacts. The firearms and swords will be discussed separately:

muskets	stove pipes for military hospitals
pistols	horsemen's caps
horsemen's swords	wagons
scabbards	barrels

[239] In 1776 Hunter sold three cannon, but he probably purchased them from another manufactory and acted only as a middle-man. The few surviving references to cannon and Rappahannock Forge involve naval contracts in which James Hunter provided all manner of products necessary to ship building. Unlike most other iron products, the casting of cannon required the use of mineral coal and was a highly specialized occupation. Hunter's estate inventory included 1,000 bushels of pit coal in the anchor shop and pieces of coal have been found on the site.

saddles	traveling forges[240]
bridles	flour and meal
bridle bits	wool cloth[241]
cooking utensils	cotton cloth
ships' anchors	spurs
axes of various types	spades
harness	shovels
camp kettles	ships' hardware[242]
wheels	knives
coarse and fine wire[243]	boots
sawed lumber	whiskey
horse shoes	canteens[244]
claw hammers	cannon balls
leather goods	camp stoves
mattocks	tool handles
shoes	fuller's shears[245]
pig and bar iron	brass hardware
files	chain
curry combs	nails[246]

---

[240] These were invaluable military supplies. They were essential for making horseshoes, which were required by the thousands of horses engaged by the military. The traveling forge also made possible repairs of metal items right in the field. There are no know surviving examples of Hunter's traveling forge, but it may be assumed that they were relatively small, perhaps no larger than 20" square. They would have had some type of bellows attachment, which might have been a standard wood and leather arrangement or even a scoop-bladed fan with a crank and reduction gear that kept the fan running after the blacksmith had stopped turning the handcrank. Forges of this type were also in use following the Revolution during the Indian Wars. Mann Page (1719-1781) wrote to Jefferson saying of the forges, "Upon the Receipt of your former Favour respecting the traveling Forges I immediately applied to Mr. Hunter, who engaged to have them finished with the utmost expedition. I for some time made frequent Enquiries into their Progress, & found that the Workmen were employed about them, but a late fit of sickness has prevented me from knowing whether they were finished or not. I will to Day make a farther Enquiry, & when they are finished, will cause them to be sent over to you." (Executive Communications, Letter, Mann Page).

[241] If Hunter was finishing his wool cloth, which is safely assumed from the fact that he offered it for making uniforms, then he had to have had a fulling mill on site.

[242] Included all manner of cleats, hooks, iron rings to link lengths of masts and spars, chain, bolts, nuts, washers for securing iron fittings, pierced strap iron for anchoring standing rigging, iron rudder posts, pulley wheels and axles, nails, etc.

[243] Wire was used in a variety of ways. Fine wire was used to wrap leather handles on swords and to make wool and cotton cards. Coarse wire was used for chain, bolts, nuts, etc.

[244] Camp kettles and canteens were probably made with tinned iron fashioned in the plating mill.

[245] Used to cut heavy wool cloth.

[246] For most of the 18th century, nails were extremely difficult to obtain, most being imported from England. Because of the shortage of nails, buildings were often pegged together, thus reducing the need for large quantities of nails. Although English law strictly forbade the construction of slitting mills in colonial America, economic demand motivated several entrepreneurs to ignore regulations and build them anyway.

When a planter determined that the tobacco had depleted his farmland, he simply moved on to another tract, built a new house and started anew. Typically, he burned his previous house and outbuildings as a means of recovering the nails used in their construction. Because nails were produced individually in a labor-intensive process, blacksmiths simply could not keep up with demand.

James Anderson, 18th century blacksmith in Williamsburg, Virginia estimated that eight boys could make 25,000 nails a week. To make the nails the smith started with a strip of iron several feet long, about ¼ inch wide and of the same thickness. This strip had been produced in a slitting mill. This square rod first had to be drawn down, meaning that it was heated and beaten out to thin and lengthen in. After drawing

steel	bullet molds
bayonets	carbines
gin falls and blocks	rolled hoops
salt pans[247]	wool cards, cotton cards and heckles
fascine knives	masts
spars	bowsprits
chests or trunks	spoons
plows of various types	forks
bullets of various types	buttons
scythes	small hand tools
cutlasses	pikes
barrel hoops	hoes of various types
griddles	knapsacks
firebacks	mallets
mauls	scrapers
grindstones	crosscut saws
whip saws	hand saws
compass whip saws	sawfiles of various types
augers	sledges
crow bars	rules
spoke and standard gimlets	compasses
saw sets of various types	carpenter's squares
chisels of various sizes	carpenter's gouges
spikes	rudder irons

No doubt Rappahannock Forge also included a store just as Accokeek had done. Sadly, the author has, as yet, discovered only one receipt. On Sept. 7, 1769 Adam Wayland[248] (c.1725-1781) paid for 6 ½ bushels of salt and paid an advance on "eight plates" (Wayland family papers).

As there are but a handful of surviving orders for weapons produced at Rappahannock Forge, determining production output is difficult. Charles Dick estimated that "when not interrupted and with a proper supply of money" his Fredericksburg Manufactory could produce one hundred muskets a month. While he never attained this level of production, it is evident that Rappahannock, operating under similar hardships but with a much larger work force and greater production capacity, could have produced many more. Ebenezer Hazard visited the Fredericksburg Gunnery in May 1777 and recorded the output of that facility at about twenty muskets per week produced by about sixty employees. The price paid for Fredericksburg muskets was £4.10 while Hunter was paid £8 per musket. Over the years many of the official papers generated during the colonial period have been lost or destroyed. The minutes of the Council of Virginia record some, but clearly not all, orders and payments for goods produced at the forge. Payment seems to have come, not as cash, but in the form of warrants, similar to modern purchase orders. In theory, these should have been redeemable for cash or tobacco. In reality, by 1781 there was little or no money in the State treasury with which to pay the warrants.

One of the earliest surviving orders was dated Mar. 27, 1776. At this time the Committee of Safety agreed "to take of Mr. James Hunter so many Falling axes, spades, shovels and mattocks as will be necessary to supply 4 regiments, and as many Picks as will be necessary for the third Regiment"

---

out, a chisel was used to cut the nails to the desired length. Finally, each nail was heated and beaten from the end to flatten and fashion a head.

[247] "On the Representation of Mr. Richard Parker manager of the Salt Works in Northumberland County it appearing to be necessary to procure a Salt pan for the said works and that the same may probably be had in a Short Time of Mr. James Hunter near Falmouth leave is accordingly given to Mr. Parker to engage with Mr. Hunter for a pan to contain about one thousand Gallons" (McIlwaine, Journals, vol. 1, p. 185).

[248] Adam Wayland was a farmer and miller who lived in Culpeper County. He established a mill there that was later operated by his son John Wayland (c.1754-1841), then grandson Simeon Wayland (1786-1847).

Adam married first Elizabeth Blankenbaker (c.1725-1775) by whom he had six children. He married secondly Mary Finks (c.1750-1830) and had two children.

(McIlwaine, Journals, vol. 2, p. 469). On May 4 the Council issued a warrant for £132.3 to Fielding Lewis for the use of Hunter "for intrenching Tools for the use of Col. Mercers Regiment, also for the use of James Hunter assistant to Richard Richards, for £5 for a Gun furnished the 3rd Regiment" (McIlwaine, Journals, vol. 2, p. 496). On June 7, the Committee of Safety ordered "that a warrant issued to Mr. James Hunter for One hundred & fifty pounds for twenty five stand of Arms, and two hundred and fifty pounds, on account, to purchase Arms, Iron, and intrenching Tools" (McIlwaine, Journals, vol. 1, p. 7).

On June 17, 1776 Hunter was paid "eighty four pounds for fourteen musquets and bayonets and ten pounds three shillings and nine pence for Ferriages" (McIlwaine, Journals, vol. 1, p. 25). On Aug. 8 he was issued a warrant for "Three hundred and Sixty one pounds five shillings for sixty muskets, 2 pair Bullit Moulds & three chests. Also one hundred & three pounds five shillings for Iron supplied Anderson the public Armorer, and for Waggonage of the Iron and Musquets" (McIlwaine, Journals, vol. 1, p. 113). He was also paid £50 for gunpowder for proving guns "manufactured for the publick" and £253.11.0 for spades and shovels "furnished for the use of the Publick Salt Works" (McIlwaine, Journals, vol. 1, p. 113). At the same time the Council ordered:

> that a Warrant issued to Mr. James Hunter for Sixty two pounds Seventeen Shilling and two pence for a Ballance due on his account for Arms and Ferriages, and for Barr Iron, Cannon Balls and Flour furnished for the Navy. Also thirty seven pounds, nineteen shillings and four pence ½ for three six pound Cannon furnished the publick (McIlwaine, Journals, vol. 1, pp. 114-115).

The next day the Council paid him £62.17.2 "for Barr Iron, Cannon Ball and Flour furnished for the Navy," as well as £37.19.4 ½ for three six-pound cannon (McIlwaine, Journals, vol. 1, p. 114).

Another order that has survived is dated Mar. 13, 1777 and was generated by the Board of War. It states, "That a letter be written by the president to Col. Stephen,[249] of the 10th Virginia regiment, ordering him to purchase 80 stand of arms, or as many more as can be procured, of Mr. James Hunter, of Fredericksburg, in Virginia the Colonel to be empowered to draw for the amount on the deputy pay master general of the southern department" (American Memory, Mar. 13, 1777). The military was woefully short of arms, this same document reporting that "Congress being informed that there are in the public armoury, in the State of Virginia, a number of arms, to the amount of about three hundred stand, belonging to the said state. Resolved, That a letter be written by the president to the governor and council of Virginia, requesting them to furnish the battalion commanded by Colonel Stephen, now at Fredericksburg, with the arms now in that place, and as many more as can be procured, for which they shall either be paid, or have the same number, equally as good, returned, at their election" (Virginia Board of War, Journal, Dec. 23, 1779-Mar. 25, 1780).

In addition to firearms and naval stores, Hunter also supplied the Continental troops with bar iron and several references to this commodity are present in the surviving records. Among these is an order for iron to be sent to James Anderson, master armorer at Williamsburg. On June 3, 1777 the Board of War ordered William Aylett[250] (1743-1780) to "send a Vessel to Mr. James Hunter at Fredericksburg for four Tons of such sized Bars as the said Anderson may want" (McIlwaine, Journals, vol. 1, p. 424).

---

[249] Col. Adam Stephen (c.1718-1791) was a Scottish physician who served in the British navy before coming to America where he settled in Berkeley County. In 1754 George Washington gave him the rank of general. He was Washington's second in command in the Virginia Regiment during the French and Indian War.

Adam represented Berkeley County at the second Virginia convention in the spring of 1775 and was chairman of that county's committee of safety. He was commander of the Berkeley County militia and, in February 1776 was appointed colonel in the Continental army. He made brigadier general the following September and major general in February 1777. In November 1777 he was dismissed from the service for bad behavior.

[250] William Aylett was a planter and merchant in King William County. He served in the House of Burgesses from 1772-1776 and was a member of the Virginia conventions from 1774-1776. On Apr. 27, 1776 Congress elected him deputy commissary general for supplying Virginia troops. On June 18, 1777 he was named deputy commissary for purchases. Two years later he was accused of corruption and resigned his office, but vigorously defended himself in the newspapers.

John Strode's letter of Feb. 19, 1777 to Patrick Henry had urged a ten-year contract between the government and the forge with a provision for compensation should the government default. The closest the state came to a contract was on June 25, 1777 when the Council minutes recorded:

> This Board taking under their consideration a Resolution of the House of Delegates directing them to agree with Mr. James Hunter for all the muskets completely fitted which he can make within twelve Months from this Time.
>
> Resolved that this Board will take all such Muskets that Mr. Hunter can Manufacture within the Time aforesaid and allow him the price of eight pounds for each providing they shall be as well filled [filed] & finished as those formerly purchased by this Board of the said James Hunter (McIlwaine, Journals, vol. 1, p. 440).

This was followed on Dec. 19, 1777 with a warrant to Hunter for £300 "on account of the Gun factory" (McIlwaine, Journals, vol. 2, p. 49).

On July 15, 1778 the Continental Congress ordered payment of "a warrant issued on John Gibson, Esqr.,[251] auditor general, in favour of the Committee of Commerce, for 417 35/90 dollars, to enable them to pay James and Adam Hunter, of Virginia, the balance of their account, dated 29 May last; the said Committee to be accountable" (American Memory, Journals, July 15, 1778).

Another large order is referred to in George Washington's orders, written at Valley Forge on May 1, 1778. He wrote, "By a letter from Colo. Moylan a few days ago, I find that his Regiment and Sheldon's[252] will want Arms, swords and pistols in particular, and as they are not to be obtained to the Northward, I beg you will engage all that you possibly can from Hunter" (Fitzpatrick, vol. 11, pp. 338-339).

On Mar. 28, 1777 George Washington wrote to Col. George Baylor[253] (1752-1784) of Caroline County regarding the outfitting of his militia unit. Washington, well aware of Hunter's manufacturing capabilities, wrote, "Surely Mr. Hunter can furnish Pistols as fast as they are wanted" (Twohig Revolutionary War Series, vol. 9, pp. 1-2). On May 1, 1778 Washington again wrote to Col. Baylor informing him of the shortage of arms and directing him to "engage all that you possibly can from Hunter" (Fitzpatrick, vol. 11, pp. 338-339). Baylor followed his instructions providing yet another surviving order this one spanning the period of Sept. 27, 1777 to June 19, 1779. Baylor's ledger includes orders for 32 sets of bridle bits, 15 swords and scabbards, 16 pairs of pistols, 1 bell, 1 sword blade, and the "hilting and polishing &c of an Of[ficer's] Sword." Bits cost 3 shillings and 12 pence per set, a sword and scabbard sold for 20 shillings, and a pair of pistols for 24 shillings. Many of the listed items were ordered for specified officers, including Adj. Victor,[254] Captains G. Lewis,[255] Gray,[256] Jones, Stith,[257] Jamison,[258]

---

[251] John Gibson (c.1725-1791).

[252] Maj. Elisha Sheldon (1740-1805) was born in Lyme, Connecticut. His reputation for wild behavior resulted in his being put out of his local church for "lascivious conduct and breach of the Sabbath." Sheldon commanded the 5th Regiment of Light Horse and served with the Continental Army during the summer and fall of 1776 as commander of a corps of Connecticut volunteer horse. On Dec. 12 of that year Congress appointed him lieutenant colonel commandant of a cavalry regiment, the 2nd Continental Dragoons. This was reorganized in 1780 as a legionary corps and he was made a brevet brigadier general Elisha Sheldon died in Vermont.

[253] From 1775-1776 George Baylor was a member of the Caroline County Committee of Safety and from 1775-1777 was aide-de-camp to George Washington. He was the commanding officer of the 3rd Regiment Light Dragoons. In the fall of 1778 Baylor's regiment was sent to Old Tappan, New York for the purpose of harassing Cornwallis and von Knyphausen on the Hudson River. On Sept. 28 British Gen. Charles Grey victor at Paoli the previous year, approached the American camp by nightfall and, in a silent and vicious attack, surprised Baylor's dragoons. Thirty Americans were killed, 50 were captured, and George Baylor was severely wounded in the lungs. Although he was eventually exchanged, Baylor never fully recovered from his wounds and he was never again physically able to command. On Nov. 9, 1782, his regiment was consolidated with the 1st Continental Dragoons under the command of Col. William Washington. On Sept 30, 1783 he received a commission as Brevet Brigadier General.

[254] John Victor (1757-1817) was born in Virginia. He married Sarah Tankersley.

[255] George Lewis (1757-1821) was an officer in George Washington's personal guard during most of 1776 In December of that year he became a captain in the 2nd Continental Dragoons and the following January

Lieutenants Randolph, Barret, W. Baylor,[259] Maj. Wolfin, and Colonels Washington,[260] Dade, and Baylor.[261]

Gen. Washington wrote to Col. Stephen Moylan[262] (1734-1811) on Apr. 29, 1778 discussing Col. Baylor's order with Hunter:

> I am as much at a loss as you can possibly be how to procure Arms for the Cavalry, there are 107 Carbines in Camp but no Swords or Pistols of any consequence. General Knox[263] informs me, that the 1100 Carbines which came in to the Eastward and were said to be fit for Horsemen were only a lighter kind of Musket. I believe Cols. Baylor and Bland[264]

---

transferred to the 3rd Continental Dragoons. In February 1779 he resigned his commission. George married Catherine Dangerfield.

256 Probably William Gray who served as a captain in the 1st Continental Dragoons in 1781.

257 On Mar. 19, 1776 John Stith (1755-1808) was appointed a 1st lieutenant in the 4th Virginia Regiment. He made captain on Mar. 12 the following year and was taken prisoner at Charleston on May 12, 1780. On Sept. 30, 1783 he was appointed Brevet Major and served until the end of the war.

258 John Jameson (1752-1837) was from Culpeper County. On June 16, 1776 he was appointed captain of a Virginia regiment of dragoons. He made the rank of major of the 1st Continental Dragoons on Mar. 31, 1777 and transferred to the 2nd Continental Dragoons the following month. In January 1778 he was wounded near Valley Forge. On Aug. 1, 1779 he was made a lieutenant colonel and served until the end of the war. John married Nancy Hayter.

259 Walker Baylor (1760-1822) was George Baylor's (1752-1784) brother and, in June 1777, was commissioned a lieutenant in his brother's regiment. In February 1780 he was promoted to captain and resigned from the army in July of that year. Walker married Jane Bledsoe.

260 Col. William Washington (1752-1810) was the son of Bailey Washington (1731-1807) and Catherine Storke (1722-1804) of Windsor Forest. He was a cousin to George Washington. In February 1776 William was commissioned as a captain in the 3rd Virginia Regiment. After being wounded at Trenton, he was appointed a major of the 4th Continental Light Dragoon Regiment on Jan. 27, 1777. In November of that year he was named lieutenant colonel commandant of the 3rd Continental Light Dragoons. They were sent to the southern department in the spring of 1779 and remained there until 1782. This group was later reorganized and renamed the 3rd Legionary Corps. William was wounded at Cowpens on Jan. 17, 1781 and on Mar. 9 Congress voted him a silver medal for his conduct there. He was wounded again and captured at Eutaw Springs on Sept. 8, 1781 and was a prisoner on parole until the end of the war. After the Revolution, he settled in South Carolina where he married Jane Riley Elliott.

261 Account ledger, Col. George Baylor's account with James Hunter, Sept. 27, 1777 to June 19, 1779, Alderman Library, University of Virginia, Charlottesville, Virginia.

262 Stephen Moylan was born an Irish Catholic and was educated in Paris. He spent three years in the shipping business in Lisbon before settling in Philadelphia in 1768 where he became a prominent merchant. On Aug. 11, 1775 he was appointed Continental mustermaster general and was responsible for maintaining the muster rolls for the Continental army as well as inspecting troops and equipment. In the fall of 1775 he assisted in outfitting several armed vessels for Continental service. On Mar. 6, 1776 he became one of George Washington's aides-de-camp. In June of that year he was named Continental quartermaster general with the rank of colonel. He resigned from service in September 1776. In January 1777 he was commissioned to raise a regiment of light horse and served as a cavalry officer for the remainder of the war.

263 Henry Knox (1750-1806) was a bookseller from Boston who spent his pre-war years reading extensively about military science, though he had no formal military training. He joined with Col. Joseph Waters in planning and overseeing the building of fortifications at Roxbury and on Nov. 17, 1775 became colonel of a Continental regiment of artillery that he commanded throughout the war. In March 1782 he was promoted to major general and in December 1783 succeeded George Washington as Commander-in-Chief of the army. He served as Secretary of War from Mar. 8, 1785 until Dec. 31, 1794. He was always a large man, but by the end of the Revolution weighed about 300 pounds.

264 Col. Theodorick Bland (1742-1790) was the son of Col. Theodorick Bland (1708-1784) of Cawsons, Prince George County, Virginia and was a physician and planter. On June 13, 1776 the Virginia convention appointed him captain of a troop of light horse. The following December the General Assembly named him

have procured Swords from Hunter's Manufactory in Virginia, but I do not think it will be possible to get a sufficient Number of Pistols, except they are imported on purpose. I long ago urged to Congress the necessity of importing a large quantity of Horse Accoutrements from France, but whether the order was ever given, or whether they have miscarried the passage I do not know (Fitzpatrick, vol. 11, pp. 338-339).

On Jan. 19, 1781 George Muter presented the carter, W. Nuttall, with orders:

You will proceed to Westham[265] & receive from Mr. Boush as many Muskets as the two Waggons can conveniently carry, which you are to deliver to Charles Dick Esq: at Fredericksburg, or to him & James Hunter Esq: as Mr. Dick shall direct. Should Mr. Dick think it proper to lodge the Arms at some distance from Fredksburg [sic], you will have it done, but still you are to proceed to Fredksburg with the Waggons, and to bring from thence in the Waggons to this place (Richmond) the articles Mr. Hunter will deliver you in consequence of the order you will carry from Mr. Armistead to him, for them.

I must recommend the most carefull attention to the Articles committed to your charge, & that you will be a expeditious as possible (War Department # 035514).

The Act for the Encouragement of Iron Works was rescinded in 1780 and, by mid-year, most of Hunter's skilled craftsmen had been drafted into military service. Those who avoided the initial draft call now crossed the river to join the Fredericksburg Manufactory. Because Fredericksburg was a public facility, its workers were exempt from the draft. By late 1780, Hunter had lost so many workmen that he was forced to suspend his small arms manufactory for want of skilled workers. Additionally, British raiding parties were wrecking havoc on gun makers. On Jan. 11, 1781 Baron von Steuben wrote to Washington warning, "An attempt might be made at Williamsburg. The next great object for the enemy being Hunters Iron Works and the Stores at Fredericksburg" (Sparks, Correspondence). At Richmond, military supplies were removed to Westham, though the British discovered them anyway. They razed the foundry and destroyed or damaged many of the muskets stored there. However, a number of these weapons were deemed repairable and Col. George Muter[266] (died 1811), the Virginia Commissioner of War, ordered them sent to Fredericksburg for repair.

The safety of Rappahannock Forge was of grave concern. On Jan. 10, 1781 Jefferson wrote to Hunter stating:

---

major commandant of the state's six troops of light horse. In early January 1776 his regiment joined George Washington at Morristown, New Jersey. He was promoted to colonel on Mar. 31, 1777 and served as commandant of the Convention Army's barracks near Charlottesville from April to October 1779. In December 1779 he resigned from service. Theodorick was a member of the Continental Congress from August 1780 to October 1783 and was a member of the U. S. Congress from March 1789 until his death in June 1790.

[265] Westham Foundry was located in Henrico County.

[266] In April 1780 the state capital was moved from Williamsburg to Richmond, creating great confusion at a time when it was least needed. The board of war was not moved at this time, but simply ceased functioning. The governor and Council assumed the duties of the board until George Muter, commander of the State Garrison Regiment, took over in June. At this time a governmental reorganization also resulted in the boards of trade and the navy being replaced with commissioners. Muter was criticized for his handling of the position, especially strong words and actions coming from Baron Friedrich von Steuben who appealed to the Assembly to have Muter fired. Steuben claimed that ordnance and other supplies were in a "disorderly situation." The ensuing investigation revealed that large numbers of muskets were unusable for want of repair and many cannon were lying useless because they weren't mounted in carriages. The Assembly determined that "the whole Business of the War Office appears to be entirely deranged [and]...the present Commissioner of War is not qualified to fill that important Office and ought to be discharged therefrom." Col. William Davies replaced Muter and served in that capacity until the end of the war.

The importance of your Works to the operations of War will doubtless point them out as a proper object of destruction to the desolating Enemy now in the Country. They are at this time at Westover and will possibly embark there. Their next expedition we cannot foresee; lest it should be to demolish your Works. I write to Genl. Weedon to take measures for protecting them: In the mean time I would advise you to move directly off into the Country every thing movable. Should you not be able to effect this by your own and hired Waggons, I hereby authorize you to impress Waggons, teams and drivers for that purpose, only placing yourself instead of the public in point of responsibility to the Owners (Boyd, vol. 4, pp. 329-330).

James Mercer[267] (1737-1793) also expressed his opinion of the value of Hunter's works. In a letter to Gov. Thomas Jefferson he wrote:

I shall therefore without appollogy proceed to inform your Excellency, that for my own knowledge of the country & the uniform opinion of all I have conversed with on the subject there is not in this State a place more deserving of public attention than this Town and its appendage Mr. Hunter's Iron Works—I am sure I need not tell you that it is from Mr. Hunter's Works that every Camp Kettle has been supplyed for the continental and all other Troops employed in this State & to the Southward this year past—that all the anchors for this State & Maryland & some for continent have been procured from the same works: that without these works we have no other resource for these articles, and that without the assistance of the Bar Iron made there even the planters heareabouts & to the Southwards of this place wou'd not be able to make Bread to eat (Executive Communications, Apr. 14, 1781).

On Jan. 19, 1781 Col. Muter advised Charles Dick that the guns from Westham were enroute to Fredericksburg with a carter named W. Nuttall. He urged that they be repaired with all possible speed and that care be taken to prevent their capture. Muter wrote:

The necessities of the State are such that the utmost expedition in repairing arms is most essentially necessary at this time. Therefore if it is possible to get a few of the Arms repaired at Mr. Hunter's Works, I cou'd wish it to be done. I have inclosed a Letter for Mr. Hunter which (if you think it proper) I wish you to be so good as to send him, along with as many of the Muskets as you think necessary to direct Mr. Nuttall to deliver at his works (Revolutionary War Manuscripts, Jan. 19, 1781).

On Feb. 12, 1781 Muter wrote to William Armistead advising him of materials requested from James Hunter:

Two Waggons are now ready to sett off for Fredericksburg for Nail rod & Camp Kettles & wait only for orders. Mr. Rose will call on you or send some person to whom you will please to give orders to Mr. Hunter for as much Nail rod & as many Camp Kettles as they can possibly carry. You will please to order it so, that the waggons shall bring as many Kettles as possible & only make up the loads as to weight with Nail rod (War Department Collection #035548).

---

267 James was the son of John Mercer (1704-1768) of Marlborough. He graduated from William and Mary and in 1756 commanded Fort Loudoun in Winchester. In 1762 he was elected to the House of Burgesses, a position that he held until 1775. James was also a member of the Committee of Correspondence and helped organize the Provincial Convention of 1774. He served as a member of the Virginia conventions of 1774, 1775, and 1776 and was on the Committee of Safety from 1775-1776. This body appointed him to the Virginia Committee of Safety. In 1779 he was selected as a delegate to the Continental Congress and from 1779-1789 was a judge of the General Court of Virginia. James was a judge on the first Virginia Court of Appeals, a position he held from 1786 until his death. In 1772 he married Mary Eleanor Dick (died 1780), the daughter of Charles Dick.

In March 1781 William Davies replaced George Muter as Commissioner of War. On the 26th ol that month Davies wrote to Hunter requesting a large order of entrenching tools:

Sir—

It is of the greatest importance to the public that a number of entrenching Tools should be immediately procured. I am informed by the Governor that you have a considerable number on hand for the Continent. The pressing applications from the Marquis and Baron to have these articles immediately forwarded, induces me to send Express to you on the subject. 800 spades, 400 common axes, 200 Broad & Grubbing Hoes, 100 pick axes, 300 fascine knives[268], & small hatchets, 6 crosscut saws, Carpenters tools of all kinds for the hands, and nails of the larger sort.

These are the articles that are wanting and not knowing how far it may be in your power to furnish them, I have mentioned them all. Whatever can be furnished I beg you to spare us, & forward them on by the first opportunity that can be procured, as the supply admits of no delay, it is so necessary to our operations below. I beg your answer by this Express, and am

Sir, very respectfully,
Your most hble sevt.
William Davies

(War Department Collection #035676)

This was followed by another letter to Hunter regarding the repair of arms and other equipment:

Sir—

You will be pleased to give Capt. Read such assistance as may be in your power toward the repairing the different equipment of his cavalry as expressed in the enclosed paper, as also toward procuring the several articles necessary for his troops as therein expressed. As he has met with much assistance from you formerly we hope for the same kind attention in the present instance.

I am Sir
Yours &c
Wm. Davies

(War Department Collection #035675)

At Rappahannock, James was struggling with a multitude of problems. As the damaged muskets were enroute from Westham, Hunter received a hasty letter of warning from Jefferson urging him to remove his equipment to a safer location. Having raided Westham, it appeared that the British troops might be on their way to destroy the works at Fredericksburg and Falmouth. Col. Muter also wrote to Hunter informing him that the damaged Westham muskets were enroute to the Fredericksburg Manufactory and to Rappahannock Forge and urging him that they might be "repaired at your works as quick as possible" (Revolutionary War Manuscripts, #035511). On Jan. 25, 1781 Hunter wrote Jefferson thanking him for the warning and informing him that he had carried out the directive to remove the equipment (Executive Communications, Jan. 25, 1781).

James was facing other problems as well. When the Westham muskets arrived at Rappahannock, Hunter was unable to carry out the repairs. His letter to Jefferson said, "But it gives me concern to acquaint your excell'y that it is not in my power to repair any of the arms sent me by Col. Winter whose letter accompanying them came to hand yesterday; my workmen in that branch having all left me, and the

[268] A large knife similar to a machete that was used to cut sticks. Fascines were used in fortifications and consisted of a fagot or bundle of rods or small sticks of wood, bound at both ends and in the middle. They were used in raising batteries, in filling ditches, strengthening ramparts, and in making parapets. They were sometimes dipped in melted pitch or tar and were used to set fire to the enemy's lodgments or other works.

manufactory of small arms being of consequence discontinued; but the orders before given by the state for other matters are going on with all possible expedition" (Palmer, vol. 1, pp. 463-464).

Some of Hunter's shops continued operating as usual. On Feb. 20, 1781 James again wrote to Jefferson stating, "I beg leave to inform you that the order given last summer for 1000 [camp kettles] to be furnished for the particular use of the state is now fully completed, and the greater part of them, including in the number the 200 received by Col. Towle, by virtue of your excellency's order to that effect, have at different times been delivered" (Boyd, vol. 4, pp. 666-667). He also expressed frustration at not having been able to complete work sent him previously. James wrote,

> And on this occasion, I cannot forbear to testify to your excellency the great regret I felt, that I was unable to render fit for service the muskets sent to this place some weeks ago; at a time too when they were so much wanted; but the making and repairing of small arms, once prosecuted to so considerable an extent at my works has been for some time past discontinued for want of workmen, all those employed in this service having left me principally because by an act of the legislature, they are restored subject to militia duty, draughts, &c from what they had always enjoyed an exemption; altho' if the privilege could again be restored I have no doubts that the works in this branch might be resumed to the great benefit of the state (Boyd, vol. 4, pp. 666-667).

In the same letter Hunter informed Jefferson that he was "engaged in fabricating a parcel together with some other things by the direction of Gen. Greene,[269] and for the use of his army; nevertheless the order your excellency has last given for 1000 additional arms to be immediately got ready for the separate service of the state, shall be particularly regarded; all possible dispatch used for its completion, and your excellency made acquainted with our progress in the execution of it; be assured no exertions within the compass of my power shall be wanting, where the public good is concerned."

On Mar. 26, 1781 Richard Claiborne (1757-1818) wrote to Thomas Jefferson explaining how the cutbacks at Rappahannock Forge impacted him. "I beg likewise to intreat of your Excellency an order for the exemption from Malitia duties all the men that Mr. James Hunter shall employ for the Public at his Iron works near Fredericksburg. Many of them have left him and gone to the Gun factory for protection, by which means I am deprived of the articles he has agreed to furnish me with, and I have some time since advanced him the money to enable him to complete them as soon as Possible" (Boyd, vol. 5, pp. 240-241).

The threat of British raiding parties continued into the spring. On May 29, 1781 Lafayette wrote to Anthony Wayne warning that Banastre Tarelton was rumored to be approaching Fredericksburg with the intention of destroying both armories. He informed Wayne that the English had repaired Bottom's Bridge, crossed it and were camped at Hanover Courthouse. He asked Wayne to immediately proceed toward Fredericksburg and meet him somewhere along the way. Lafayette wrote, "Should Enemy destroy Hunters Works before We meet, it will be a heavy blow to these Southern States. A few hours may perhaps decide a great deal in the fate of this Warr, and I shall not loose a Moment to forward the movements of this Army towards the wished for junction" (VCRP, Papers Relating).

On the English side, Cornwallis was faced with a decision. On June 30, 1781 he wrote to Sir Henry Clinton informing him of his decisions regarding troop movements. Cornwallis reported that after crossing the James River at Westover, he moved on to Hanover Courthouse and crossed the South Anna. He was well aware of being flanked by Lafayette and knew the Americans believed the British army to be

[269] Nathanael Greene (1742-1786) was considered the best American military strategist of the Revolution. He was born to a Quaker family in Warwich, Rhode Island and worked as a young man in his father's iron foundry. In September 1773 Nathanael was dismissed from his Quaker meeting for attending a military parade. Two years later he was commissioned as the youngest brigadier general of the Continental Army and was given command of three Rhode Island regiments. In August 1776 he was promoted to major general. George Washington made him quartermaster general of the army in February 1778 and he replaced Horatio Gates late in the latter part of 1780. This appointment was in response to Gates' defeat at Camden after which he fled the field. Nathanael fought three major battled in the Carolinas, namely Guilford Courthouse, Hobkirk's Hill, and Eutaw Springs. Though he lost all of these engagements, the effort so weakened the British army they nearly collapsed.

headed for Fredericksburg. Cornwallis, however, had received some unspecified information regarding Hunter's works and wrote, "From what I could learn of Hunter's Iron Manufactory, it did not appear of so much Importance as the Stores on the other side of the Country and it was impossible to prevent the Junction between the Marquis and Wayne." Cornwallis chose instead to detach Lieutenant Colonels Simcoe and Tarleton "to disturb the Assembly then sitting in Charlottesville, and to destroy the Stores there, at old Albemarle Court House" (VCRP, Papers Relating). Tarleton's troops were said to have destroyed "1000 new muskets made at the Public Gun Works *near* [italics added] Fredericksburg" (Reilly 39). It is unclear whether these weapons were made at Fredericksburg or Rappahannock.

Hunter struggled to keep some elements of his factories productive while removing, then returning, valuable gunmaking equipment. On May 30, 1781 Hunter wrote to Gov. Jefferson that, "Tarleton with 500 Horse is reported to have been at Hanover Court yesterday & last night within five miles of Bowling Green, on his way to destroy my works—unless my Sword Cutler and Artificers, that could make the swords, are returned on furlough, it is impossible to furnish them. At present I am removing my tools and a total stoppage of everything" (Boyd, vol. 6, p. 41). Gordon Barlow, noted historian, believes that Hunter's equipment was moved to the Stevensburg[270] Armory just outside Culpeper. He also notes that when Hunter returned the equipment to Rappahannock Forge, he didn't re-open some of his shops. At this same time, Stevensburg (opened c.1780) acquired some state-of-the-art equipment. Mr. Barlow suspects that by this time Hunter was growing weary of the struggle to keep his works open and simply wanted to be done with it. John Strode had been gone nearly two years, the government wasn't paying him for goods already produced and, by result, he was unable to pay his workers. The repeal of the Act for the Encouragement of Iron Works was resulting in the loss of many of his craftsmen who could go across the river to Fredericksburg to avoid the draft. According to payment vouchers in the collection of the Virginia State Library, in 1783 workers at the Fredericksburg Manufactory included manager William Grady, William Moxley, David Partlow, and Leonard Young. Others probably left for lack of pay, the critical shortage of skilled gunsmiths making employment readily available in other armories.

The enemy's rumored approach proved to be just that. Shortly thereafter, Hunter was given some relief from the draft, was able to procure more arms makers and resume work, though on a limited scale. A letter from Hunter to Col. Oliver Towles[271] (1736-1825) dated Nov. 22, 1781 stated, "Since the Assembly extended the like indulgence to my workmen as those employed in the Public Factories, I have resumed the manufactory of small arms—Workmen are putting the machinary for grind'g & boring &c. in order." He had on hand and ready for use "one traveling Forge with its utensils and Harness complete," approximately 500 bridle bits, a quantity of curry combs, and 1,000 horsemen's swords "to the pattern forwarded from General Green's Army by order of Colonel Washington." As to his receiving payment for completed goods, Hunter wrote, "I trust thro' Col: Davies, will be put on such a footing, as I may receive something solid to pay wages & buy Provisions (my own being deliver'd the Public) & to pay the first cost of the materials wanted, if my labour & prop[erty] is with held" (Palmer, vol. 2, p. 618). The pattern mentioned was a sword captured at the battle of Guilford Courthouse in North Carolina and which had been sent to Rappahannock in March 1781 so that it could be reproduced and issued to American soldiers.

James' wartime expenditures (and frustrations) were not confined to the forge. Military regiments were free to impress whatever supplies they needed as they passed through an area. Property owners were forced to surrender food, livestock, wagons, guns, pasture, hay and grain for horses, or anything else required by the soldiers. In 1781 James Hunter submitted a public claim for goods taken from his

---

[270] Though little more than a crossroads today, Stevensburg was once a busy area. The little village was located at the intersection of the old Carolina Road (State Route 663) and Kirtley Road (State Route 600). Near the present site of the Stevensburg Methodist Church was a Quaker meeting house called Southland that was established as early as 1782. This served the Quaker community around Stevensburg and Mt. Pony.

[271] Oliver Towles resided in Spotsylvania County. He studied law under Edmund Pendleton (1721-1803) and, at the outset of the Revolution, had a large practice. On Feb. 16, 1776 he was commissioned a captain in the 6th Virginia Regiment and made major on Aug. 15, 1777. Oliver was taken prisoner at Germantown and not exchanged until late 1780. On Feb. 12, 1781 he was commissioned Lt. Colonel of the 5th Virginia Regiment and retired from military service on Jan. 1, 1783. He married Mary Chew Smith, the daughter of Larkin Chew (died 1770) of Spotsylvania and the widow of John Smith of Rickahock.

Spotsylvania plantation including 106 pounds "jesuits bark[272] and cask" valued at £106.81, 101 pounds of bacon, 1,218 pounds of flour, and "106 tight casks." The following year he submitted a claim from Stafford for ten beeves and corn and fodder valued at £8.19 (Abercrombie, vol. 3, 859, 869, 873).

Payment for Completed Orders

In January 1781 Hunter wrote to Gov. Thomas Jefferson asking to be paid for "the warrant granted me on the treasury sometime ago, for £50,000" as well as for certificates from Gen. Gates[273] for materials totaling some £130,000. Also included on the "Return of Provisions" dated Sept. 30, 1777 and included among the papers of the Continental Congress is a receipt documenting James' contribution of 711 pounds of beef and 2,919 pounds of flour (Papers of the Continental Congress, Return of Provisions).

Virginia authorities were quick to contract with Hunter for the goods that he could produce. They were notoriously slow to pay, however, and Hunter (like Fielding Lewis and Charles Dick[274]) invested a good deal of his own money to keep the works afloat financially. In February 1777 at Gen. Washington's demand, Hunter received an order for "so many arms as may be sufficient to arm a regiment" (McIlwaine, Journals, vol. 1, p. 332). This meant that the general needed at least 680 muskets. While there is no surviving documentation proving that Hunter delivered the order, it would have been highly uncharacteristic had he not done so. Nowhere in the surviving official records of Virginia is there mention of his having been paid for so many weapons.

James seems always to have been short of cash at the forge, a problem that was exacerbated by dealing with the government. As early as April 1774 James complained, "I am realy much strained for Money to procure Pig Iron to keep my Forges working" (Hunter Papers, Apr. 13, 1774).

The problem with the Board of War's refusal to pay for officer's equipment had come up between Hunter and Baylor previously. On Oct. 12, 1779 Hunter wrote to George Baylor enclosing a copy of the account and a copy of a letter he had received from the Board of War in Philadelphia (Hunter Papers, Oct. 12, 1779). The Board had informed Hunter that "Your Draft on Richard Peters Esq.[275] for £7056.5 Virg. Currency which was presented this morning & payment refused. Col. Washingtons warrants were left with the Board sometime ago and the reason now given for refusal is that the articles specified were chiefly for Officers. By some late regulations a fixed sum is allowed for the Equipment of Officers in lieu of their Accoutrements in necessary arms &c for Privates are paid by the Board of War" (Hunter Papers, Oct. 12,

[272] Jesuits Bark comes from a species of tropical trees of the Chinchona family that grow in South America, specifically in the Peruvian Andes. The bark of this tree contains quinine and was used by Jesuit priests as early as the mid-17th century to treat malaria. The priests took large quantities of the bark back with them to Spain, France, and Italy where it became known as Jesuits Bark. The inner lining of the bark was stripped from its woody outer covering, dried, ground into powder, and made into a bitter-tasting tea.

[273] Horatio Gates (c.1728-1806) was born in Maldon, Essex, England, the son of Robert and Dorothy (Parker) Gates. He served as a captain in the American army through much of the French and Indian War and was seriously wounded while serving with General Edward Braddock's (1695-1755) troops. In 1773 he bought a plantation in the lower Shenandoah Valley where he built a home called Traveler's Rest. This is located near modern Kearneystown, Jefferson County, West Virginia. George Washington called Horatio back into service and he commanded the American forces at the victory at Saratoga (1777), but was soundly defeated in 1780 at Camden, South Carolina. Horatio married first in 1754 Elizabeth Phillips. He married secondly in 1786 Mary Vallance.

[274] On Jan. 23, 1781 Charles Dick sent a letter to Gov. Jefferson expressing his concern over the lack of state support for the gunnery. He wrote, "I never intended to resign the Business of the Factory, as I had such a considerable share in its formation &c, and the pleasure of its thriving to such a Degree for the public Benefit; could I been even but allow'd a bare Maintenance for my fidelity and Services, But True it is! I am now providing and contracting for the proper Materials and Provisions for the Factory on my own credit and Interest, for no body will trust the Public a farthing, and shall have more Honour to be under your Excellency and Counsil's Appointment and Directions, not doubting but I shall be supported with Money to carry on the Business with pleasure and benefit to the Public" (Boyd, vol. 4, p. 430).

[275] Richard Peters (1743-1848) was from Pennsylvania and served as Secretary of the Board of War.

1779). Hunter informed Col. Baylor that "the payment of my Ball. for Arms delivered for the use of your Regiment appears at greater distance than I expected and as I have suffered much for want of the Money, shall be greatly obliged to you to come up, if you can possibly do it with any convenience for it is not in my person at present to wait on you & hope you excuse and put me in some Tract to obtain payment & if you can make it convenient between & next Sunday we might write Missrs. Fitzhugh and Mercer on that score, who I dare say will do all they can to facilitate the payment with such instructions & certificates as you think proper to forward." As with the other orders, there is no indication of whether or not James received payment for these items. He seems, however, to have suffered as we do today with ever changing government regulations.

Despite Patrick Henry's declaration of the value of the iron works, by 1780 Hunter's business was in serious straits. John Strode had retired to Culpeper for health reasons. Not only was Hunter not receiving timely payment for his goods, but many of the forge's wagons had been pressed for military use and his workers were leaving for the Fredericksburg Manufactory where they were exempt from the draft.

The drafting of Hunter's workers was in every way counter-productive to the war effort and, at various times forced him to close his works. On Apr. 16, 1777 Richard Henry Lee wrote to George Washington asking him to consider exempting iron workers from the dragt. Lee wrote:

> Conversing lately with Mr. James Hunter of Fredericksburg, whose labors have benefitted the public greatly, I find that the indispensable article of iron has been greatly affected, and its production injured, by the constant practise of inlisting the Laborers in those works, and pressing the Teams belonging to them. There are few things more capable of throwing distress among the people, and injuring public affairs, than such a proceedure. I would therefore submit it to your consideration Sir, whether (until the Legislatures can provide compitent laws) it will not greatly remedy the evil, if you were, by order published in all the papers [to] forbid all Continental Officers from inlisting persons engaged with, and actually serving in any iron works within the United States, or from pressing any horses, teams, or Carriages of any kind belonging to such works (Twohig, Revolutionary War Series, vol. 9, pp. 178-179).

Washington had a different view on protecting iron workers from military service. On Apr. 24, 1777, he answered Lee and expressed his views on the subject:

> You are not aware of the evil consequences, that would follow a general exemption of all persons concerned in iron-works from military duty; they are very numerous, and in this part of the country form a great majority of the people. Besides, why should the ironmaster carry on his trade without restriction, when the farmer, equally useful for the support of the war, ...shoemaker, and other manufacturers, ...may have their servants and apprentices taken from them at pleasure? (Twohig, Rev. War Series, vol. 9, p. 257)

There does seem to have been some attempt on the part of Virginia authorities to pay Hunter for the materials he was producing. One particularly interesting effort towards this end was made on June 29, 1780 when the Council ordered the governor "to direct that the tobaccoes collected in the counties of Stafford and Spotsylvania for the publick, under the Act of Assembly for laying a tax payable in certain enumerated commodities be delivered to Mr. Jas. Hunter; part of which he is to take at £50 __ P Ct—in part paymt of ye drafts of Congress in his hands—for 123,077 45/90 Dollars and the residue Mr. Hunter is requested to keep in his possession until further instructions from the board" (McIlwaine, Journals, vol. 2, p. 263). It is unknown if James ever received this payment or not and, if so, what became of the "residue." However, the order does show an honest intent by authorities and provides modern researchers with a hint of the extent of operations and production at Rappahannock.

In late 1776 the Council issued several warrants to Hunter as payment for a variety of supplies provided by Rappahannock Forge. On Nov. 11 he received a warrant for £1,347.12.1 ½ that "being the amount of his Account for Arms, Camp Utensils, Waggonage and other Necessaries, as settled by the Commissioners" (McIlwaine, Journals, vol. 1, p. 235). A few days later he was issued a warrant for £175.15 for 25 muskets and bayonets and for " Chest for Package also for Bar Iron, Steel and Waggonage" (McIlwaine, Journals, vol. 1, p. 248). On Dec. 18 he was paid £956.5 it "being the amount of fifteen hundred Bushells of Salt purchased by the Governor and Council" and £122.5 for muskets, bayonets, and waggonage (McIlwaine, Journals, vol. 1, p. 288).

On Dec. 4, 1779 Abner Vernon placed a notice in the *Virginia Gazette* seeking artisans and offering iron products for sale. The notice read:

> Wanted at Hunter's iron works, to whom suitable encouragement will be given, a masterly hand in a wire mill; also nailers, card makers, and other artists in iron manufactures; where may be had, bar iron, nail rods, rolled hoops, plates and sheet iron, anchors, and a variety of other articles in the iron branch, with pots, kettles and other castings.
>
> Abner Vernon
>
> The best price will be given for old brass or copper, by applying to Robert Nicholson or William Hunter, in this city.

On Nov. 1, 1779 the War Office directed William Finnie[276] (1739-1804), Deputy Quartermaster General "to pay all such Articles of the above account as were furnished for Non Commissioned Officers & Privates & charge it in his account with Col. B. Flower[277] CGMS. The Board are not authorized to give any orders about paying for Articles furnished Officers." William Finnie added another note dated Williamsburg, December 13, 1779 stating that "in Obedience to an Order of the Hon. Board of War at Philadelphia...I have settled Mr. James Hunter's Account for sundry Military Stores &c furnished Col. Baylors Regiment of Light Dragoons amounting to £727.7 Virginia Money, that he has received £10161.12 at sundry times from Col. Baylor and that there appears due him a Ballance of £7056.5 from which I have deducted £1235.3 for sundry articles charged therein for Swords, Pistols & Bridle Bits delivered the Officers of said Regiment, the Board being of Opinion they have no authority to pay for the arms furnished officers and that I have given this certificate to enable Mr. Hunter to demand that sum from the Commander of the Regiment" (Virginia Board of War, Letter Book, June 30, 1779 - Apr. 7, 1780).

On Mar. 29, 1781 Maj. Richard Call[278] wrote from Petersburg, Virginia to Gov. Thomas Jefferson regarding the manufacturing of swords at Rappahannock and the precarious financial situation of the arms manufacturers:

> I have received by Express from Lieut. Coll. Washington, which he directs me to send to Mr. Hunter as a pattern & have swords made for the men—but the great injury every Mechanic who has done work for the Cavalry sustained by being kept out of his money I am afraid will prevent Mr. Hunter from undertaking to make them unless he could be certain of getting paid when they are done. If Your Excellency will be pleased to give him some such assurance it will be greatly encouraging to the Cavalry and finding themselves equally armed with the enemy will give confidence knowing that bravery will then ensure success—the sword is the most destructive and almost only necessary weapon a Dragoon carries. Our mounted men at present have swords but the generality of them are much inferior to the British (Governor's Letters Received, Mar. 29, 1781).

---

[276] William Finnie was from Williamsburg. In the summer of 1775 he commanded an independent company of militia guarding the public magazine at Williamsburg. He was named quartermaster general of the colony's forces in October of that year. On Mar. 28, 1776 Congress appointed him deputy quartermaster general of the southern department and he served in that capacity with the rank of colonel until the close of the war. He served as mayor of Williamsburg from 1783-1784.

[277] Benjamin Flower (1748-1781) was born in West Hartford, Connecticut and was a commander of Washington's Flying Camp formed in the summer of 1776. This unit was intended as a mobile reserve for the anticipated battles of New Jersey and New York. In Jan. 1777 he was reassigned as colonel of a regiment of artillery artificers. In this post he created a special unit of skilled craftsmen in the areas of carpenters, blacksmiths, wheelwrights, harnessmakers, coopers, farriers, and nailers. This regiment acted as ordnance workers that tested and maintained artillery. Flower was wounded in action and died in 1781. He was buried at Christ Church in Philadelphia.

[278] On June 4, 1776 Richard Call was appointed 1st lieutenant in the 1st Continental Dragoons. He made captain on Dec. 4 of that year and served in George Baylor's regiment through his appointment as major in 1782. He retired from service in 1783.

On Jan. 25, 1781 Hunter wrote to Jefferson asking to be paid for goods already delivered. He wrote, "I have sent by the bearer Mr. [Charles] Dick, the warrant granted me on the treasury sometime ago, for £50,000; of which I have not yet received payment, but beg I may now as the reasons for obtaining it and which I had the honour to lay before your excell'y still subsist with undiminished force; and as I have also certificates from Gen. Gates of the delivery of sundries the manufactures of my works, furnished during the course of the last summer, for the use of the army under his command to the amount of at least £130,000,[279] a considerable portion of which is due for workmen's wages, provisions, &c." (Boyd, vol. 4, pp. 448-449). This was but a partial amount owed Hunter by the state and reflects the magnitude of his business in Stafford. Hunter ended his letter by offering "a parcel of coarse woollens, sufficient for the clothing of 150 or 200 men, which I would deliver if the state has occasion for them; and receive in payment tobacco, inspected at the warehouses of Fredericksburg and Falmouth at the price of £50 per hundred weight." Hunter was never fully reimbursed for materials manufactured at Rappahannock, a fact that contributed heavily to the massive debts that plagued his estate. By 1779 there had been a marked depreciation in the currency and Virginia found it difficult to pay her bills. Prior to the Revolution, John Robinson was treasurer of the Virginia colony. For some years Robinson secretly loaned public funds to his debtor friends, many of the loans being unsecured, leaving a shortfall of some £100,000 in the treasury (Watkins 55). As a result, the colony was severely short of funds as she entered the Revolution and, essentially, was bankrupt by 1781.

As was reflected in John Mercer's letter written during the winter of 1766/67, Hunter's financial difficulties actually began almost as soon as he began building Rappahannock Forge. The Lawson vs Tayloe suit includes a letter written by Thomas Lawson to John Tayloe and written in October 1771 in which Lawson mentioned that Adam Hunter had paid him a visit at Occoquan inquiring as to when Thomas planned on shipping more pig iron to Rappahannock. Lawson told Adam that he was unable to send more iron because James owed him a considerable sum for previous purchases. Thomas also informed John Tayloe that he had recently heard that Hunter had sold a number of slaves in order to meet the demands of some of his creditors. Lawson was quite unhappy because he hadn't known anything about the sale and none of the proceeds had been used to pay off Hunter's debt to Occoquan. Thomas continued by mentioning that Hunter also owed Capt. Hatton[280] £893.19 with interest that was also long overdue. In another letter from Lawson to Tayloe, the former said of Hunter and his debts, "[Hunter] sent me a Bill of Exchange lately for £100 in part payment for the Pig Iron. Mr. Strode had last Fall, and which ought all of it to have been paid for long ago; this Bill I must carry to Baltimore with me, and try to get some Winter cloathing for our people" (Fredericksburg Circuit Court Records, "Lawson vs Tayloe"). The system of credits and accounts used during this period was often the undoing of many businesses, as they could not pay their debts until they were paid themselves.

Hunter's letter of January 1781 reflects the close connection that existed between Rappahannock Forge and the Fredericksburg Manufactory. Fielding Lewis and Charles Dick shared Hunter's precarious financial situation. On April 5, 1781 Charles sent a request to Governor Jefferson for a warrant of £100,000 saying "it is not near equal to one thousand pounds in good times." Still, the requests went unanswered. What is clearly evident is that Hunter, Dick, and Lewis were all linked by an unwavering patriotism and commitment to the cause of independence, investing most if not all of their personal fortunes to keep their respective armories in operation. Their belief in the ideals of independence caused them to contribute all they had and stake their personal credit to assure the successful operation of their works despite clear signs that they would never be reimbursed for their efforts.

---

[279] In an 1810 deed of trust between John Strode and James Madison, the conversion between pounds and dollars was given. At that point, a pound was worth approximately $3.35. Using those numbers (which are rough at best), Hunter's vouchers of £180,000 converted to about $603,000.

[280] This may have been Walter Hatton (died c.1781) who, from 1760-1775 was Collector of Customs in Accomac County. At the outbreak of hostilities, Walter and his family returned to England and, based upon a claim dated June 2, 1777, was granted £80 in lieu of his official position in Virginia. Walter attempted to return to Virginia with Lord Dunmore in 1781, but was shipwrecked on the voyage. On Jan. 10, 1787 his widow, Ellen submitted a claim for the loss of income as a revenue officer and an estate and personal goods valued at £3,140. On Mar. 15, 1784 his widow, Ellen, submitted another claim for the loss of William's office of £200 per year and other property valued at £3,148.

Fielding Lewis suffered more from this than did Dick or Hunter. By 1780 the inflation rate was a staggering 1000%. Lewis invested nearly all of his personal fortune in the Fredericksburg Manufactory and borrowed huge sums as well. On Feb. 9, 1781 Lewis wrote to the state treasurer informing him that had it not been for his personal advances of money, "the factory must have been discontinued, as no money could be had at the treasury or, so little, that the business must have suffered." Fielding had been instructed to borrow for the use of the state all the money that could be obtained and he wrote that he believed he could raise between 30,000 and 40,000 pounds, "seven thousand of which I lent the State being all that I had at that time on hand." He lamented that this grave situation left him unable to carry on his own business or even pay his taxes for lack of funds. He closed his letter with an uncharacteristically strong statement, "Can it be expected that the State can be well served when its best friends are used in the manner I have been treated" (Executive Communications, Feb. 9, 1781)? The strain took a toll on Lewis. By the fall of 1781, his health was failing. He left Fredericksburg and died soon after.

Yet the need for military supplies continued. In February 1781 Jefferson had written to Hunter asking him to make 1,000 additional camp kettles. As previously mentioned, Jefferson noted the loss of state records that made it impossible for him to ascertain how many of the previous orders for kettles had been filled. He continued, "We shall be glad however, whatever they be, that you now furnish us with one thousand for the separate use of the State and to be called for by my order only." He added a post script, "We will send 2 waggons a week hence for camp-kettles and continue them going backward and forward on the same business if you can make them fast enough til we get the whole number" (Boyd, vol. 4, p. 551).

On Mar. 1, 1781 Abner Vernon wrote to Richard Young[281] (1740-1815), quartermaster in Fredericksburg, to report on the progress being made at Rappahannock:

> I just received a Letter from William Armistead Esqr.[282] directed to Mr. Hunter with Two Waggons for Loads of Camp Kettles, the Governour some time ago ordered a Thousand to be made wch. all finished, & delivered except seventy four & them I now send down to the Ferry to save trouble of the Publick Waggons coming over the River. I also return 50 Kettles which I borrowed from Mr. Wilson in your absence, & no entery is to be made for them—in all sent today is 124. The 74 I shall charge as delivered to you & will call for Certificate when convenient. Please give particular directions about Loading the Waggons as the drivers are both Boys. I expect if convenient youll forward as many more Kettles as will make up full Loads—am busy at present or would wait on you as the Waggoner tells me he brot. no orders to you. I inclose Mr. Armisteads Letter for your perusal wch. please keep till I see you.[283]

By late 1781 Hunter seems to have reopened his gun manufactory. In November of that year, James wrote to Gen. Oliver Towle thanking him for protecting the works during Tarleton's invasion. He added:

> Since the Assembly extended the like indulgence to my workmen as those employed at the Public Foundries, I have resumed the manufactory of small arms. Workmen are putting the machinery for grinding & boring &c in order & artificers collecting, which may be improved to the benefit of the State...I have also on hand a Thousand Horsemen's Swords to the pattern forwarded from General Greene's Army by order of Colo. Washington (Palmer, vol. 2, pp. 618-619).

Payment from the state to Hunter was never forthcoming. Nor were Charles Dick or the heirs of Fielding Lewis paid for their efforts, or even reimbursed for the thousands of pounds from their personal fortunes that they had willingly advanced to keep the armories open and productive. Giles Cromwell wrote,

---

[281] Richard Young was assistant deputy quartermaster and a resident of Fredericksburg. He married Mary Margaret Moore and died in Kentucky.

[282] William Armistead (1750-1799) was the son of Col. John Armistead (died 1779) of New Kent County. During the Revolution, William served as an agent of the State charged with providing the military with arms, clothing, and other necessities. He married first ___ Latimer and secondly Mary Latham Curle.

[283] The original of this letter is in the collection of Mr. Tim Garrett of Fredericksburg.

"[Virginia's] neglect to stand behind the iron industry after the Revolutionary War undoubtedly caused the bankruptcy of the Fredericksburg Armory, the Westham Foundry, and Hunter's Iron Factory, all within two years after the war" (Cromwell 150). Of course, the fact that Virginia had essentially bankrupted in 1781 was exacerbated by the near bankruptcy of the fledgling United States. By 1783 the total debt of the fledgling United States exceeded $42,000,000 (Thomas Ruston Papers). For a list of warrants and payments to Hunter see Appendix A.

Years of financial stress and the difficulty of keeping skilled workers took a toll on Rappahannock and, eventually, Hunter was forced to cease manufacturing materials for the state. On Apr. 22, 1782 he wrote that he "would very readily render the Public any assistance" but the little attention given his works had forced him to discharge all his workmen. He noted that this had followed considerable expense for additions and repairs to his plant. This marked the end of James Hunter's valiant attempts to arm the State, though the forge remained in operation for a considerable period following the Revolution.

Contained in the surviving Stafford County court records is a single receipt of payment to Hunter from the State of Virginia. This entry reads, "To Sundries furnished the Army Oc: from 9th Sept. 1780 to 10th April 1782 p acct & Vouchers £645:5:4 ¾; to Sundries on hand Vizt 1 Traveling Forge with a full set of Smith's tools £83.2.10; 32 Horsemens Swords 60/; 135 Curb Bridle Bitts @ 8/ £54.0.0; 21 pr Stirrup Irons @ 7/ £7.7.0; 116 Bridoons @ 4/ £23.4.0; 1 large Rifle Gun with Moulds Wiper &c: Carries 4 Oz. Ball £30." [284] An interesting coincidence that will likely never be explained is the match between the items on this list and the estate inventory of items contained in the "Counting Houses &c" (see Appendix B).

Rappahannock Forge Weapons

In addition to a wide variety of military supplies and naval stores, Rappahannock also manufactured swords and various types of guns. Hunter's artisans produced four models of firearms of which we are aware: pistols, muskets, carbines, and amusettes. Nathan Swayze, an antique arms expert, spent two years researching the weapons produced at Rappahannock Forge and located only 10 pistols, 4 muskets, and 4 amusettes that could be attributed to that works (Swayze 7, 24, 31). Recent research has since located a fifth amusette, five muskets, and one carbine.

Although Rappahannock Forge was a major producer of arms during the Revolution, there are few remaining examples of their firearms. The explanation for the scarcity of examples of these arms may be quite simple. Arms produced during the Revolution were pressed into immediate and extremely hard service; many were severely damaged or destroyed. Firearms contained moving parts which, over time, became damaged, worn out, or out-dated. Hunter's guns left Rappahannock clearly marked with "Rapa Forge" on the lock plate and "I. Hunter" on the barrel. As they became unserviceable, these weapons were taken to armories for repairs. A broken lock was replaced. A damaged barrel was removed and a new one fitted in its place. Broken stocks were repaired or replaced. Many of these weapons saw service later during the War Between the States and the process of repair and upgrade continued as guns were converted from flintlock to percussion. Parts from some of Hunter's guns, damaged during the Revolution, were removed, stockpiled, and used on other weapons. Additionally, Hunter's muskets were .80 caliber which, by the 1860s, was not a standard size. The answer to this problem was to simply replace the barrels. As locks and barrels were replaced and/or altered, manufacturer's marks disappeared. The majority of the 19 known surviving Rappahannock Forge firearms show signs of having been altered or repaired, though they all had enough of the original factory features to make them identifiable as having come initially from Rappahannock Forge.

The artisans at Rappahannock made their guns from pig and bar iron either produced on site or purchased from other furnaces and forges. The 1797 edition of Encyclopedia Britannica notes, however, that the best iron for guns came from another source. "For the materials, the softest iron that can be procured is to be made use of. The best in this country [England] are formed of stubs, as they are called, or old horse-shoe nails; which are procured by the gunsmiths from farriers, and from poor people who subsist by picking them up on the great roads leading to London. These are sold at about 10 s[hillings] per cwt [hundredweight] and twenty-eight pounds are requisite to form a single musket barrel" (Encyclopedia, vol

[284] Stafford County Court Record Book, 1784-1785, p. 271.

VIII, p. 241). Because the horseshoe nails had endured excessive pounding, not only in the process by which they were manufactured but in daily use on the horses' feet, they were relatively free from impurities that would make them brittle. Thus, they were soft and ideal for making gun barrels.

The wooden gun stocks were made from black walnut, an indigenous Virginia tree the wood from which was noted for its fine grain. Although Rappahannock had a brass furnace, presumably built with the intention of using the brass for gun mountings, most of the surviving weapons have iron fittings.

Muskets:

The 1797 edition of Encyclopedia Britannica described a musket as "a fire-arm borne on the shoulder, and used in war, formerly fired by the application of a lighted match, but at present with a flint and lock. The common musket is of the caliber of 20 leaden balls to the pound, and receives balls from 22 to 24; its length is fixed to 3 feet 8 inches from the muzzle to the touch-pan" (Encyclopedia, vol. XII, p. 554). These were smoothbore weapons, that is, without rifling on the interior of the barrel.

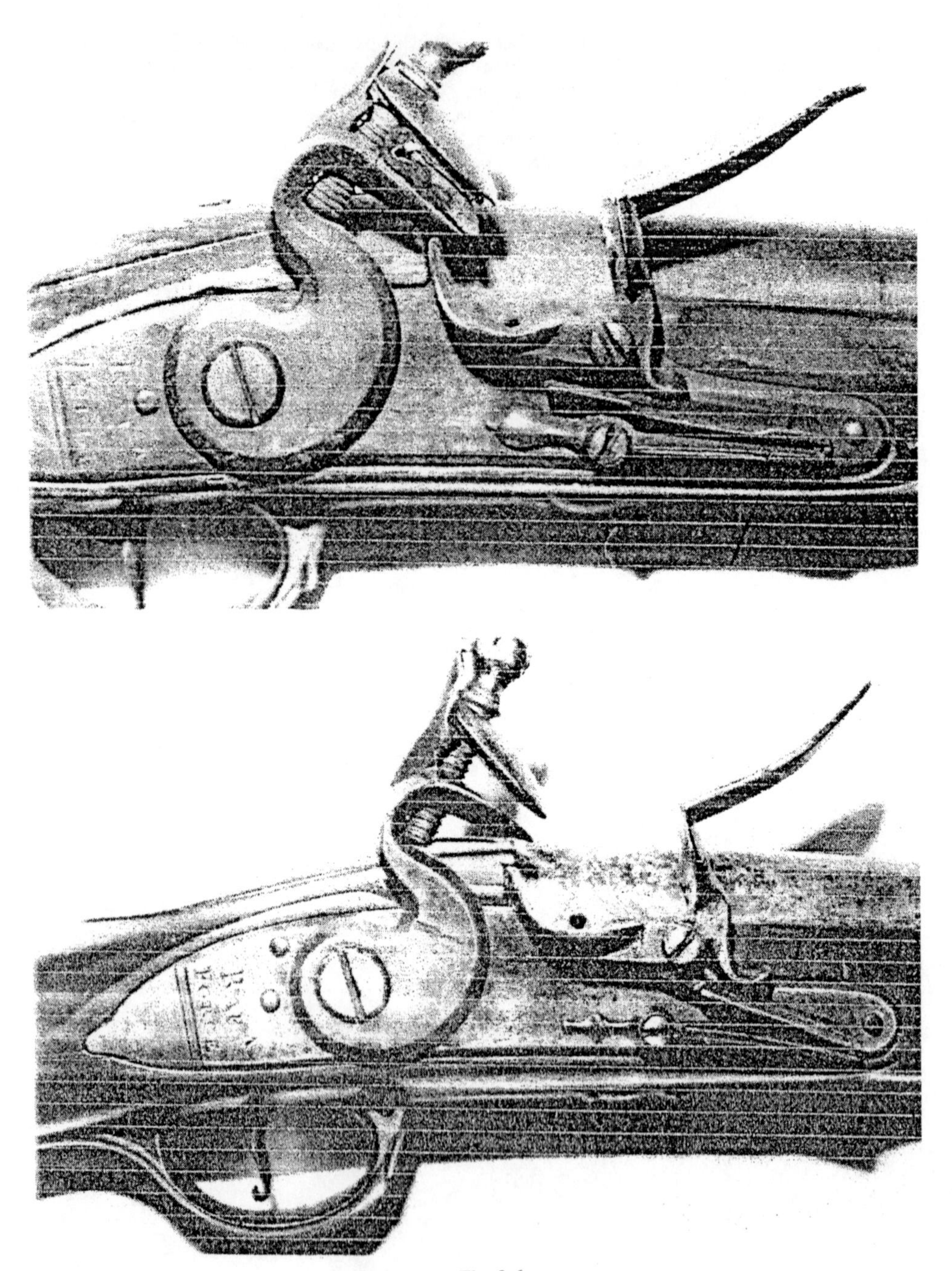

Fig. 2-6
Rappahannock Forge lock on a musket.
(Reproduced with permission of the American Society of Arms Collectors).

Based upon the extant orders for Rappahannock Forge weapons, Hunter produced more muskets than any other type of firearm. On June 6, 1776 Hunter delivered several well-made muskets to the Virginia Convention of Delegates. This seems to have been his first venture into manufacturing arms for the state, though his workers probably had been making guns for some time for domestic use. The delegates ordered the Committee of Safety "to engage with Mr. James Hunter, of Fredericksburg, for as many good muskets, with bayonets, sheaths, and steel ramrods as he can manufacture from this time, at the rate of six pounds for each stand; and that they allow the same price to any other person who shall manufacture arms within this colony, of equal goodness with the sample now produced by the said Hunter" (Palmer, vol. 8, p. 167). The following day, the Committee issued a warrant or payment voucher for 25 muskets, probably representing the 25 sample muskets that Hunter had presented to the Convention to demonstrate his manufacturing capabilities (McIlwaine, vol. 2, p. 49). A stand consisted of a musket, ramrod, bayonet, and shoulder strap. Rappahannock Forge manufactured hundreds of these weapons, one order alone accounting for at least 680 muskets (McIlwaine, vol. 1, p. 332). On June 17, 1776 Hunter was paid for 14 muskets, for 60 on Aug. 8, for 25 on Nov. 22, and an unspecified number ordered and paid for on Dec. 8 (See Appendix A for a list of all known orders and payments).

Rappahannock Forge muskets were closely modeled on the popular long land pattern commonly called the "First Model" British Brown Bess. Unfortunately, only five of Hunter's muskets are known to exist, four of which are in private collections. The extant examples of Hunter's muskets average 57 to 60 inches in length and are all about 80 caliber. Three of the five are engraved with the customary "Rapa Forge" behind the cock and "I Hunter" on the barrel. The fourth musket, a part of the Colonial Williamsburg collection, does not have "I Hunter" on the barrel. The fifth musket, owned by Mr. Charles G. McDaniel of Fredericksburg, bears neither of these identifying marks on the exterior, but was authenticated in Williamsburg based on manufacturer's marks on the inside of the lock. This weapon, along with a musket made at the Fredericksburg Manufactory, is on display in the Fredericksburg Area Museum and Cultural Center.

Due to the loss of records, Rappahannock's production capability is unknown, though production figures recorded by Charles Dick at the Fredericksburg Manufactory may be used as a rough guide. In September 1781 Charles noted that, given money, workmen, and supplies, his factory was capable of producing 100 stands of muskets per month. Hunter had far more craftsmen and could easily have surpassed that figure.

The standard powder charge for this weapon was about 135 grains or approximately 1/3 ounce. Powder charges varied greatly, however, and it is difficult to estimate exact powder usage. The British tended to use far more powder than did the Americans who faced a perennial shortage of gunpowder. Like those made at the Fredericksburg Armory, Hunter's muskets were fitted with stocks fashioned of American black walnut. Flayderman's Guide values these guns from $7,500 to $20,000 depending upon condition.

Before being shipped out to waiting regiments, each firearm was inspected and tested. The inspector of arms was required to be a master gunsmith in his own right. On at least one of the surviving muskets, the inspector's initials, "IP" are engraved on the inside of the lock plate. During this period, the letter "I" was interchangeable with "J." This engraving then may be translated to read "J P." Although many of the names of the workers at Hunter's forge are forever lost, the name of Joseph Purkin, a master gunsmith, is known and he may well be the inspector who inscribed his initials in the known surviving examples of Hunters muskets, pistols, and amusettes. An Englishman by birth, Joseph was employed as a master artillery artificer at Rappahannock Forge from c.1776 to c.1780. Upon leaving Rappahannock, Perkin worked as a gunsmith in Philadelphia from 1780-1793. He then moved on to Harper's Ferry, Virginia (now West Virginia) where, from 1793-1797 he was engaged in arms repair and served as superintendent from 1796-1810 (Hartzler 75). He served simultaneously as superintendent of the U. S. Arsenal at New London, Virginia, a position he held from 1799-1802. Joseph Barry described Perkin as "an English Moravian...an amiable, unsophisticated man, and tradition still tells of his simplicity of dress and deportment. He was accompanied by a Mr. Cox, "who had followed him to Harper's Ferry from southern Virginia where Mr. Perkins had formerly resided (Barry 15-16). Upon retiring from Harper's Ferry, Joseph returned to Philadelphia where he later died.

Pistols:

Many collectors of antique guns believe that the first American single-shot military pistol was the "North and Cheney" manufactured in 1797. In truth, the first such weapon was made at least twenty years earlier at Rappahannock Forge. According to Nathan Swayze and Flayderman's Guide, the pistols manufactured at Rappahannock are considered as the first or earliest American military handguns. Most of these were actually made under contract for the state of Virginia during the Revolution and are among the best marked of any early American-made pistols. They appear to be good, sturdy weapons that could stand up to much use, though 18th century pistols were so notoriously inaccurate that they actually made better clubs than guns.

Most mounted and naval personnel as well as most officers carried pistols. Those issued to the navy were often fitted with a belt hook. Officers' pistols were often mounted with silver and tended to be rather ornate, such as those made at Rappahannock for Gen. Charles Lee.

The Rappahannock Forge pistol was modeled directly upon the gun used by the British Light Dragoons. This weapon was introduced into the colonies c.1759, so there would have been numerous examples in existence by the time Hunter began manufacturing his handguns. Production of these guns at Rappahannock commenced sometime prior to June 21, 1776 when John Strode advertised for workers skilled "in the manufactory of small arms, a number of hands who understand the file, also anchor smiths, blacksmiths, and nailors, to whom suitable encouragement will be given (*Virginia Gazette*, June 21, 1776).

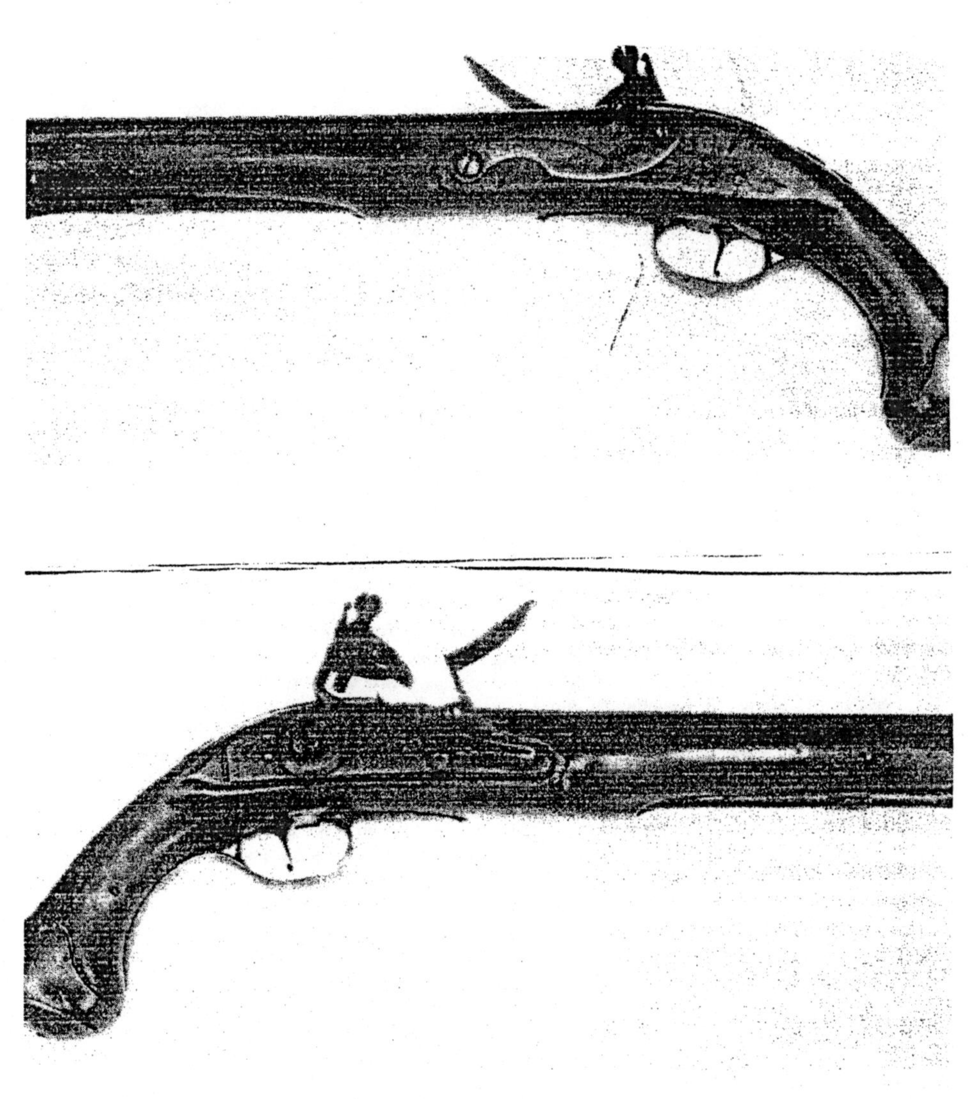

Fig. 2-7
A Rappahannock Forge pistol.
(Reproduced with permission of the American Society of Arms Collectors).

There are 10 known surviving Rappahannock Forge pistols, most of which are in private collections. While dimensions differ slightly, the average overall length is 15 inches, with barrels of 9 inches. Calibers vary from 66 to 69. Flayderman's Guide values these guns from $7,500 to $30,000 depending upon condition. Specimens of Rappahannock guns are so rare as to encourage forgers to make replicas and sell them as authentic. Charles Chapel warns gun collectors, "Do not buy a pistol of this description at any price unless three experts of national reputation examine it and pronounce it genuine!" (Chapel 56).

Amusettes:

Massive semi-shoulder arms designed to fill the gap between small artillery and shoulder weapons date from as early as the 16th century. Over the years, these guns have been known by such names as amusettes, arquebus a croc, wall guns, and rampart guns. As the latter two names suggest, these weapons could be propped on a fortification wall or on the railing of a ship or skiff and employed somewhat like cannons. Serving as a fill-in between a shoulder arm and a cannon, the amusette had a much greater range than an ordinary musket. Being vastly lighter in weight than a cannon, it was adaptable for use on boats, in quickly erected forts, and in situations where it was not practical to transport cannons.

On Feb. 4, 1776 Fielding Lewis wrote to George Washington saying, "I propose making a Rifle next week to carry a quarter of a pound ball. If it answers my expectation, a few of them will keep off ships of war from our narrow Rivers, and be usefull in the beginning of an engagement by land" (Abbot, Revolutionary War Series, vol. 3, p. 246). If Fielding Lewis was suggesting that amusettes be manufactured at the Fredericksburg Manufactory, there is no evidence that they ever were. By February 1776, Fredericksburg was not yet capable of producing such a weapon. They still lacked the equipment as well as the craftsmen to accomplish such a feat. The project was transferred, instead, to Rappahannock Forge where a gun of unequaled range and accuracy could be made.

Throughout the history of the wall gun, most versions were smooth-bored flintlocks. While an unknown gunsmith did venture to rifle an American-made amusette during the first half of the 18th century, the process was perfected at Rappahannock Forge (Darling 2-5). The truly remarkable feature of Hunter's wall gun was its amazing accuracy, a result of the rifling. Even with today's technology and equipment, it would be difficult to duplicate what Joseph Purkin accomplished with hand tools and a careful eye. Hunter's amusettes were rifled with 13 lands and grooves that spiraled up the length of the barrel. This unusual type of rifling was known as "polygroove" and had been used by gunsmiths for several centuries. Modern guns are rifled by carefully calibrated, computer-monitored equipment and normally have two to four grooves. Barrel rifling causes the bullet to spin as it flies towards its target, thus dramatically improving accuracy. However, if the grooves are not absolutely even and correct in their placement up the barrel, the bullet will "wobble" in flight, making it exceedingly difficult to hit a target.

The 13 grooves in the amusette didn't necessarily make the gun more accurate than a similar gun with less rifling. One of the drawbacks to rifled long guns was the difficulty of loading them. In a smoothbore weapon, the powder, patch, and ball were dropped down the muzzle and tapped into place with the ramrod. The interior diameter of the barrel was slightly larger than that of the ball; the cloth patch filled the space between the two. In a rifled gun, the ball fit tightly inside the barrel and had to be forced down with the ramrod. Patches were unnecessary because there was no extra space between the ball and the inside of the barrel. The extra rifling in the amusette may well have increased loading efficiency by reducing the amount of friction between the barrel and the ball.

Although Hunter's amusettes do not seem to have been widely utilized during the Revolution, they were nonetheless important. Their amazing accuracy was attested to in a letter from Gen. Charles Lee[285]

---

[285] The Virginia Lees are quick to point out that Gen. Charles Lee was no relative of theirs. English born, he was the youngest son of Charles Lee. Seemingly born with a love of the military, he was commissioned in his majesty's service at age 11. He was described by J. T. Headley in Washington and his Generals as "Fiery, impetuous, and headstrong, the young officer from this point starts on a career so wild and irregular, and adventurous—now flashing up in splendor, and now sinking in darkness—that his life seems a strange romance rather than a reality." Charles served in Braddock's Campaign of 1755, then went to the Mohawk Valley where he purchased a commission as a captain. There he was adopted by the Mohawks and married the daughter of a Senaca chief. He served with distinction under John Burgoyne (1722-1792) in Portugal

(1731-1782) to George Washington dated May 10, 1776. Lee wrote, "I am likewise furnishing myself with four ounces rifle amusettes, which will carry an infernal distance the two ounced hitting a half a sheet of paper 500 yards distant" (Abbot, Revolutionary War series, vol. 4, pp. 257-259).

Ebenezer Hazard also made note of the amusette during his visit to Hunter's works in 1777. He wrote, "Saw at Mr. Hunter's Gunnery two Rifles, which carry a four Ounce Ball, they are to be fixed in the same Manner with a Swivel Gun, & are made for the Congress: several of them have already been sent to the Camp" (Shelley 418). It is unknown how many of these weapons were made at Rappahannock; only one known receipt survives, that record being amongst the few surviving Stafford County court records (Stafford Deed Book S, p. 271).

In design, the amusette was somewhat similar to the early German rifles, having a squatty, thick appearance with a relatively straight stock. The influence of Hunter's Pennsylvania Quaker gunsmiths is also evident in the accentuated curl at the rear of the brass trigger guard and the brass side plate.

James Hunter's amusettes were about 61 inches in length and weighed approximately 53 pounds, making them far too heavy to be carried long distances in the field. Typical of weapons of the period, the stocks were of native black walnut, the hardware of brass and iron. Three of the five known surviving amusettes have the letters "I P" engraved behind the cock, the initials of the previously mentioned Joseph Perkin. He was certainly employed at the forge during the time at which these weapons were being manufactured and, based upon the positions he later held, was eminently qualified to build Hunter's wall gun. These guns are also marked with "IH" and/or "Rapa. Forge" on the barrel or behind the cock. Three of the surviving amusettes have round barrels and two have octagonal barrels. The caliber of the barrels was 1 3/16 inches (1.1875" or 30.16 mm). The standard powder charge for this weapon was probably about 200 grains or a little more than ½ ounce, though by increasing the charge the range would also increase. Various sizes and types of shot were used in the amusette. A 1 3/16-inch round lead ball weighs about one ounce. Contemporary accounts refer to the use of 2 ounce and 4 ounce shot, indicating that this weapon was also being fired with elongated projectiles.

A metal prop attached to the stock enabled the gun to be supported on a solid base. There are two types of props represented on the surviving amusettes. One is a forked support that would have been ideal for use on a fence, wall, or tree stump. The second type is a swivel closely resembling an ore lock. It appears from the design of this swivel that it was intended to be slipped into a metal sleeve or hole, probably on board ships. This would have been eminently useful on ships as cannon were only mounted on the sides of the boat, requiring that the boat be turned in order to aim and fire. Use of the amusette provided much more flexibility and improved defenses should the ship be attacked from the fore or aft, neither of which were armed with cannons.

On the right side of the stock of the Rappahannock Forge amusette is a breech box. This is a small compartment with a sliding cover and was used to carry small cleaning instruments that screwed onto the end of the ramrod. The design of the compartment is a distinctive feature on Hunter's amusettes and reflects the craftsmanship of his Pennsylvania gunsmiths.

After the Revolution, West Point in New York became the major depot for the storage and repair of the remaining weapons. As the various pieces became obsolete and were discarded, examples of each type were reserved to create a teaching collection for the academy. This collection was not formally cataloged until c.1873. A notation on the paperwork accompanying one of the two amusettes in the Rock Island Arsenal states that the U. S. Marine Corps captured the weapon in China during the Boxer Rebellion. No further explanation is recorded for how an 18th century wall gun found its way to China.

In June 2001 Walter V. Roberts examined the amusette housed at Springfield Armory.[286] This weapon is one of two of the surviving wall guns made with octagonal barrels. The barrel was produced

---

and retired on half-pay in 1763. In 1765 he joined the Polish army, returned to England, and came to America in 1773 where he immediately became involved in the Revolutionary cause. He purchased land in western Virginia and in 1775 Congress appointed him a Major General. At the siege of Boston, "his dirty habits and obsenity gave offense" but he was "endured for what he was supposed to know." Rappahannock Forge made a matched pair of pistols for Charles Lee; these are now on display at the Smithsonian in Washington.

[286] Prior to its transfer to Springfield, this weapon was housed at the Watervliet Arsenal in West Troy, New York. The following pencil notation appears inside the barrel channel, "Cleaned by Paul Jefferson, May 5, 1876" (Springfield Armory Catalogue Record 1155, Accession 02).

with polygroove rifling, the grooves making one complete turn in 53 ½ inches of barrel length. The octagonal barrel and polygroove rifling are typical of Pennsylvania-made rifles of the pre-Revolutionary era. Roberts speculates that the octagonal-barreled weapons may have been prototypes produced from Fielding Lewis' suggested design of February 1776. If this is true, these prototypes would have closely resembled the weapons Hunter's Pennsylvania gunsmiths were accustomed to making. After testing, some changes may have been recommended, including the substitution of a simpler round barrel in place of the more labor-intensive octagonal one. Further support of these being prototypes is indicated by the lack of Roman numerals and other markings on the metal parts that were used in the modified assembly line process to keep together parts for specific guns. As prototypes, the marking of individual parts would have been unnecessary.

The frizzle on the Springfield amusette shows a few striations indicating only light use. The muzzle is slightly flared. The ramrod is of iron, and the accompanying paperwork indicates that when the gun was transferred from Rock Island to Springfield, the hammer was missing. Springfield sent the amusette to Williamsburg where a new hammer was made and installed.

There are five known surviving examples of the Rappahannock amusette, one each at the Springfield Armory (Mass.), the Smithsonian Institution (Washington, DC), West Point Arsenal (New York), and two at the Rock Island Arsenal (Illinois). Rock Island, which at one time had three of Hunter's amusettes, transferred one of them in 1958 to the Smithsonian Institute in Washington, DC.

Carbines:

Rappahannock Forge also produced carbines, only one surviving example of which is known to exist. This weapon was discovered in 2002 in the corner of a Florida junk shop where it had been cast with with an extra-long, church candlesnuffer and a baseball bat. It is now in a private collection in Massachusetts.

The 1797 edition of Encyclopedia Britannica described this type of weapon as:

> A carabine is a small sort of fire-arm, shorter than a fusil, and carrying a ball of 24 to the pound, borne by the light-horse, hanging at a belt over the left shoulder. This piece is a kind of medium between the pistol and the musket, and bears a near affinity to the arquebus, only that its bore is smaller. It was formerly made with a match-lock, but of late only with a flint-lock (Encyclopedia, vol. IV, p. 156).

When Mr. Tim Wider found his Rappahannock Forge carbine, the stock was badly weathered but otherwise was in fair condition. The surviving carbine is 53 inches in length and the lock is marked with the characteristic "Rapa Forge." The lock is 6.5 inches long and is held in place by a 5.5-inch long S-shaped brass side plate with 2 iron screws. It has a brass trigger guard and brass butt plate that is marked 11G/27/3RT. At the time of its discovery, the only missing part was the hammer; all other components were intact. Approximately 1.5 inches of the barrel had been removed at a previous time, though the bayonet lug was still in place. This weapon is 65 or 66 caliber, considerably smaller than Hunter's muskets. This is consistent with the use of the same smaller rounds in both cavalry pistols and carbines.

Many carbines were shorter in length than standard muskets and were often issued to mounted troops. However, the use of carbines was not restricted to horsemen. Officers often chose carbines because they were lighter in weight than muskets and easier to carry on a march. Although mounted troops did not often find themselves in need of bayonets, manufacturing specifications of the period required all military contract weapons be fitted for a bayonet. Thus, the presence of the bayonet lug on this firearm is not unusual.

This weapon has been inspected by a noted authority on 18th century firearms and certified as genuine. The owner has had the original barrel length restored and the stock repaired.

All of Hunter's weapons used black powder. This was easily made given access to the necessary ingredients. Most gunpowder was made locally and the Fredericksburg Manufactory complex included a powder magazine. Gun powder is a mixture of salt peter, sulfur, and charcoal, "usually granulated; which easily takes fire, and when fired, rarefied or expands with great vehemence, by means of its elastic force." The 1797 Encyclopedia Britannica provides directions for making this versatile explosive:

> Dr. Shaw's receipt for this purpose is as follows: Take four ounces of refined saltpeter, an ounce of brimstone [sulfur], and six drams of small-coal; reduce these to a fine powder, and continue beating them for some time in a stone mortar with a wooden pestle, wetting the mixture between whiles with water, so as to form the whole into a uniform paste, which is reduced to grains, by passing through a wire sieve fit for the purpose; and in this form being carefully dried, it becomes the common gunpowder (Encyclopedia, vol. VIII, pg. 236).

Swords:

Questions abound concerning Hunter's manufacture of swords. Based upon newspaper notices, he seems to have begun making swords at about the same time that he began fashioning guns. John Strode advertised for "a number of hands who understand the file, also Anchor Smiths, Blacksmiths, and Nailors, Wire Drawers, Card Makers, File Cutters, Sithe and Sword Cutlers, to whom suitable Encouragement will be given." He continued by offering "Tradesmen of every Kind, who incline to settle may have Half Acre Lots of Land at a small Ground Rent for ever, and may be assisted in building, Provisions, and Materials, and the Produce of their Labour taken in payment for the same" (*Virginia Gazette*, July 4, 1777).

Gordon Barlow and Giles Cromwell, noted firearms experts, believe that Hunter may never have produced any swords at all. No production figures have ever been found, though several orders for swords survive, one of which was for 1,000. Referred to in letters and documents as "horsemen's swords," these weapons featured a wire-wrapped leather handle and a slightly curved blade. While infantry swords were usually straight, the horseman's sword was curved and designed to be used with a sweeping downward stroke from horseback. Although the blades were not especially sharp, they could still deliver the enemy a stunning blow. Infantry swords had sharp points, straight blades, and were designed for impaling.

On March 30, 1781 Gov. Jefferson wrote to Maj. Richard Call of the 3rd Dragoons, "We mean to direct the Commercial Agent to order 500 Horsemen's Swords to be made immediately by Mr. Hunter on Account of the State (Boyd, vol. 5, p. 285). Accompanying that order was a sample sword, captured from Tarleton's troops at the Battle of Guilford Courthouse in North Carolina along with the information that Col. William Washington (1752-1810) wanted Hunter to use this sword as a model for his production. The following May Jefferson doubled the order, instructing Hunter to employ every hand he could spare to produce the swords. Unfortunately, this request came at a time when Hunter was responding to the rumor that Tarelton was headed towards Fredericksburg. At Jefferson's insistence, Hunter dismantled and removed most of his equipment and had few workmen left on site. Tarleton never materialized and, by November 1781, Hunter was able to write, "I also have in hand a thousand horsemen's swords to the pattern forwarded from General Green's army by order of Col. [William] Washington."

There do not seem to be any surviving swords that can be unquestionably credited to Rappahannock Forge. Of course, collectors and historians have been searching for swords engraved with the characteristic "Rapa Forge" or "I H" that appeared on Hunter's firearms. However, there are many surviving swords with the letter "H" stamped on the back of the knuckle-bow. Harold Peterson's research has revealed three other known swordmakers of the period who might have marked their products with the letter "H." George Heighlerger manufactured swords in Philadelphia before and after 1781. George Hinton also produced swords in Philadelphia from 1787-1790 and Ezekiel Hopkins made swords in Rhode Island during the Revolution. Mr. Peterson also notes that the numerous surviving swords stamped with "H" were manufactured from 1775-1781 and were quite standard in design, with a wide iron stirrup hilt and pierced counter guard. The leather grips were wrapped in fine wire. Peterson writes, "Its strong, simple lines and the rough forging of blade and hilt bespeak its manufacture in America during a period of emergency." These swords average 1.9 pounds each and are 41 inches in overall length. The blade is 36 ¼ inches long by 1 ½ inches wide at the hilt (Peterson American Sword, 292). One is left with the tantalizing possibility that at least some of the swords marked with the letter "H" actually originated at Rappahannock Forge. It must also be noted, however, that the great majority of swords manufactured during this period had no production markings at all, regardless of the maker. Not until the War Between the States did manufacturer's marks become commonplace on swords.

Rappahannock Forge After James Hunter

Despite a severe shortage of funds, Rappahannock continued operations through the close of the Revolution and into the early years of the 19th century. Johann Schoepf[287] who visited Falmouth in 1783-84 said that the forge was still a major industry. He wrote, "Above Falmouth, near the falls of the Rappahannock, is one of the finest and most considerable iron-works in North America. More than 6-800 tons are worked there yearly, it is said. My. Hunter is the owner. These works are distinguished besides a rolling and a slitting mill, and of this there are only two or three in America, the former British government having prohibited the setting up of mechanisms of that kind" (Schoepf, vol. 2, p. 42). Production at the forge continued until at least 1803, possibly as late as 1820.[288]

The newspapers carried no announcement of James Hunter's death. The man who had done so much for the cause of American liberty was buried behind what is now Union Church in Falmouth and eventually faded from most memories. The only known reference to his death is contained in William Allason's (1731-1800) account book in which he stated that James died on Nov. 18, 1784 (Allason ledger). The exact reason behind the family's not placing an obituary in the paper is unknown, but it may be speculated that by not doing so they hoped to delay creditors' claims on his estate.

Part of the difficulty in researching Rappahannock Forge is that the company books and papers have disappeared. The last surviving executor, Patrick Home[289] (died 1803), still had possession of these documents when William Waller Hening used them in 1802 to reject a British mercantile claim on Hunter's estate. The Rappahannock ledgers and papers must have constituted a tremendous mass of documents. By the time Patrick died in 1803, the courts were familiar with the estate; in fact, they had assumed administration of the estate a year earlier. Patrick died unmarried and it is uncertain who cleared the house and settled his affairs. He had no immediate relatives in the area and Dr. Alexander Vass[290] (died 1814) acted as his administrator. Whether the records were returned to someone associated with the estate or were retained by the court is unknown. Actually, the whereabouts of these records was an issue not long after Patrick's death. In early 1837 James Logan placed a notice in the local newspaper that read:

> Whereas Patrick Home, of Rappahannock Forge, died in 1803, leaving a variety of valuable documents, which unforeseen circumstances that have occurred since, have thrown into the hands of parties unknown to his Representatives, and as these documents are valueless to the holders, it would be esteemed a great favor if they would forward the same to Mr. James Logan, at the Farmers' Hotel, Fredericksburg, who would willingly afford a reasonable recompense (*Fredericksburg Political Arena*, Jan. 6, 1837).

It is unknown if Logan was successful in his quest. In all likelihood, James was the nephew of Patrick Home whose sister, Helen, married Maj. George Logan of Scotland. The Overton County, Tennessee court records include a record for the appointment of Alexander Vass, Robert Dunbar, and Robert Patton as their attornies to look after their interests regarding Patrick's estate.

---

[287] Johann David Schoepf (1752-1800) was a physician and son of a German merchant. From 1777-1784 he resided and traveled in America, assisting with the Revolution and keeping a journal about his experiences.

[288] The 1809 Adjutant General's Report included an inventory of industrial buildings. Listed on this inventory was Strode's massive armory at Rappahannock. An iron furnace was built in Fredericksburg in 1830, perhaps because Rappahannock had ceased production and the area was without a local source of iron.

[289] Patrick Home was of the Hume family of Wedderburn Castle in Duns, Scotland, his mother having been James Hunter's sister. Patrick was closely related (possibly a nephew) to George Hume (1698-1760) of Wedderburn, the exiled Jacobite of Spotsylvania and Culpeper counties.

[290] Little is known of Dr. Vass. In September 1805 he removed from Falmouth to Stevensburg "or his plantation not far from it, where he means to practice Physic, Surgery & Midwifery as heretofore." He offered to sell his house and lot "on an elegant and healthy situation adjoining Falmouth." (*Virginia Herald*, June 21, 1805). Alexander married Mary Thornton (born 1773), the daughter of Francis Thornton (1737-1794) of The Falls, Spotsylvania, and Anne Thompson (c.1743-1794).

At the time of his death, James Hunter owned over 6,000 acres in Stafford County[291] plus land in Culpeper, Fauquier, Fredericksburg,[292] and Spotsylvania.[293] The 1785 personal property tax records for Stafford list James' estate with 72 slaves at the forge, 45 at Stanstead, and 83 at Accokeek. This latter figure indicates that, well after the Revolution, there was substantial work being carried on at Accokeek. The following year Adam Hunter was named as the taxpayer but the locations of the slaves were not specified, though their total number had decreased only by 23. James' personal estate was inventoried in April 1785 and included property on his lands in Stafford, Culpeper, and Fauquier (See Appendix C). The commissioners valued his personal estate at £12,735.16.4. Of the 253 slaves listed, 76 were "at Mansion House and lower Plantation," 32 were at "Stanstead Plantation," 10 remained at Accokeek, and 66 at the forge. It is unclear why there was such a discrepancy between the numbers of slaves listed on the personal property taxes and in the inventory, both of which were made in 1785.

James Hunter's will was recorded in Stafford County on Apr. 11, 1785 (see Appendix B). He named Adam and Abner Vernon co-executors and devised half of his estate to his brother. One fourth, "that shall remain after my debts, Legacies, and other demands be discharged," was to pass to each of his two nephews in Scotland, one the son of his deceased brother, John,[294] and the other to Patrick Home (died 1803), son of his sister, Jane.[295] James also asked that Patrick come to Virginia to help settle the estate. According to William W. Hening's reports in the British Mercantile Claims, Patrick was orphaned at an early age and had spent much of his life in France with his uncle, Archibald Hunter (1734-1796). He returned to Scotland shortly before coming to Virginia in 1791 and qualified as an executor the following year. James also left several specific bequests including £1,000 to each of his cousins, James, Jr. (1746-1788) and William (1748-c.1773),[296] "upon condition [they] not bring any Suit against my Estate." Finally, James released "my wench, Aggy[297] and her children to be freed from all service." He asked Adam to give her a parcel of land from his Culpeper tract and provide her with two slaves, several cattle and hogs, a mare and two colts, three other horses, corn and pork for the ensuing year as well as "corn sufficient for feeding

---

[291] The 1800 Stafford land tax records listed James with 4,976 acres (Rappahannock Forge, Stanstead, and Rocky Pen Plantation), 870 acres (the Baxter tract), and 200 acres (Accokeek).

[292] On June 9, 1760 Alexander Wright, merchant of Fredericksburg sold James Hunter and George Frazier (died 1765), merchants, a 2/3 interest in lots 23, and 24 in Fredericksburg.

[293] On Dec. 8, 1767 James paid £358.15 for 350 acres in St. George Parish, Spotsylvania. He purchased this land from Thomas Reeves and his wife Sarah of Spotsylvania and the deed described the tract as being opposite the falls of the Rappahannock. That would have placed the property across the river from the forge.

[294] There is no further mention of this devisee and it is unknown if he received his portion of the estate or not.

[295] Jane Hunter married in Scotland Patrick Hume, Sr. Both of them must have died early as young Patrick was raised by his uncle Archibald (1734-1796).

[296] These were the sons of James' uncle, William Hunter (died 1754) whose will was recorded in Spotsylvania County (Spotsylvania Will Book B, p. 185). William left to son James the "lots I now live upon, called the Ferry lots," with the benefit of the ferry, a tract of land adjoining the town of Fredericksburg, and a tract at Fall Hill known as Silvertown Hill. Son William inherited 300 acres in Orange County and 400 acres in Culpeper that his father had purchased from George Hume. Daughter Martha was to receive £1,000 and the boys were to be educated at William and Mary. James Hunter of the iron works was named executor of the estate and there was considerable dispute amongst family members over his handling of that estate.

On Aug. 2, 1758 William Taliaferro (1726-1798) posted a £12,000 guardian bond as guardian of James, William, and Martha Hunter, infant orphans of William Hunter (Spotsylvania Will Book B, p. 187). On Dec. 2, 1766 James Hunter posted a £3,000 guardian bond as guardian of his nephew, William, who was still under age. John Glassell and Adam Hunter were James' security.

James, Jr. died in Georgia and was the man with whom John Banks had had disastrous business dealings. William moved to Essex County, married, and became the progenitor of that branch of the Hunter family.

[297] On Feb. 7, 1801 Samuel Slaughter and Richard Young Wiggenton (c.1767-1807), commissioners of the high court of chancery, were directed to sell part of James' Culpeper land in order to satisfy Hunter's old debt to Richard Corbin's estate. This was done "subject however to the devise made by James Hunter to his mulatto woman Aggy."

the two slaves, and horses allotted her." Adam was to see that she was provided with "houses on the said Land, where she shall choose, fit for her Use, and to her liking." She was also given James' old chair and "all the furniture now in the house she sleeps in."

Long-standing debts, illnesses, and disputes between executors and heirs complicated the settlement of the Hunter estate. Patrick's arrival on the scene caused an immediate split between Adam and Abner and drew John Strode back into forge affairs. Patrick teamed with Adam against Abner Vernon who called upon his old friend Strode in an effort to keep the forge operating in the post-Revolution recession. Although Patrick initially appears to have been a divisive influence who severely hampered settlement of the estate, the reality may have been quite different.

For five or six years following James Hunter's death, production at the forge continued under the leadership of Adam Hunter and Abner Vernon. The latter was descended from a long-established and highly esteemed Quaker family in Pennsylvania who had accompanied William Penn from England to the New World. In 1682 three Quaker brothers, Randal, Thomas, and Robert Vernon (died 1710) emigrated from Stanthorne, County Palatine, Chester, England and settled in Chester County, Pennsylvania. Thomas and Randal were active in their local Monthly Meetings and their names appeared frequently in those records. Robert didn't take such an active part in his community or church and, consequently, little is known of him (Futhey 755). All three brothers left issue and Abner may well have been a grandson of one of them. It is unknown exactly when Abner came to Virginia. He was employed as the bookkeeper for some 30 years before accidentally drowning in 1792, suggesting that he may have come to Virginia with John Strode in the 1760s. Surviving records in Stafford and Culpeper counties reveal very little information about Abner. He was listed in the tax records as a resident of Stafford and seems to have lived somewhere in Falmouth.[298] He also acquired a considerable landed estate in Culpeper, though there is no evidence that he ever resided there. On Mar. 10, 1781 John Strode and his wife, Ann, sold Abner three adjoining tracts totaling 374 acres. According to the deed, Abner paid £10,000 for the land (Culpeper Deed Book K, page 218).[299] Abner added to his estate on July 8, 1786 with the purchase of 319 acres excepting ½ acre sold to Francis Hume for a burying ground (Culpeper Deed Book N, page 297, July 8, 1786). In 1790 he paid Thomas Strode £400 for another 400 acres adjoining John Strode's land (Fredericksburg Court Records, Deed Book A, page 262). Abner's property was in the southeastern part of Culpeper, not far from the Rappahannock River.

Upon his arrival in Virginia in 1791 Patrick immediately assumed leadership of his uncle's business and estate. He called upon the assistance of George Wheeler,[300] a gunsmith who may have served in some administrative function at the forge.

---

[298] Abner was listed in the 1783 Stafford personal property tax books with 4 slaves and 1 horse.

[299] This seems an excessive amount of money to pay for a relatively small quantity of land, though the court confirmed the purchase price in 1836 (Fredericksburg Circuit Court, Clarke vs Finnall, LC-H/66-06/1836).

[300] It is unclear where George Wheeler originated, but he made his first appearance in the Culpeper records on July 18, 1790 when he purchased 302 acres on the north side of Baker's Mountain and the west side of Wolf Mountain. The purchase was made possible by the default of Joseph Strother, deputy sheriff, who had failed to turn in money collected for "Publick Taxes for the year 1784." He had worked for James Barbour, "late Sherif of said County" and, unable to produce the money the residents had entrusted to him, his property was sold at public auction. George acquired his plantation for £35.

From 1794-1804 George Wheeler operated a gun manufactory in Culpeper (Palmer, vol. 9, pp. 12, 192, 203, 204, 214). By 1800 he was in partnership with Patrick Home. George had varied business interests and, on Sept. 15, 1794, advertised in *Bowen's Virginia Sentinel* for a hatter. By 1800 George was experiencing financial difficulties and advertised his tobacco, wheat, and corn farm in Culpeper (*Fredericksburg Courier*, July 21, 1800). He also tried to sell his manufactory, an ad being placed in the *Virginia Herald* on July 31, 1798. This notice provides a detailed description of a gun-making facility of the period:

> For Sale. My Tilt Hammer in Culpepper county, with the land thereunto annexed. There is on the land sufficient quantity of wood to work here for several years; some cut and hauled, the rest very convenient. As this property is in an agreeable and healthy part of the country, and the Commonwealth wanting to contract for arms, a person with capital sufficient to carry it on, may make almost what he pleases. A Grinding House, with a

During this period, George and Patrick also formed a partnership in Wheeler's Manufactory at Stevensburg in Culpeper where muskets, rifles, and pistols were made on contract for the state. George's advertisement for the sale of the gun factory and his subsequent partnership with Patrick Home may have been the result of a pressing need for cash that was met by a loan from Patrick. On Oct. 22, 1798 George executed a deed of trust in which he conveyed to Patrick 39 acres on Rush River in Culpeper "on which is erected a house called the tilt hammer house with all its fixtures and tools, one house with the water wheel, called the Gun Factory with all its fixtures and tools &c. This house is two story high the upper is for the purpose of stocking muskets" (United States Circuit Court Records, "Hunter's exts vs Home's admr, 1805).[301] This property secured a £1,000 loan from Patrick.

George made his earliest known contract with the state in 1798 (Palmer, vol. 8, p. 506). On Sept. 24, 1799 Wheeler and Home wrote to the governor and "proposed to manufacture arms at their factory in Culpeper. One thousand stands in fifteen months from date of contract on Cook's plan, the gun and Bayonet compleat at fifteen Dollars, equal to patern furnished, with the American Eagle engraved on the plate of the lock" (Palmer, vol. 9, p. 49).

When Patrick died in 1803, Caleb Morrison[302] became George's partner. Wheeler's Manufactory was still producing arms when George died in 1809, though the quality of the weapons was never what state authorities had hoped (Palmer, vol. 9, p. 295).

George Wheeler assisted Patrick in the settling of Hunter's Culpeper estate. Records there contain a deed and mortgage in which George, acting as "attorney and agent for Patrick Home, executor of James

---

separate water wheel, double geared, almost finished, and in a few weeks will be ready to grind and bore gun barrels by water, which will save much labor, and at the same time, do all its parts more perfect, and with more expedition than can be done by hand or any other mode. She is calculated for 3 welding fires, two breechers, and 6 stockers. The bellows work from the loft by water and are so constructed that by drawing down a rod at the fire, the blast of each can be stopped without stopping the wheels. The whole [is] exceedingly strong. The Harness frame, where the drawing hammer works, is secured by sills sunk deep in the earth, 17 by 22 inches square, dovetailed and keyed. It will employ 13 hands in the different branches, which is sufficient to make 36 muskets per week. For Terms apply to George Wheeler, Fredericksburg.

Wheeler again offered to sell the manufactory, placing a notice in the Apr. 19, 1804 issue of the *Virginia Herald*. This, too, was unsuccessful as he still had possession of the property at the time of his death.

On Oct. 30, 1800 Wheeler and Home placed an advertisement in the *Maryland Herald* seeking gunsmiths for Wheeler and Home's Manufactory in Culpeper. They were in need of "one good Mounting or Bayonet Forger, two good Filers and one or two good Polishers, to such they will give generous wages." At the end of the notice, they listed seven gunsmiths already in their employ, including Francis Dowler, George Brenise, Thomas Patton, John Resor, John Kayler, Michael Nichol, and Peter Link. While it may seem strange to list those already in the employ of the forge, gunsmiths were a closely-knit and very peculiar lot who had strong opinions of those with whom they worked. By listing the workers already employed at the forge, the subscribers probably hoped to attract others familiar with their work.

In 1794 George married Lydia Calvert in Culpeper and his will was recorded in Culpeper on Apr. 17, 1809. He mentioned his wife Lydia and children George, Elizabeth, Polly, Margaret, and Ambrose. He wrote, "Whereas I have entered into a contract with the United States (in part with Caleb Morrison) for the manufacture and delivery of a certain number of Muskets it is my will and desire that the same may be completed as soon as possible and the profits arising therefrom after supporting my family to be laid out for their benefit."

[301] George Wheeler died before repaying the loan. In 1805 Hannah Hunter, widow of Archibald Hunter, sued Patrick's administrator claiming that George had owed her husband money. She also claimed that Patrick had owed Archibald $3,000 "for divers sums of money paid and advanced by the said Archibald to and for the use of the said Patrick." Hannah sought to force the sale of the gun factory. This effort must not have been successful as the factory continued in operation until sometime after 1809.

[302] In 1808 Caleb Morrison married Sally Browning in Culpeper.

Hunter," conveyed to Philip Lightfoot 111 ½ acres belonging to Hunter's estate, part of the Roote's tract and known as Hunter's Quarter" (Culpeper Deed Book U, pages 452 and 455).

At first glance it might appear that Adam and Abner were the best choices to manage James' estate and the continued operation of the forge. They had been involved in the business from the beginning and were intimately familiar with it. Certainly the few records surviving in Stafford and Fredericksburg indicate that Adam and Abner were managing quite well until Patrick Home's arrival in 1791. The reality, however, may be quite different. It must be remembered that James Hunter specifically asked in his will that Patrick come to help with the estate. The author speculates that James did so because he was fully aware of how difficult the settlement would be and that Adam and Abner were not necessarily up to the task. Within a month of James' death, Adam suffered the first of several strokes, the last leaving him totally incapacitated.[303] Abner was well advanced in years and may not have been in the best of health. James called upon Patrick because he trusted his nephew's judgment.

Shortly after Hunter's death, Adam and Abner began trying to collect debts due the estate. These ranged from as close as Falmouth and Fredericksburg to as far away as Maryland and Jamaica. On June 13, 1785 Adam and Abner appointed Samuel[304] and Robert Purviance,[305] merchants of Baltimore, as their attorneys to collect debts due in Maryland (Stafford Deed Book S, p. 255). A week later, they named William Hylton of Jamaica as attorney there "to demand of John Dixon, Esqr...in the said Island of Jamaica...all sums due James Hunter, dec." (Stafford Deed Book S, page 356). The Jamaican debts amounted to over £7,317 and had been due since 1762. Hylton was authorized "on failure of payment to take all Legal Methods" to obtain the money and to sell "the estate called Salem in Hanover, Jamaica." There are no known records that reveal how these proceedings fared.

Adam and Abner also published notices including one that appeared in a Richmond newspaper:

> Stafford, June 15, 1785—All persons indebted to the Estate of Mr. James Hunter, late of Stafford, deceased, are requested to make payment to the Subscribers, his Executors—The situation of the estate induces them to urge the debtors to be speedy in their

---

303 One lawsuit brought by Hunter's executors included a deposition from Robert H. Hooe regarding Adam's physical condition. According to Hooe, on Dec. 13, 1784 Adam was "seized with a Paralysis stroke, which deprived him for some time of the use of his Limbs and greatly debilitated his intellectual Organs, and on the twenty sixth of Novr. 1787 he was revisited with a similar stroke which totally deprived him of the use of Speech and power of voluntary locomotion, in which situation he still remains." This same suit included a copy of a letter Adam wrote to Thomas and John Backhouse, Jr., sons of John Backhouse of Liverpool. He wrote, "My infirmity still retards my progress, but as my Brother Mr. Archibald Hunter from Dunkirk promises to join me next April and remain with me I shall by his assistance be enabled to put matters in such tr__ as will answer the best purposes" (U. S. District Court Records, Hunter's exors vs Backhouse's admx, 1797).

Another letter from Adam Hunter to Thomas Ridout (1754-1829), merchant of Alexandria, and dated Dec. 20, 1784 stated, "In future, will be necessarily confined, as my duty as an Executor requires, to the adjustmt [sic] of my late Brothers Estate, Which added to my bad state of health, Being now laid up by a severe paralytic shock, By which I am deprived of the use of one side, and obliged to use a substitute in writing, Induces me in future to decline all connection in Business" (Hunter, Adam, Letter).

304 Samuel Purviance (c.1740-c.1788) was from Donegal, Ireland. He came to America c.1754 and settled in Philadelphia where he was a merchant. While in Philadelphia Samuel was an outspoken opponent of the Stamp Act. In 1763 he and his brother, Robert, established a mercantile store in Baltimore, but Samuel did not move to that town until 1768. Included with the store was a distillery and wharf. In November 1773 Samuel was appointed trustee of the poor in Baltimore County. In 1788 Samuel was a member of a group exploring the Ohio River. Their boat was captured by Indians and some of the party escaped, though Samuel did not. He was never heard from again (Purviance, n.p.).

305 Robert Purviance (1733-1806) was born at Castle Fin, Donegal Ireland. His grandfather was a Huguenot driven from France in 1685 who took refuge in Ireland. Robert came to America in 1763 and established himself as a merchant in Baltimore. During the Revolution, he was a naval officer in Baltimore and, in 1794, Collector of the Port. In 1768 his brother, Samuel, joined him in Baltimore and the two engaged in a lucrative mercantile business there.

payments. All Claimants on said estate are also requested to send in a state of their claims as expeditiously as possible.

Adam Hunter
Abner Vernon

N.B.—Sundry Negro Tradesmen with or without families, property of the said estate, to be sold, apply to the Executors.
(*Virginia Gazette or American Advertiser*, July 9, 1785).

The *Virginia Gazette* of July 19, 1785 carried two notices pertaining to the settlement of the estate. In one of these Adam announced that the partnership of James and Adam Hunter had been dissolved by his brother's death. The second notice was under the names of Adam Hunter and Abner Vernon, executors, who requested all those indebted to James' estate make prompt payment of their debts. They also informed the public that "Sundry Negro tradesmen with or without families are to be sold."

James Hunter's estate was burdened with a number of debts that the executors were unable to pay in part because of the many unpaid debts owed to the forge. Typical of the period, individuals as well as businesses operated on a bewildering system of credit. Throughout the colonial period there was a perennial shortage of cash. Customers, who rarely had any cash in their pockets, maintained accounts with businesses, which they paid when they were able. Payment was usually made with tobacco notes or with a combination of notes and cash. There was a general understanding that payments on accounts were due annually, but a bad tobacco crop often forced extensions on accounts for an additional year or more. Because merchants and businessmen were unable to pay their suppliers until they received payment from their customers, the suppliers were also forced to extend credit or risk losing their market altogether. This unwieldy system proved the downfall of many businessmen who turned to the courts in an effort to force payment of long-standing debts, usually accomplished through the sale of land. Despite the suits, however, many debtors successfully avoided paying their creditors, resulting in an untold number of bankrupted businessmen.

Without access to the business ledgers, it is impossible to determine how many people owed money to the forge. Considering the size of the operation at Rappahannock, there were undoubtedly quite a large number of outstanding debts. Records in Fauquier County reveal that in 1788 Adam and Abner brought suit against Rev. Rodham Kenner[306] who owed the forge £7.1.4 ¾ for "Work Labour & Service done & performed" on two mill gudgeons he had had steeled (Fauquier Mill Records, Adam Hunter, 1788-001). Long-lost Stafford records may have contained similar suits.

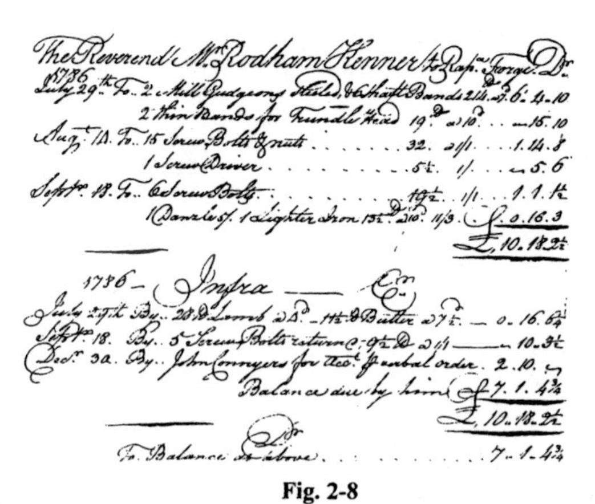
The Reverend Mr. Rodham Kenner to Rapp. Forge Dr.
1786 July 29th To 2 Mill Gudgeons Steeled, & [illegible] Bands [illegible] 6. 4. 10
2 Thin Bands for Trundle Head 19 [illegible] 15. 10
Aug. 14. To 15 Screw Bolts & nuts 32. 1/1 1. 14. 8
1 Screw Driver 5½ 1/ 5. 6
Septr. 18. To 6 Screw Bolts 19½ 1/1 1. 1. 1½
1 [illegible] 1 Lighter Iron 15½ [illegible] 11/3 £ 0. 16. 3
£ 10. 18. 2½

1786 Infra Cr.
July 29th By 28 [illegible] Lamb 2d. 14½ [illegible] Butter 7½ 0. 16. 6¾
Septr. 18. By 5 Screw Bolts returned 9½ [illegible] 1/1 10. 3½
Decr. 30. By John Conyers for [illegible] order 2. 10.
Balance due by him £ 7. 1. 4¾
£ 10. 18. 2½
Dr.
To Balance as above 7. 1. 4¾

Fig. 2-8
Rodham Kenner's account with Rappahannock Forge, 1786 (Fauquier Court Records).

[306] Rodham Kenner (1740-1803) was the son of Howson Kenner of Fauquier. Rodham was born in Northampton County and died in Kennersley, North Carolina. In 1763 he married Elizabeth Plater, the daughter of George Plater (1744-1803).

From time to time Adam and Abner sold small groups of forge slaves in an effort to raise cash. In 1787 they ran the following:

> To be sold at public sale on Wednesday the 7th of February next if fair, otherwise the next fair day, before Mr. Benson's Tavern in Fredericksburg, ABOUT One Hundred SLAVES, consisting of Men, Women and Children, in families and single, the property of the estate of JAMES HUNTER, Esq., deceased, amongst whom are several valuable tradesmen, viz, Blacksmiths, Tanners, Ship and House Carpenters, Wheelwrights, a Bricklayer[307], Miller, and Labourers. Twelve months credit will be allowed the purchasers, giving bond and approved security, to bear interest from the date, if not punctually paid when due. Ten per cent discount will be allowed for ready-money.
>
> ADAM HUNTER, ABNER
> VERNON, Executors, Stafford
> County, Jan. 4, 1787
>
> (*Virginia Journal and Alexandria Advertiser*, Jan. 4, 1787)

The periodic sale of slaves continued for some years, notices of those sales being placed in the local newspapers. On Jan. 25, 1799 Joseph Ennever, acting as either forge manager or Adam Hunter's executor, offered to sell or hire out "Two excellent Negro Ship Carpenters" (*Virginia Herald*, Jan. 25 1799). On Feb. 18, 1803 Francis Thornton,[308] Robert Patton[309] (1750-1828), and Robert Walker[310] (1766-1808) commissioners and administrators, announced the sale of slaves, "late the property of James Hunter esq. dec." to satisfy a debt to James Mills[311] to whom they had been mortgaged (*Virginia Herald*, Feb. 18 1803). Another notice appeared in the *Virginia Herald* advertising the sale of 50 or 60 "Men, Women, and Children, amongst whom are some valuable Tradesmen, such as Blacksmiths, Nailers, Forgemen, House Carpenters, and a ship Carpenter" (*Virginia Herald*, May 24, 1803). At this same sale Robert Patton intended to sell "some household Furniture, Plantation utensils, a Waggon and team, some Horses and

[307] On Feb. 4, 1787 George Washington wrote to Richard Henry Lee regarding the sale of Hunter's slaves He wrote, "it is not my wish to be your competitor in the purchase of any of Mr. Hunters tradesmen especially as I am in a great degree principled against increasing my number of Slaves by purchase and suppose moreover that Negroes sold on credit will go high—et if you are not disposed to buy the Bricklayer which is advertised for Sale, for your own use, find him in the vigour of life, from report a good workman and of tolerable character and his price does not exceed one hundred, or a few more pounds, should be glad if you would buy him for me...If he has a family, with which he is to be sold, or from whom he would reluctantly part I decline the purchase, his feelings I would not be the means of hurting in the latter case, nor at any rate be incumbered with the former" (Abbot, Confederation Series, Vol. 5, pp. 10-11)

[308] This was probably Francis Thornton (1767-1836) of Fall Hill, Spotsylvania County. He was the son of Francis Thornton (1737-1794) of Fall Hill and Ann Thompson. In 1790 Francis the younger served as a justice of Spotsylvania. He married Sally Innes, the daughter of Judge Harry Innes of Kentucky.

[309] Robert Patton was born in Ayr, Scotland. A factor for merchant William Cunningham, around 1769 he settled in Culpeper and oversaw Cunningham's business in both that town and in Falmouth. By 1770 he had permanently established himself in Fredericksburg. In 1792 he married Anne Gordon Mercer, the daughter of Gen. Hugh Mercer (1725-1777). On part of the old Fall Hill tract Robert built a fine home he called White Plains. In 1802 he was elected to the vestry of St. George's Episcopal Church but became disenchanted with that denomination. Six years later he and his wife were members of Fredericksburg Presbyterian Church.

[310] This was probably Robert Walker (1766-1808), a merchant in Fredericksburg.

[311] Mills and Hunter had been in business during the 1760s and this long-standing debt to Mills probably dated from that period. Hunter had mortgaged to James Mills, "sundry Lands and Slaves, and among others a Negro man Slave named Cyrus upon the Conditions in the said deed mentioned, which said Slave has since been sold by Adam Hunter and Abner Vernon Executors of said James Hunter, to Thomas Camp." (Culpeper Deed Book Q, p. 494). Overton Cosby (died 1834), executor of James Mills, deceased, agreed to convey his interest in Cyrus to Camp. Cyrus was included in the inventory of Hunter's personal estate made following his death (see Appendix C).

Cattle, and a quantity of blacksmith's Tools." For whatever reason, that sale was not successful for the property was advertised again on Dec. 3 (*Virginia Herald*, Dec. 3, 1803).

The estate was also due debts from Fredericktown, Maryland where Frederick Duvall was appointed to collect payments and settle affairs at the 8,151-acre Fielderia furnace tract. The Library of Virginia has a collection of six letters written by Abner Vernon to Baker Johnson (1747-1811) of Frederick County. These documents relate to the settlement of Hunter's estate and the division and sale of the old Fielderia tract. A writ of possession, issued on June 20, 1791, described the various tracts comprising the 8,151 acres of Fielderia. Although James Hunter had foreclosed on Gantt in 1767, the latter remained on the property and had to be forcibly ejected by court order (Vernon, Letters).

On Apr. 8, 1791 and for several weeks thereafter, Patrick Home and Adam Hunter advertised the sale of the Fielderia tract. The notice provides an interesting description of the tract:

> Valuable Lands for Sale. To be Sold on Monday, the 13th Day of June next, at Mrs. Kimball's Tavern, Upwards of Seven Thousand Acres of Land, formerly the Property of Fielder Gaunt.
>
> On these Lands there are a considerable quantity fit for meadow; several valuable streams of water, on one of which there is a Grist and Saw Mill, and there has been a Furnace, to which an ore bank is very convenient. Part of these lands lay near that strong and flourishing place called Frederick-Town; and the rest is near the river Potowmac, the navigation of which it is expected will soon be accomplished.
>
> The whole of the Land will be laid off in lots, to contain from one to 300 acres, and the location of the lots will be such as to render the purchase of two or more entirely convenient.
>
> Bond, with approved security, will be required, and the payments made in the following manner, viz.—One third of the purchase money to be paid on the day of sale, the other two-thirds in two equal annual payments thereafter, without interest. Conveyances will be made on the entire payment of the purchase money. A plat of the lands may be seen, and information had, by applying to Mr. Samuel Duvall.[312]
>
> Adam Hunter,
> and
> Abner Vernon
> Executors of James Hunter, deceased
>
> Frederick county, Maryland, 22d April, 1791
> (*Virginia Herald and Fredericksburg Advertiser*, Apr. 8, 1791)

Business continued at Rappahannock Forge long after the Revolution though the extent and variety of production during this period is uncertain. On Dec. 17, 1788 James Hunter, Jr. wrote to his brother-in-law, John Russel Spence, regarding anchors for two ships. He wrote, "I hope you may contrive them from the Forge, but there can be no doubt they make them there...Iron mentioned for bolts should come on immediately" (Hunter-Garnett Papers). The personal property tax records also indicate ongoing activity at Rappahannock long after James Hunter's death. In 1789 Adam paid taxes on 49 slaves and 14 horses at the forge, well below the quantity on site during the Revolution, but a considerable number nonetheless (Vogt, Tithables, vol. 2, p. 467). In fact, Adam, as representative of his brother and Rappahannock Forge, remained the largest slave owner in Stafford until his death in 1798. Well after James' death, Thomas Jefferson corresponded with representatives of Rappahannock regarding the finishing of a gun for his friend, Philip Mazzei.[313] This correspondence, which involved a dispute over the mounting of a gun barrel,

---

312 Samuel Duvall (1748-1811) was born in Frederick County, Maryland and was the son of Capt. William Duvall (1723-1810) and Priscilla Prewitt. In 1744 Samuel married Priscilla Ann Dawson (1756-1836). He was a surveyor and represented Frederick County in the Maryland House of Delegates from 1781-1783.

313 Filippo Mazzei (1730-1816) was born in Poggio a Caiano and died in Pisa, Italy. He received the degree of Doctor of Medicine from the University of Florence and practiced for several years in Smyrna, Turkey. From 1755-1773 he was the agent of the Grand Duke of Tuscany in London and was engaged in commerce. In December 1773 he came to Virginia with several other Italians to introduce the cultivation of grapes, olives, and other Italian agricultural products. At that time he resided in Albemarle County where he

continued for some time and on Dec. 12, 1792 Jefferson wrote to Abner Vernon complaining of the charge of seven guineas for the "mounting of Mr. Mazzei's gun barrel."[314]

Newspaper notices advertising services and goods at the forge indicate that activities continued there until at least the early part of the 1800s. The Fredericksburg Circuit Court records include part of a ledger from 1793-1797 listing a number of purchases made at the forge. These included such activities as drawing wire, "sundry smithwork," making carriage wheels, nails, cart boxes, cutting wood for carts, tanning and dressing hides, sawing lumber, hauling and waggonage, making horseshoes, and carpenter's work (Fredericksburg Court Records, Hume vs Hooe). It appears that the administrators continued to operate the most productive shops, generating much-needed income.

Following Hunter's death the executors sold some of his real estate, though final settlement of the estate did not begin until 1802 when the federal court assumed administration of his estate. Although the executors ran several notices for the sale of Hunter's 1,200-acre Marsh tract in Fauquier County, it took some 14 years before it finally sold. One early newspaper notice described the property:

> By virtue of the authority given us by the Will of James Hunter, esquire, deceased, will be Sold, on Wednesday the 20th day of May next,
>
> A valuable
> Tract of Land,
>
> lying in the county of Fauquier, on the Great-Marsh, containing about eleven hundred acres, near five hundred acres of which are capable of being made excellent meadow, and the rest fit for making tobacco, small grain, &c—A farther description is thought unnecessary, as it is presumed any person inclined to purchase will view the land.
>
> The said Tract will be sold altogether, or divided into parcels, as may best suit the purchasers. The sale to be on the premises. One third of the purchase money must be paid on the day of sale---six months credit will be given for another third---and for the valance twelve months credit; the purchaser giving bond with approved security. Upon payment of the money an indisputable title will be made by,
>
> Adam Hunter,
> Abner Vernon, Ex'rs.
>
> April 20th, 1789
> (*Virginia Herald*, Apr. 30, 1789)

The tract did not sell at that time and was advertised again in October 1789 (*Virginia Herald and Fredericksburg Advertiser*, Oct. 1, 1789). Again no purchaser was found the property was not sold until 1803 when the remainder of Hunter's property was auctioned.

The settlement of James' estate was complicated by the demands of his numerous creditors. Several suits survive in Fredericksburg and Richmond and there are indications that papers used to be on file in Stafford as well. The earliest surviving court records regarding Hunter's estate date from Nov. 18, 1794 and are on file in the Fredericksburg Circuit Court. This case provides an interesting view of the early years of this estate. According to the suit:

---

became a personal friend of Thomas Jefferson. Filippo, or Philip as he was called here, was an active supporter of the movement for independence. From 1779-1783 he was an agent in Europe charged with obtaining army stores and loans for the State of Virginia. He returned to Virginia in 1784. Later, he was charge d'affairs in Paris, then Privy Councillor of the King of Poland. In 1802 he received a pension from the Emperor Alexander of Russia. Philip published the first French-language history of the American Revolution in 1788.

[314] Some time prior to this correspondence, Mazzei had purchased a rough gun barrel from Rappahannock and left it there to be finished, stocked, mounted, and a lock made for it.

he the said Abner Vernon and your Orator Adam Hunter (being the only persons nominated in the testament to be Executors thereof who were then in Virginia) only took upon themselves the burthen of Executing the same, and your Orator Adam Hunter being in a short time after he had taken that burthen upon himself deprived totally of his speech and almost totally of the use of his limbs, he by that means became incapable of taking an active part in the Execution of his office, and was compelled from necessity to intrust the whole management and direction of the Estate of his testator to the management of the said Abner Vernon his Co-Exor and he did continue to manage it from that time until the arrival of your orator Patrick Home in Virginia in the year [1791], and for some time afterward, at length your orator Patrick Home qualified himself to act as a Exor of the same Testament, and being desirous of becoming acquainted with the true Situation of that Estate, which by that qualification he had acquired a right to manage...as well as fearful that much waste had been committed thereof by the said Abner Vernon, he required him to account for his Executorship so far as he had gone therein, but could not prevail on him so to do. Your orator then commenced suit against him in this honourable Court in order to compell him so to account but even before a Bill was filed therein the said Abner Vernon was drowned in fording a River, and the same suit was thereby abated (Fredericksburg Court Records, Hunter's Exor vs Vernon's Admr).

Patrick was unable to obtain from Abner Vernon any formal accounting of his uncle's estate. After Abner's unexpected death in 1792, Patrick commenced a suit in the Stafford court against John Strode.[315] He not only demanded an account of the estate's debts and assets, but also wanted Strode "to pay what upon such account, and the Decree of this Court thereon, the said Abner Vernon might appear to be indebted to your orator, as his Co-Exor." The Stafford justices agreed that Patrick had a right to know how the estate had been managed and ordered Strode to provide the account. Strode and Vernon had been close friends for some thirty years. They had been instrumental in the establishment and success of Rappahannock Forge and they both deeply resented Patrick's intrusions. On Feb. 24, 1794 Strode, acting as Abner's administrator, commenced a suit in Stafford against Patrick Home and Adam Hunter "laying his damages at fifteen hundred pounds and indorsing his writ 'for Labour and Services done by the Intestate' and subsequent to the decree to wit, on the first day of July in the same year he commenced another action against them also on the case, laying his damages at three hundred pounds, and indorsing his writ 'for money paid and advanced by the intestate and his admr for the Defendants.'" Adam and Patrick claimed that Strode's actions were "brought only to vex and harass them...It is their firm opinion that the said Abner Vernon died greatly indebted to their testator's Estate" (Fredericksburg Court Records, Hunter's Exor vs Vernon's Admr). Adam and Patrick claimed they were unable to afford the costs of this suit until they received a judgment on their previous suit against Strode. Adam's failing health only added to Patrick's burden. By 1787 Adam was totally incapacitated, the full responsibility of settling Hunter's estate falling solely to Patrick Home. He wrote of the situation, "I am bringing matters to a termination, as fast as the nature of the case will admit; my Uncle's helpless situation is a considerable ban on my progress to that wished for event" (VCRP, Letter, Patrick Home to James Home).

Abner died intestate late in 1792. The first notice of his death read:

> We feel much concern in relating a melancholy accident which happened on Wednesday evening last. Mr. Abner Vernon, of Stafford, in attempting to ford the Rappahannock at the falls, was thrown from his horse and drowned. He was a man of most amiable temper, cheerful and kind to every one, sturdy in his principles, pure in his morals; esteemed and respected by his numerous friends for his great integrity and usefulness, and who, now sensible of his worth, lament his loss with sincere and general sorrow (*Virginia Herald*, Dec. 13, 1792).

Abner's body was not quickly recovered. A second newspaper notice read, "The body of Mr. Abner Vernon, who was drowned in this river the 5th December last notwithstanding the most diligent search was not found until the 13th instant. We understand that the papers &c. which he had in his pockets, and which were of confidential value, were not materially injured" (*Virginia Herald* of Feb. 21, 1793). According to William W. Hening's report, Abner "had had the chief management of the affairs of his

[315] The Stafford records pertaining to this suit have been lost. References to it are contained in the Fredericksburg Circuit Court Records, Hunter's Exor vs Vernon's Admr, CR-DC/V/560-95/1799.

testator, [and] his untimely death occasioned them to be left in great confusion" ("British Mercantile Claims," vol. 28, p. 223). On Feb. 11, 1793 his widow, Jane, renounced her right of administration of his estate in favor of John Strode (Stafford County Scheme Book Court Orders, 1790-1793, p. 381). In accepting this responsibility, John was drawn back into forge affairs and forced to deal with Patrick Home. The settlement of Abner's estate resulted in several court suits between John and Patrick and, no doubt, caused both men a good deal of worry.

Abner's personal property in Culpeper was advertised in the Dec. 11, 1794 issue of the *Virginia Herald.* His estate there included "a large stock of Horses, Cattle, Sheep and Swine; plantation utensils, and a great variety of Household Furniture.[316] The sale to continue, night excepted, until the whole be disposed of; at the close of which will be hired out, several very valuable slaves."[317]

On Mar. 3, 1797 Abner's family granted John Strode a power of attorney authorizing him to act in their interests with regard to Abner's estate. This document suggests that most of the family remained in Pennsylvania. Those relatives named included Sarah Asbridge "late Vernon Widow," Joseph Vernon, and Thomas Taylor, all of the township of West Town, Chester County; Samuel Vernon of Delaware County, Pennsylvania; and George Vernon of the township of East Whiteland in Chester County. They empowered John Strode "to lease, lett, sell or dispose of either by public or private sale...all such lands tenements and herediments and all and every part of such real or landed estate whether in fee simple or otherwise which our beloved relation and friend Abner Vernon late of the State of Virginia was seized at the time of his decease" (Culpeper Deed Book X, page 9).

Strode brought suit against Patrick and Adam claiming that the forge owed Abner's estate back wages due for Abner's services as bookkeeper and manager of the works "and other Estates of the said James Hunter before and since his decease." Strode also sought to recover £300 which he claimed was due Abner from the purchase of "a quantity of Pig Iron for the use of the estate of the said James Hunter" for which Abner had paid £444.12 from his own pocket. Vernon purchased the iron from William Beale, "agent for John Tayloe Esq.,[318] at his Iron Works." Abner's repeated requests prior to his death had failed to motivate Adam and Patrick re- pay him. According to the writ, "Vernon could not obtain Pig Metal for carrying on the forge of said James Hunter...on any other terms, and which Pig Metal was as this defendant has heard and verily believes worked up into bar Iron and sold for the benefit of said Testators Estate and Patrick Home one of the Complainants confessed to be true and promised to pay the said sum of £___ for the said Pig Iron to the said Beale" (Fredericksburg Court Records, Strode vs Home &c). The difficulty of procuring iron on anything except a cash basis was likely a result of the enormous debts belonging to Rappahannock Forge and a 30-year history of the company's inability to pay its bills.

This case was continued month after month through 1797, though the pig iron issue was not settled until 1801. These monthly continuances were due in part to the absence of Strode's attorney, Robert Mercer[319] (1764-1800). He instituted the case in Stafford Court but "the week before the next Court on which he considered it necessary to file his declaration, was called off among the militia which were ordered to the Westward in defence of the General Government and from that sudden call arose the

---

[316] John placed a notice in the *Virginia Herald* of Apr. 4, 1793 for the stud services of Abner's stallion, Bucephalus, standing at the Vernon farm in Culpeper. The fee for this outstanding horse was "Five Dollars the season—pasturage gratis."

[317] Included in the Fredericksburg Circuit Court records is a deed dated Dec. 28, 1790 in which Thomas Strode conveyed to Abner Vernon of Stafford 400 acres in Culpeper adjoining John Strode's land. Abner paid £400 for the property (Fredericksburg Deed Book A, page 262).

[318] John Tayloe (1721-1779) of Mt. Airy was the son of John Tayloe (1687-1747) and Elizabeth Gwyn (1692-1745), the daughter of David Gwyn (died 1703) and the widow of Stephen Lyde (died 1711). John, Jr. was educated in England and built the fine home in which he lived in Richmond County. In 1744 he was a signer of the Treaty of Lancaster, a member of the Ohio Company, was a trustee of the town of Dumfries in Prince William County, a breeder of fine racehorses, and owned Neabsco and Occoquan iron works. In 1747 he married Rebecca Plater (born 1731), the daughter of Col. George Plater (1695-1755) of St. Mary's County, Maryland. In 1757 John was appointed to the Colonial Council under Lord Governor Dunmore John Murray (1732-1809) and, upon the reorganization of the state government, was elected to the first Council of State under Governor Patrick Henry. He resigned this position on Oct. 9, 1776.

[319] Robert was the son of John Mercer (1704-1768) of Marlborough. In 1791 he represented Stafford County in the Virginia House of Delegates.

misfortune of nothing more than the incipiture appearing on the Records of the County Court of Stafford." John Strode claimed that he only brought the suit against Adam and Patrick because he feared a suit by Beale against Vernon's small estate would "tend to ruin it and greatly distress his widow." Regarding Abner's paying for the pig from his own pocket, John noted, "the debt for which the said Vernon made his own estate liable for the said Pig Iron, [was] a benevolent and friendly act in him and more than he as Exor was bound to have done."

On June 10, 1797 the Stafford Court heard the case as a plea of trespass (indebtedness). "John Strode...complained of Adam Hunter and Patrick Home Surviving Exors of the last will and testament of James Hunter deceased...whereas the said Adam Hunter and Patrick Home on the 25th day of March [1792] in the lifetime of the said Vernon at the County aforesaid, was indebted to the said Abner Vernon in the sum of £222.6...[having been] laid out and advanced to and for the use of the said Defendants as Executors...[who] undertook and then and there faithfully promised the said Abner Vernon that they would well and truly pay him...the said sum of £222.6." The case was again continued, not being heard by a Stafford jury until Mar. 15, 1798 at which time they found in favor of Strode for £261.5.7 (Fredericksburg Court Records, Strode vs Home &c.).

True to form, Adam and Patrick appealed the decision to the higher court in Fredericksburg. This court also found in favor of Strode and ordered Patrick and his co-signer, Robert Dunbar, to take out a bond for £600 to guarantee payment. The court records state, "Whereas upon an appeal prayed by said Patrick Home and Adam Hunter surviving Exors of James Hunter from a judgment obtained against them by John Strode adm of Abner Vernon in the County Court of Stafford, for the sum of Two hundred and sixty one pound five shilling, and seven pence Damages and Costs, Judgment did on the 14th day of the present month past in the Circuit Court of Fredericksburg for the said Strode it was considered that the said Judgment should in all things be affirmed, and that the said Strode should recover of the said Home Damages according to law for retarding the Executors" (Fredericksburg Court Records, "Hunter's Exors vs Vernon's Admr). Patrick appealed the decision to the Court of Appeals in Richmond. The case wore on until July 15, 1801 when Patrick and Robert Dunbar signed bonds to John Strode for £489.12.2. Of this £237.3.5 was to be paid in cash and the remainder was obtained by the sale of four slaves in front of Stafford Courthouse on Aug. 2, 1801 (Fredericksburg Court Records, "Strode vs Home &c.).

Despite the court orders, payment to Abner's estate was not forthcoming. John continued to press the issue, bringing suit in the federal court against the commissioners appointed to settle Hunter's estate. This case was finally determined in favor of Abner's estate, but not until 1832, some 12 years after Strode's death. John Strode claimed that Abner's estate was due £2,415.13.0 for services rendered at the forge. Patrick and Adam disagreed claiming that Abner's estate actually owed money to the forge. Commissioners Robert Hening and John W. Green were appointed to settle the accounts between the parties. They issued their report on June 5, 1811 and determined that both estates were in debt to each other. The commissioners found that Abner had sold one of James Hunter's certificates for $11,961 or £3,588.5.6 ½, although unauthorized to do so. Hunter's estate owed Abner £240 per year as forge supervisor and an additional £60 per year from 1785-1792 as one of James' executors. However, the commissioners found that Abner had allowed over £2,555 in debts due Rappahannock Forge to go uncollected or be made without security, several of these receiving the commissioner's comment, "lost by your neglecting to collect it." In the end the commissioners decided that Hunter's estate was indebted to Vernon's estate £2,361.1.0 (U. S. Circuit Court Records, Strode vs Patton and als, 1832).

After her husband's death, Abner's widow Jane married William Ball[320] of Stafford. Apparently, William Ball and John Strode had some differences of opinion as to William's rights to property Jane had acquired from Abner's estate. Suits brought in the District Court of Fredericksburg and in the County Court of Culpeper resulted in an agreement between the two. The dispute involved "in some instances a claim which said Ball sits up as his own right in property—purchased or claimed to be purchased at public sales by the said Jane during her widowhood and in other instances respecting his claim to her Dower in the said Estate real and personal." The agreement with Strode specified that William Ball "shall have and hold in love of his wife [sic] Dower in the said Estate and in lieu of all other Claims against the said Estate or against the said John Strode...the following property to wit: Armistead by Trade a Black Smith, Aaron, Matt, and Reubin labourers, four male negro Slaves: the smith tools which said Armistead works with and

---

[320] This was probably William Ball (died 1815), the son of William Ball (died 1782) and Martha Brumfield. William, Sr. was born in Pennsylvania but later settled in Virginia.

all the wearing appearel of the above mentioned deceased, a silver watch, Clock, Bedding and house hold furniture, a riding Carriage, Horses, cattle, Stock of every kid, Plantation Utensils; including every Article more or less which the said Jane purchased on the 26th day of December 1794 at the sale of the personal property of the estate, which said enumerated Slaves and personal property are to be and remain the property of the said William Ball, his heirs and assigns forever, also the House and tract of land about 300 acres thereunto belonging whereon he now resides together with all its appurtenances for and during the natural [life] of his the said Jane all which Estates Slaves and personal property He the said William Ball shall have and keep peaceable possession of and enjoy in manner and form aforesaid without hindrance interruption or molestation from the Heirs of the said Estate or from the said John Strode in his private capacity as Administrator of the said Estate or as Attorney in fact for the said Heirs and further all charges which the said John Strode may have in his Books of account either for himself or for the said Estate against the said Ball or against his wife while a widow of what nature soever be the same more or less shall stand as pair and forever discharged." Ball agreed to drop his claim to every other part of Abner's estate and promised to deliver to Strode by the end of the year James and Milly, two mulatto servants, and Andrew and Winny "all four of which reasonably well Cloathed who are in possession and are part of the Slaves for which he now stands sued." Ball agreed to "receive the hire or wages of the last mentioned four slaves for the present year" and John was to receive the hire of Armistead to the end of the year, then give him up to William Ball along with his tools (Culpeper Deed Book X, page 9, May 18, 1802).

Adam and Patrick fought John Strode every step of the way, even going so far as to publish notices in the newspapers stating, "All persons indebted to the estate of James Hunter, Esquire, of Stafford, deceased, are hereby forewarned from paying any monies to any person except the subscribers who are the surviving executors" (*Virginia Herald*, Dec. 26, 1792 and Jan. 3, 1793). Apparently, Patrick and Adam were concerned about people paying Strode money due Hunter's estate.

Patrick and Adam also engaged in animated squabbles with their neighbors near the forge. Around 1793 William Richards[321] (1765-after 1803), owner of Richards' Hill to the immediate east of the forge, decided he wanted to build a gristmill on his property. Europe's ever-growing demand for American flour was a welcome relief to businessmen in debt-ridden America and caused a sharp, though short-lived renaissance in Falmouth. Richards' proposed mill would have been the third such structure built just to the west of Robert Dunbar's bridge between Falmouth and Fredericksburg. When Patrick and Adam made renovations to the forge complex in the early 1790s, they also constructed a small gristmill on the east end of the tract. This later became known as the Falmouth Mill. William Richards' proposed mill was intended to stand just a few feet downstream from the new mill built by Patrick and Adam. Building a new mill required the blessing of the county court which, by law, was required to hear complaints from those above and below the proposed mill site. Patrick immediately objected to Richards' mill, perhaps fearing the competition, and claiming that the land upon which William planned on building the mill actually belonged to Hunter's estate. Patrick asked that William be required to produce proof of ownership of the land (Fredericksburg Court Records, Home Exor of Hunter vs Richards). Even though William was able to produce a clear record of title back to Thomas Vicaris' patent of 1690, Patrick appealed the Stafford court's decision to allow the building of the mill. In the end, William Richards' was allowed to build his mill (Fredericksburg Court Records, Home Exor of Hunter vs Richards).

---

[321] William was the son of John Richards (1734-1785) and Susannah Coleman (died 1778), the daughter of Robert S. Coleman. William was the grandson of William Byrd Richards (born c.1697) of Caroline County. William Byrd's brother was Mourning Richards, architect, carpenter, and contractor of Aquia Church. Their sister, Catherine, married the Rev. Robert Innes (1720-1799) of King and Queen County. Robert was a close friend of the Rev. John Moncure (1710-1764), first rector of the present Aquia Church.

In 1775 John Richards came to Falmouth from King and Queen County. Here he bought 300 acres above Falmouth from John Dixon (c.1740-1791). Initially called Richard's Hill, this plantation was later known as Ingleside. The Richards family was long associated with tobacco export and sales, warehouses, mercantile operations, estate management and administration, and public affairs.

William's sister, Elizabeth (born 1760), married Falmouth merchant Daniel Triplett (1753-1818). In 1787 his sister, Mildred or "Milly," married William Scandrett Stone (1765-1827), the son of George Stone (1741-1771) and Mary Scandrett of Maryland. William Richards married Ann, the daughter of John and Ann Blackwell. Following the Revolution, William was deeply involved in the milling industry in Falmouth.

At about this same time, William built a canal beginning about one half mile below Rappahannock Forge to a millpond where Falls Run flows into the river. This canal was intended to be part of a much longer canal beginning at Greenbank carrying on through Falmouth and incorporating Hunter's old canal. Although only intermittent parts of this canal were built, William Richard's section became the foundation of the flour milling industry that briefly breathed life back into post-Revolution Falmouth.

On May 11, 1795, in the midst of their dispute with William and his mill, Adam and Patrick asked the Stafford Court for permission to build another gristmill on their property between the forge and Falmouth. They claimed that the land upon which their dam would abut was either theirs or the Commonwealth's. William Richards "who conceived himself to be affected by the building of the said Mill," requested that the applicants produce evidence of their title to the land. The sheriff was ordered to impanel a jury to hear the case. To the undoubted delight of William Richards, on Oct. 12, 1795 the court ordered that Adam and Patrick not be allowed to build their mill, as they were unable to prove ownership of the land (Fredericksburg Court Records, Bet Richards and Hunter's Exors.).

This argument continued through 1797. On Feb. 21 of that year sheriff Hancock Eustace[322] (1768-1829) was ordered to summon Adam and Patrick for yet another hearing regarding Richards' Mill. The summonses were not delivered and deputy sheriff Enoch Mason[323] (c.1769-1828) added a wonderfully amusing note to the bottom of the document. He wrote, "Mr. Adam Hunter one of the persons mentioned in this writ did some time ago escape from the Gail of Stafford County and I have not since been able to find him, either with this summons or with an escape warrant which is in my hand." There is no explanation as to why Adam was in jail to begin with, or how he had managed to escape when he was so severely handicapped. Several times during the 18th century, the sheriff complained to the justices that the jail was in need of repairs and that escapes were likely. The fact that a paraplegic could escape from a nearly new jail provides new insight into the condition of the jail and the level of security maintained there. Enoch Mason had to appear before the justices to explain why he had not produced Adam's body on court day. A writ was then given to deputy sheriff William Mason to serve on Patrick Home. Equally unsuccessful in his search for the summoned, William returned the writ to the clerk without serving it (Fredericksburg Court Records, Hunter's Exors vs Richards).

Some months later when Adam and Patrick finally appeared in court, Adam claimed that he could not possibly present his case, being unaware that the hearing was to take place that day. Attorney Benjamin Both claimed that "the said Home because he is debarred of the benefit of the said Adam Hunter by reason of want of notice, and the said Adam Hunter because he is from the same cause an equal sufferer with the said Home, and they do therefore jointly and severally protest against all further proceedings of this Court in this session" (Fredericksburg Court Records, Hunter's Exors vs Richards). The case was continued.

Patrick's dispute with William Richards didn't die easily, and his ongoing suit provides an interesting account of the roads around the forge. In a case heard in Fredericksburg William asked permission "to open a new road from the main County road (which leads from the counties of Culpepper and Fauquier to the Town of Falmouth) to his water gristmill on Rappahannock River to leave the said main road a small distance below the house now occupied by Benjamin Bussell and to go from thence the most elegible way to his said water gristmill" (Fredericksburg Court Records, Hunter's Exors vs Richards). On July 14, 1800 the court-appointed several "viewers of the road" who suggested that a road be "opened from the main county road at its junction with the Forge road a little below the house of Benjamin Bussell and passing down the said Forge road within a small distance of a locust Tree leaving the said Tree to the right and passing down a small path by Beck Howards until it crosses a ditch which is the dividing line between the applicant and the Estate of James Hunter thence round the said applicants wheat field fence into the forge road which leads to Falmouth, thence down said road a small distance below the applicants gate then

---

[322] Hancock was the son of Isaac Eustace (died c.1795) and Agatha Conway (1740-1826) of Fauquier County. Hancock served in the French and Indian War (1755-1763) under the command of Col. George Washington. On June 4, 1789 he married Tabitha Henry (died c.1840), the daughter of the Hon. James Henry (1731-1804) of Fleet's Bay, Lancaster County. Hancock resided at Woodford, part of which is now Meadowlark Subdivision on Garrisonville Road (State Route 610).

[323] Col. Enoch Mason was the son of John Mason (1722-c.1796) and Mary Nelson (172_-c.1801) of Aquia. On April 28, 1796 Enoch married Lucy Wiley Roy of King George County. They resided at Clover Hill, now Rose Hill Farms Subdivision in Stafford.

leaving the Forge road and passing down the hillside to the right of the applicant's Mill" (Fredericksburg Court Records, Patrick Home ads William Richards; Patrick Home vs William Richards).

The surviving court documents present Patrick Home as a "rabble-rouser" who did little but bring confusion and discontent to people who had long worked well together. However, it is risky to make judgments about a person's character based solely upon public documents. Patrick seems to have made a sincere effort to sell the forge and settle his uncle's estate. Further, he does not seem to have profited financially by his twelve-year effort to do so. The Virginia Colonial Records Project contains several letters written by Patrick to his family back in Scotland. These documents present a far different picture of this interesting man. On Mar. 5, 1796 Patrick wrote to James Home, probably an uncle or cousin. Of James Hunter's estate Patrick wrote:

> I am and have encountered severe struggles in the prosecution of my business in this Country, the Estate of my late uncle I found encumbered with a heavy Load of Debt, that circumstance added to the mismanagement committed since his Death, has involved me in trouble, anxiety, and ingratitude. I had likely to have been overwhelmed, in the commencement of my career, but a firm and resolute intention buoyed me up, and rendered me invulnerable to the malevolence of faction.
>
> I wish you to make yourself acquainted with the character and stability of Mr. Forbes, Nassau Street, Soho, London, and without delay communicate the result of your enquiries to me. He solicits me to take a concern in the Forge, and with him, or a person he may nominate. Previous to an undertaking of a thing of such magnitude as this, and the responsibility in which I should be implicated, I am desirous of obtaining the best and fullest information, in order to enable me to form some idea of the Grounds on which I may establish my safety in the Connection (VCRP, Letter, Patrick Home to James Home).

In a letter to merchant and kinsman James Hunter,[324] Patrick wrote, "I am now on contract with a James Forbes Esqr. for sale of the Forge. Should I obtain the price I ask for it, Eighteen thousand Pounds Sterling, a very handsome legacy will be our Lot, and my fatigue and labour will be amply remunerated" (VCRP, Letter, Patrick Home to James Hunter). Known surviving records contain no further references to James Forbes or to this proposed partnership.

Under pressure from creditors, Patrick continued to seek buyers for his uncle's property. Beginning in May 1798 and continuing until the end of September that year, he advertised the forge property in the newspapers. These advertisements describe the shops remaining at Rappahannock:

> For Sale,
> The Iron-Works & Mills
>
> Belonging to the Estate of James Hunter, Esq. of Virginia, deceased, most delightfully situated on the Falls of Rappahannock river, two miles from the town of Fredericksburg, and one mile from the town of Falmouth and tide water: Consisting of a Forge 128 feet by 51, eight fires and four hammers, a Coal House 80 feet by 40; a Merchant Mill, 70 feet by 36, with two pair of French bur stones, and every other necessary apparatus for making flour, constructed upon the modern improved system;[325] also a Gristmill, 20 feet by 18,

[324] It is unclear how this James Hunter was related to James Hunter of Virginia.

[325] The "modern improved system" refers to modernizations made according to recommendations by Oliver Evans (1755-1819) of Newport, Delaware, a brilliant millwright, inventor, and engineer who automated the old water-powered mills. Evans' improvements included rolling screens or grain cleaners, conveyors for moving materials laterally, elevators that raised grain and meal to upper floors, chutes for lowering materials between floors, and a "hopper boy" in which freshly ground meal was cooled. Once the grain entered the mill, via a water-powered wench that hoisted the heavy sacks to the upper floor, the miller didn't have to handle it again until the finished flour or meal poured through a chute into a waiting bag or barrel. All cleaning, grinding, cooling, and sifting were handled automatically. Evans' innovations not only made the life of the miller easier, but also were the basis of many technological improvements that led to the Industrial Revolution. Oliver Evans also made substantial contributions to the development of the high-pressure steam engine that was later used in locomotives.

with one pair of stone; and a Saw Mill 55 feet by 11, the running gears and machinery of the same are new, most judiciously fixed and executed. Contiguous thereto are a Smiths Shop, 60 feet by 26, with three fires; and a Stable 54 feet by 27. All these buildings are of stone, of neat and substantial workmanship. There are also a Nailery, a Tanyard, Coopers, Carpenters, and Wheelwrights Shops, with tools and utensils for the several mechanical branches, and houses for the managers, millers and workmen. The water is taken out of the main body of the river, and conducted to the works by a deep canal, capable of furnishing more large & extensive improvements. The head and fall are about 19 feet. The great and various local advantages that this most valuable property possesses, cannot be accurately and fully enumerated. Few such places in America can be found, that are better calculated for man to exercise his ingenuity in the erection of the vast diversity of works which require the power and aid of water. Adjoining thereto are about seven thousand Acres of Land, on which are some farms, and a sufficient quantity of wood for the use of the iron works. There is also some good Meadow Land.

If the above property is not sold on or before the 1st November next, I will Rent the same. On application the terms will be made known by

Patrick Home,
Surviving exec'r of James Hunter,
esq. deceased.

Rappahannock Forge, 14th April, 1798
(*Virginia Herald,* May 2, 1798)

Two weeks later, Patrick's advertisement was still in the paper, but he had added a line stating that the sale would take place on Oct. 1, 1798 "at the Tavern of Mr. John Benson, in the town of Fredericksburg" (*Virginia Herald*, May 19, 1798). At the bottom of the notice appeared another offer to sell the Marsh Tract in Fauquier. It is significant that this notice contained no mention of a blast furnace, indicating that by this time iron smelting was no longer being conducted at Rappahannock Forge.

No purchaser came forward and the forge continued in operation. Following the Revolution, the American economy wavered, but the forge found a reprieve in the same European events that caused a rebirth in the town of Falmouth. The Napoleonic Wars (1793-1815) had so upset the social and economic structure of Europe that those countries were unable to feed themselves or conduct basic manufacturing. They were forced to look to America for flour and iron products. The situation at the forge was further improved by the Indian wars at home that created an additional need for iron. Increased activity in and around Falmouth boosted business, as well. To encourage American iron production following the Revolution, the government imposed an import duty of 15% on all foreign iron entering the United States after 1784. There was an additional 10% added if the incoming iron was shipped in a foreign vessel. Patrick's improvements to the mills and physical plant at the forge reflect his effort to capitalize on this welcome demand for manufactured products.

While business continued at Rappahannock, persistent creditors hounded the executors for payment of Hunter's old debts. Unable to generate the needed capital by selling the forge, Patrick placed a notice in the *Virginia Herald* stating that "All those indebted to Rappahannock Forge, whose accounts are of one or more years standing, are requested to make immediate payment to the Subscriber, or grant specialties for their respective balances, which will alone preclude the disagreeable necessity of instituting suits against them, as no longer indulgence can possibly by given" (*Virginia Herald*, Apr. 25 and May 2, 1793). As nearly everyone was short of cash during this period, the threat of a lawsuit probably did little to encourage people to pay their debts to the forge.

For nearly 11 years Adam was without the use of his arms, legs, or speech. He wrote or dictated his will in June 1787 while he was still able, but didn't die until March 1798 (see Adam Hunter's will, Appendix D and his inventory, Appendix E). His chosen administrators were his friends James Somerville[326] (1742-1798) and Henry Mitchell[327] (c.1740-1798), Fredericksburg merchants. To his brother

[326] James Somerville, Sr. (1742-1798) was the son of William and Janet (Mitchell) Somerville of Glasgow. Shortly after his arrival in Fredericksburg, he became involved with Henry Mitchell in a mercantile business. A successful businessman, James became one of Fredericksburg's wealthiest merchants and

Archibald, "now residing at Dunkirk in the Kingdom of France...it is my will, that as soon as my said Brother, or any Devisee of his, shall become a Citizen of the United American States, and be thereby capable of accepting a grant of real Estate in America, that the said James Somerville and Henry Mitchell release and convey unto the said Archibald Hunter...all the Estate I now have in America or else where." Adam specified that if Archibald died before taking ownership, the estate was to pass to Archibald's children at the discretion of the trustees. By the time Adam's will was recorded, both of the original administrators were dead. Joseph Ennever (died c.1849), David Briggs[328] (1730-1813), and Nicholas Payne[329] (died 1802) put up bond as replacement administrators and were approved by the court (Fredericksburg Will Book A, p. 3, PP. 94-96, May 16, 1798). Neither Archibald nor any of his heirs became American citizens and Adam's interest in James' estate was sold.

Shortly after Adam's death, Joseph Ennever announced in the newspaper:

> Notice is hereby given to all those who have claims against the estate of Adam Hunter, Esq, deceased, late of Stafford county, to bring them in properly authenticated that provision may be made for their liquidation. Those who are indebted to said estate are desired to make immediate payment.
>
> All those who are in possession of any Books belonging to the deceased, are most earnestly requested to return them by the last of next month.
>
> Joseph Ennever
> Administrator to the Will annexed.
>
> Stafford, May 17, 1798.
> (*Virginia Herald*, May 19, 1798)

---

owned extensive tracts in Spotsylvania, Culpeper, and Orange counties. He lived in the lower part of Fredericksburg on Lot 274 on which stood a small stone warehouse and wharf that he had purchased from Robert Johnston. He was a member of the Fredericksburg Masonic Lodge and donated the land whereon the Masonic Cemetery is located. He was mayor of Fredericksburg 1784-85, 1787-90, 1792-93 and was a member of the first Fredericksburg Hustings Court. He died unmarried in Port Royal where he had gone in an effort to recover his health. His obituary appeared in the Apr. 28, 1798 issue of the *Virginia Herald.*

Around 1795 James wrote to his brother Walter in Scotland asking that his nephew, James, Jr. (1774-1858) be sent to him in Virginia. Young James arrived shortly thereafter and entered his uncle's business. Young James inherited his uncle's Rapidan River plantation called Somervilla. In 1810 he married Mary Atwell (1778-1845) of Fauquier.

[327] Henry Mitchell was a Scottish merchant whose name appears several times in the papers of the Virginia Colonial Records Project. On Jan. 29, 1784 Henry submitted a Loyalist Claim for the loss of his salary of £100 per year as a factor in the Glasgow trading company of George McCall and Richard Smellie, £200 in expenses, £2,000 in debts and the loss of an unspecified amount of property. According to his deposition, he "left his native Country, Scotland, at an early period in life and went over to North America; that he was settled near Twenty Years at Fredericksburg one of the principal Commercial Towns in Virginia, as a Factor for and Partner with a considerable trading House in Glasgow." Tensions with England resulted in his leaving Virginia in 1777 on board the ship *Phoenix*. This carried him to New York where he engaged in trade until 1781. He returned to Virginia in 1781 in a vain attempt to recover his property. Unsuccessful in that effort, he sailed for his native Scotland, bankrupt.

[328] Like Adam Hunter, David Briggs was suspected of loyalist affections and authorities confiscated his papers. David, a well-respected merchant in Falmouth and Fredericksburg, was born in Fifeshire, Scotland. Although trained for the ministry, he chose to come to Virginia where he settled in 1752 in Falmouth. In 1771 David married Jean McDonald (c.1750-1810), the daughter of Neal McDonald, rector of Brunswick Parish. David purchased land in what is now the Hartwood area of Stafford County and built a home he called Stony Hill. His son, James McDonald Briggs (1787-1845), inherited his father's plantation.

[329] Nicholas Payne was the son of Merriman Payne (c.1713-1773) of Lancaster County. Nicholas married first c.1780 Elizabeth Towles. He married secondly Frances (Bruce) Banks (1735-1818), widow of Gerard Banks (c.1725-1787). Nicholas lived much of his life in Spotsylvania, but resided in Stafford after marrying Frances Banks.

In addition to the advertisements published in the Virginia papers, Patrick also placed notices in British newspapers. In his May 11, 1801 letter to James Hunter of Scotland, Patrick asked Hunter to place an enclosed notice in the Edinburgh papers (VCRP, Letter, Patrick Home to James Hunter, Duns). Five years of unsuccessful efforts to sell the forge had caused Patrick to reduce the asking price. He wrote, "the lowest price I would wish to take for it, is £10,000 Sterling—one half paid down in Cash, the other half payable in 12 months. The expense of this advertisement I shall pay my Brother to repay you, he seeming so anxious as well as I myself am, to return to my native Country, I have concluded to take £10,000 Sterling, which is a sum far below its value. With respect to the slaves they are perfectly distinct and not to be included at all in the above terms, & you'll please observe that the present number of them may be diminished by death & other accidents, this will be necessary to mention to the purchaser." The enclosed notice sent to Scotland for publication read:

> For Sale. The Iron Works belonging to the Estate of James Hunter of Virginia, deceased, delightfully situated on the Falls of Rappahannock River, within one mile of Falmouth & Tidewater, & two miles from the Town of Fredericksburg, where the River is navigable for Vessels of considerable burden. The Works consist of a Forge 138 feet by 51, Eight fires, & four hammers; a Coal House 80 feet by 40. A Merchant Mill, 70 feet by 36, with three pairs of Stones, French Burr, with the late improvements of Oliver Evans[330] of New Castle County on the Art of Manufacturing grain into flour, or meal, viz: for Elevating grain or meal from the lower to the upper Stories, for conveying the same from one part of the mill to another, for cooling the meal, & attending the Boulting Hoppers; all this Machinery quite new; as are also the Boulting cloths, consisting of all the four kinds, viz: Superfine, Fine, Middlings, & Ship Stuff: A Gristmill, built in the year 1798, with one pair of Stones: A Saw Mill 11 feet by 48 erected in 1798. A Slitting Mill, but of which little else than the walls are in repair. These Buildings are of Stone, of neat & substantial workmanship. The different departments are conveniently disposed at proper distances, on a capacious canal calculated to supply works still more extensive; well secured against casualties by Freshes or high floods; has its source in the main body of the River, a copious body whereof is at pleasure collected & turned in by a complete set of strong Dams, which have not given way since their formation, near thirty years ago: the head & fall of water operating on the wheels is about 20 Feet; almost all the works are in good repair, & the whole may be rendered at a small expence. To the appendages thereto are: a convenient Tanyard constructed in the year 1793; a Smith's Shop, built in the same year; a Wheelwright's & a Cooper's Shop, both built in 1796, which houses for the Managers, Workmen, &c, and a Stable 54 feet by 27, built in 1798. And about 8000 Acres of Land contiguous, well wooded including some farms & some rich low grounds lying on the river. Together with the premises will be sold the Slaves now employed in the Works, consisting of Forgemen, Blacksmiths, Nailers, Wheelwrights, Coopers, Millers, & Carpenters. These Iron Works have advantages over any others in America, particularly in regard to the sale of the produce, there being none of the kind to the Southward thereof, to most of which extensive & rich Country there is easy conveyance by water; nor is there any Forge within 90 miles.
>
> The terms of Sale will be made known by Applying to Mr. James Hunter, Merchant in Dunse near Berwick on Tweed. Many other advantages peculiar to these works will be best pointed out on the premises, which on application will be shown by—
>
> Pat. Home
> Only surviving Executor & one of
> the devisees of the said James
> Hunter Esq., decd.

[30] Robert Dunbar eventually purchased this mill. In 1812 Oliver Evans brought suit against Dunbar for ;2,000 for infringement of his patent rights (U. S. Circuit Court Records, Oliver Evans vs Robert Dunbar, 812).

It is evident that business continued at Rappahannock Forge long after James Hunter's death, albei on a much reduced scale. From 1783-1790 a considerable number of slaves were assigned to the forge though their numbers gradually declined during the period. There are no personal property tax records available between 1791 and 1811. The Stafford County personal property tax records list the following:

1783—James Hunter

Location	Slaves	Horses	Cattle
Forge	87	23	10
Stanstead	104	71	20
Total	**191**	**94**	**30**

1785—James Hunter's Estate

Location	Slaves	Horses	Cattle
Forge	72	18	42
Stanstead	45	8	27
Accokeek	83	29	28
Total	**200**	**55**	**97**

1786—Adam Hunter

Location	Slaves	Horses	Cattle
None Listed	177	16	103

1787—Adam Hunter

Location	Slaves	Horses	Cattle
Forge	53	25	23
Not Named	75	20	76
Total	**128**	**45**	**99**

1788—Adam Hunter

Location	Slaves	Horses
Forge	48	18
Not Named	26	10
Stanstead	9	5
Total	**83**	**33**

1789—Adam Hunter

Location	Slaves	Horses
Forge	49	14
Home	22	9
Stanstead	9	5
Total	**80**	**28**

1790—Adam Hunter

Location	Slaves	Horses
Forge	47	12
Home	30	14
Total	**77**	**26**

Notices in the Virginia newspapers indicate that in 1792-1793 Patrick invested money to upgrad some of the old forge shops. Not only would upgrading the facilities have generated much needed income but it would have made the complex more attractive to potential purchasers. The previously quoted notic sent to the Edinburgh newspapers provides dates of construction for some of these shops. In 1793 a notice was placed in the *Virginia Herald* "to inform the customers of Rappahannock Forge, and the public in general, that the works at that place are now carried on upon a plan by which their orders will be thankfully and readily complied with...The Blacksmiths, Wheelwright, & Tanning Business, &c. are likewise carried on at the same place, in their various branches.—The Gristmill is now in excellent order, under the direction of a good Miller, and ready for the manufacturing of flour...Wheat, corn, bacon, raw hides, &c. will be received in payment for iron and smiths work, &c. at the highest prices." To the bottom of this lengthy notice was a post script which read, "There is on hand at this time, a stock of excellent Iron, also a large Anchor, weighing upwards of 1200 wt and a wrought iron Carboose,[331] sufficiently large for any vessel carrying an anchor of that weight; which will be sold cheap for cash or produce." Thomas West signed the notice (*Virginia Herald*, May 30, 1793).

Thomas West was a gunsmith and blacksmith and the son of Edward West (1757-1827). Edward, a gunsmith, silversmith, clockmaker, and inventor was very likely employed at the forge prior to or during the Revolution. In 1802 Edward, then residing in Lexington, Kentucky, received a U. S. patent on a safety gun lock. Thomas advertised his father's lock in the *Virginia Herald*:

> PATENT GUN LOCKS. The Subscriber offers for sale, the right of making and using Edward West's improvements in Gun Locks, which is considered of the utmost importance to every person using Fire Arms, and more particularly to the Army and Militia of the United States. There are two very great advantages resulting from this improvement. The first is that no accident can possibly happen by a gun going off at half cock, as it has in that situation no connection with the dog, so frequently not with the trigger, or any other external part of the lock. It cannot possibly catch at half cock, as there is no notch that the dog can catch in; and the counter-dog is placed in the act of cocking as not to be restored, but by the cock first falling or being let down. The great security in this improvement ought certainly to recommend it to general use, to avoid the innumerable accidents that have and must unavoidably happen, from the present plan. Application made to Mr. Basil Gordon, Falmouth, will be communicated and attended to.
>
> Thomas West
>
> (*Virginia Herald*, Sept. 6, 1803)

For some years prior to his employment at Rappahannock, Thomas West had operated the Falmouth Nail Manufactory under the firm name, Thomas West & Company. His two partners in this venture were Daniel Triplett[332] (1753-1818) and Birkett Davenport[333] (1738-1817). A notice advertising their manufactory read:

---

[331] This was a large cast iron cooking stove designed for use on board ships. While it is certain that anchors were made at Rappahannock Forge, records of the Virginia Navy Board specifically mention carbooses being made at Isaac Zane's Marlborough Iron Works and being sent to Hunter for distribution to the navy. It is unknown if Rappahannock also produced carbooses.

[332] The Falmouth Nail Manufactory "formerly carried on under the firm of Thomas West & Co., is now the sole property of the subscriber [Daniel Triplett]" and offered for sale nails, brads of all sizes, spikes and ships nails. Daniel also sought to employ journeymen nailors in his factory (*Virginia Herald*, May 9, 1793).

Daniel Triplett was the son of Francis Triplett (died c.1765) of King George County. In 1777 Daniel married Elizabeth Richards (1760-1826), the daughter of John Richards (1734-1785) and Susannah Coleman (died 1778) of Ingleside. He served as a justice for Stafford in 1803. Later that year he moved to Richmond where he established a grocery store. In 1808 he established the firm of Daniel Triplett and Company and placed a notice in a local paper offering to hire 10-12 slaves to work in the coal mines near Richmond.

[333] Birkett Davenport was the son of George Davenport (died 1756) and Susanna Edmonds (1721-1771) of King George County. Birkett served as a justice of the peace in King George before moving to Culpeper. He married Eleanor Brown (1752-1790) and their daughter Susanna (1768-1798) married Philip Rootes Thompson (1766-1837).

Nail Manufactory.

The Subscribers beg leave to inform the public, that they have lately enlarged their Nail Manufactory, in the town of Falmouth, and that they will now supply merchants and others, with all kinds of Nails and Brands, on reasonable terms. Three months credit will be given to merchants, who take quantities, and a generous discount made from the retail prices. Orders to any amount under one hundred thousand, will be punctually complied with in one week after they are given in. They will take a few boys, to be bound for five years, to the Nailing business; black ones will be preferred.

Cash, Pork, Beef, and Indian Corn, will be received at the Factory for Nails, or any kind of Blacksmiths work done there.

Thomas West & Co.

Falmouth, Feb. 21st, 1791
(*Virginia Herald and Fredericksburg Advertiser*, Apr. 8, 1791)

Patrick must have made Thomas West a lucrative offer because the latter gave up the nail manufactory, the former partnership being dissolved "by mutual consent" as of Jan. 15 1793 (*Virginia Herald*, Mar. 28, 1793).

Despite what sounded like a promising re-awakening of the forge, however, the newspapers contain no further advertisements for work performed at the forge nor of Thomas West; yet activity there continued. On Dec. 10, 1802 an unsigned notice appeared in the *Virginia Herald*:

> Take Notice! That a number of the inhabitants of Culpeper, Fauquier, and Stafford Counties intend to prefer a petition to the next General Assembly of Virginia; for leave to open and extend the road from the Spotted Tavern, by the Rappahannock Forge, across the Rappahannock River to Mortimer's Island, to intersect the road leading to Fredericksburg and to build a Bridge over the said river.

It might be surmised that the petitioners wished to pass through the forge property because there was still business going on there. Traces of that road remain on the island, as do the abutments from the dam on the north side and the bridge on the south side of the island.

James Hunter's executors also spent years attempting to sell Stanstead. On July 31, 1794 they placed a notice in the *Virginia Herald* that read in part, "This seat, in point of elegance, beauty, and many other advantages, is equal, if not superior, to any in this part of the country." The mansion house and plantation were described as containing "near four hundred acres of Land, lying on the Rappahannock, opposite the town of Fredericksburg, of which it commands a complete view...the Ferry Landing on the north side of the river and Boats will be sold with this tract." Also offered for sale was an adjoining tract of 385 acres. The subscribers added, "If it is not sold by private bargain before the 1st day of January next, it will be exposed to sale by Public Auction, on the first day of April 1795, at the house of Mr. Benjamin Turner, in the town of Falmouth."

No buyers were forthcoming and a month later Adam and Patrick issued another notice announcing, "Notice is hereby given that the sale of the mansion house and plantation of the late James Hunter, Esq. deceased, by public auction, is unavoidably postponed. Any person inclinable to purchase the said property, may apply to the subscribers, living in Stafford county, who will dispose of the same by private contract" (*Virginia Herald*, May 29, 1795).

Based upon information contained in the British Mercantile Claims, Patrick purchased Stanstead for himself at this time and took up residence in his uncle's old home.[334] He also purchased the 95-acre forge tract, paying a discounted price for both. Patrick's motivation for these purchases is unclear. The property had been for sale for at least four years, perhaps longer. The buildings on both sites were in a rapid state of decline as little or no maintenance had been carried out for years due to a lack of funds. Patrick's purchase, however, raised the ire of the estate's creditors. The report in the British Mercantile Claims states that "The Mansion house and iron works, with the lands annexed, have been offered for sale

[334] Patrick employed overseers to manage affairs on Stanstead. From 1797-1799 Thomas Withers filled this position. From 1800-1803 Fielding Alexander (died 1847) served as overseer of both Stanstead and Rocky Pen plantations.

by Mr. Home and struck off to himself at a little more than £3,000, supposed to be about one-half of their original cost, though they had gone much to decay for the want of funds to keep them employed. Whether this purchase of Mr. Home's will be confirmed by a court of competent jurisdiction is yet to be ascertained. The creditors complain much of the transaction" ("British Mercantile Claims," vol. 28, p. 223). While no records survive that reveal the official outcome of Patrick's questionable purchase, in 1802 Patrick sold the forge to Robert Dunbar. Patrick died in April 1803. On Dec. 15 of that year Stanstead was divided into two parcels. Joseph Ennever bought the house and 263 acres and Thomas Collins bought the remaining 271 acres.

Hunter's creditors weren't the only ones displeased with the management of Hunter's estate. In 1797 John Brown (died c.1804), Thomas Mountjoy (1737-1840), and William Phillips (1744-1797), justices of Stafford County, brought suit against Adam Hunter, Abner Vernon, John Strode, Charles Carter, William Fitzhugh (1721-1798), Harris Hoe[335] (1736-1824), and Andrew Buchanan (c.1750-1804). According to this suit, on April 11, 1785 Adam and Abner had qualified as executors of the estate, with the latter four standing as security. These men had agreed to pay a £100,000 bond, equal to $333,333.66, for faithful completion of their duties, "to be paid to the said justices whenever demanded, with a condition to the said writing obligatory annexed...to the following effect that whereas the said Adam Hunter and Abner Vernon had that day qualified as executors in the will of James Hunter Esqr., deceased, now if they should well and truly make or cause to be made a true and perfect inventory of all and singular the goods chattels and credits of the said deceased that then had or might hereafter come to their hands, possession or knowledge, or to those of any other person or persons for them...and further should make a just and true account of their accounting and doings therein, when required by the said Court, and pay and deliver the Legacies in the said will contained and specified so far as the goods chattels and credits would [allow]." The justices claimed that the executors "did not keep or perform the said condition but ___ it in this, that they did not pay a debt due by judgment obtained in the Circuit Court of Fredericksburg by Thos. B. Morton and Gabriel Jones[336] [1724-1806] Executors of the last will and testament of Thomas Lord Fairfax for and against the said Adam Hunter and Abner Vernon Executors of James Hunter for the sum of £3,000 Virginia Currency to be discharged by the payment of £1,500 with Int. from 20 day of Apr. 1766 also 31125 lb. Tob. @ 174:95" (Fredericksburg Court Records, "ustices vs Hunter &c). The Stafford justices further claimed that the executors "did not well and truly pay to the said Rebecca Backhouse...the amount of two judgments," one awarded May 27, 1795 for $16,000 plus 220 pounds of tobacco and $24.53 in damages. The second judgment was awarded two days later for $8,238.45. This case was heard in Stafford, then in Fredericksburg, and finally in the U. S. District Court in Richmond. It was dismissed on Nov. 28, 1810 because all the litigants were dead (U. S. District Court Records, Justices of Stafford County vs Strode als, 1810).

In two separate sales carried out in 1801 and 1819 respectively the court sold Hunter's Culpeper property. Francis Corbin (1759-1821), surviving executor of Richard Corbin instituted one of the suits that precipitated these sales. Francis sought to settle an unsatisfied deed of trust between James Hunter and Richard Corbin that dated from Nov. 4, 1766. The high court of chancery appointed Samuel Slaughter,[337]

---

[335] Harris was the third son of Howson Hooe, Sr. (1696-1773) and Anne Frances Harris. He was born in Prince William County but later moved to Stafford where he served as a justice of the peace 1779-1785 and possibly other years as well.

[336] Gabriel Jones was the son of John Jones (died c.1727) of Montgomery County, North Wales. In 1720 John and his wife settled near Williamsburg, Virginia where Gabriel was born. Gabriel was trained in the law in London. He returned to Virginia and, in 1746, was appointed King's Attorney for Augusta County at the age of 22. He married Mrs. Margaret Strother Morton (1726-1822), the daughter of Maj. William Strother (c.1696-1733) of King George County and the widow of George Morton (1717-1766). Gabriel was one of the most distinguished lawyers of his day. He represented Augusta County in the House of Burgesses from 1757-1758 and again in 1771. A boundary change in 1777 made him a resident of newly-formed Rockingham County and he represented that county at the Constitutional Convention of 1788. Noted for his intelligence and integrity, his one fault was a nearly uncontrollable temper punctuated with vivid profanity. He actively practiced law from 1745 until his death in 1806.

[337] Samuel Slaughter (1774-1846) married first Frances Banks and secondly Virginia Stannard.

Carter Beverley,[338] and Richard Y. Wiggenton,[339] commissioners and directed them to sell 1,750 acres that Hunter had purchased from John Spotswood (Culpeper Deed Book V, page 513, Feb. 7, 1801). The court decreed that Patrick had been given until Dec. 5, 1800 to pay Francis Corbin £5767.3.7 ½ with interest and his costs or "the right of the said Patrick Home and all persons claiming under James Hunter deceased, to redeem the lands described and the slaves named with the increase of the females" was void. This was the same debt mentioned by John Mercer in his letter to his son George (Mulkearn 206). High bidder for the property was William C. Williams of Culpeper. Money raised by the sale was paid to Francis Corbin "subject however to the devise made by James Hunter to his mulatto woman Aggy." Williams paid £3,000 for the tract and as that amount was insufficient to meet the debt, the commissioners proceeded to sell another 151 ½ acres, it being the upper part of a tract purchased by Hunter from Col. Philip Rootes[340] and adjoining the lands of John Hilton, Richard Y. Wiggenton, and the above 1,750 acres.

The second suit was heard in 1819 and was brought by James Ross, executor of James Mills, against Robert Patton, administrator of James Hunter, John W. Green, Alice G. Williams, executor of William C. Williams, deceased, Philip Lightfoot, and John Robertson, Attorney General of Virginia. John Stannard Marshall (1818-after 1880), commissioner of the Superior Court of Chancery in Fredericksburg, was directed to sell so much of a 500-acre tract purchased by James Hunter from Philip and George Rootes[341] "as may be in the possession of John W. Green by virtue of his purchase from William C. Williams and of Philip Lightfoot by virtue of his purchase from Williams. Philip Lightfoot purchased 11 acres and John W. Green purchased 337 ½ acres for $2,500 (Culpeper Deed Book LL, pages 14 and 15, Aug. 16, 1819).

On July 28, 1801 Dr. Alexander Vass, Scottish merchant of Falmouth, purchased the previously mentioned Falmouth Mill from Patrick Home, "Executor and Devisee of James Hunter."[342] Vass, in turn, sold the mill to Robert Dunbar on Apr. 13, 1803. This sturdy little one-story mill was built of stone and had a wooden roof. In 1822 it was insured for $2,000 by then-owners Daniel Grinnan[343] and John Mundell and, at the time this policy was issued, John O'Rion[344] was the miller (Mutual Assurance Society, Dec. 16,

---

[338] Carter Beverley (1774-1844) was the son of Robert Beverley (died 1800) of Blandfield in Essex County. Carter served as a justice in Culpeper in 1799 and lived the latter part of his life in Augusta County. He married Jane, the daughter of Ralph Wormley of Rosegill, Middlesex County.

[339] Richard Young Wiggenton (c.1767-1807) was the son of John Wiggenton (1741-1825) and Elizabeth Botts (1741-1824), the daughter of Seth Botts (c.1713-1776) and Sabina Bridwell (c.1717-1785). John and Elizabeth were born in Stafford and moved to Culpeper where they lived at Greenfields. John was the adopted son of Richard Young. Richard Y. Wiggenton married Mary Jones (died 1847), the daughter of Capt. Gabriel Jones (1741-1777) and Martha Slaughter.

[340] Col. Philip Rootes (died before 1787) was the son of Maj. Philip Rootes (died 1756) of Rosewall in King and Queen County. In 1756 Philip married Frances Wilcox. He served as sheriff of King and Queen County in 1765.

[341] George Rootes was the son of Maj. Philip Rootes (died 1756) of Rosewall in King and Queen County. He removed to northwestern Virginia where he represented Augusta County in the House of Burgesses.

[342] In the Fredericksburg Circuit Court records, Deed Book D, page 292, is a deed dated July 28, 1801 in which Patrick Home sold to Dr. Alexander Vass for $2,155 the lot and mill seat adjoining Falmouth and adjoining the lot commonly called Froggetts. These had been conveyed by the Rev. John Dixon to Dr. Charles Mortimer on Dec. 4, 1773 and conveyed by Mortimer to James Hunter on June 24, 1775.

[343] In 1828 Daniel Grinnan owned a gristmill on the north branch of the Rappahannock River in Fauquier County. His descendants resided in southwestern Stafford County for many years.

[344] John O'Rion placed a notice in the Oct. 26, 1822 issue of the *Virginia Herald* that read:

> The Subscriber having rented the Corn Mill in this place, begs leave to inform his friends in town, and to those persons who have been accustomed to send their grain to it, that he intends to devote his attention entirely to this branch of business, (of which he has some knowledge,) and flatters himself that he shall be enabled to give satisfaction. As the Mill, for the last few years, has not been in good repair, he has deemed the above notice essential. He will also grind Plaster upon the lowest terms, and in the best manner. Persons sending grain to the Mill, need be under no apprehension from the Plaster getting with their meal, as ever precaution shall be used to prevent it.
>
> John O'Rion

1822). To the east of this mill stood the Thistle Mill, owned by Alexander Vass and his partners, John (born 1773) and Wright Southgate. According to the insurance policy, to the west of the mill stood "the Suburbs." Another Mutual Assurance Society policy was issued on the Falmouth Mill in 1829 when it was owned by George Curtis (c.1767-1844) and, at that time, it was valued at $1,500 (Mutual Assurance Society, Dec. 31, 1829). The grist and merchant mills were located on the east end of Hunter's property, adjoining the town of Falmouth.

John Strode's son-in-law, Mordecai Barbour[345] (1763-1846), also had a hand in keeping the Rappahannock Forge merchant mill working. On Aug. 9, 1805 he published a notice in the *Virginia Herald* regarding the purchasing of bulk quantities of wheat:

> Forge Mills. Cash will at all times be given at these mills for good Wheat—and wheat will be manufactured for merchants or farmers for toll on reasonable terms.—Any person not disposed to take the market price at the time of delivering his wheat, may at his option at any time within four months, fix a day on which he will take the current prices at which time a negotiable note will be given payable 60 days thereafter.—A Miller skilled in his business who can come well recommended will meet with employment on liberal wages.
>
> Mordecai Barbour

James Hunter was far from bankrupt when he died. His personal estate alone was appraised at over £12,700 and he owned thousands of acres in Stafford, Fauquier, and Culpeper counties; yet his debts outweighed his assets. There were numerous claims against Hunter's estate, not all of which were justified. From 1800-1802, reports were filed on the status of the many pre-Revolutionary debts owed to British merchants. These British Mercantile Claims contain two lengthy reports, dating from c.1802, pertaining to Hunter's affairs that reflect some of the difficulties in settling an estate of this magnitude. Samuel Lyde (died c.1806), surviving partner of Lionel[346] and Samuel Lyde, merchants of London, had made a claim against Hunter's estate for £2,290.[347] The agent investigating this claim wrote, "Of all the claims which have yet come under my observation, this is pregnant with the most obvious marks of an attempt to impose on the United States. From the books and papers in possession of Patrick Home of Stafford, the only surviving executor of James Hunter, (now lying before me) it appears evident that there is not a shilling due the claimants but on the contrary that they are upwards of £90 in debt to the estate of the testator." Other claims were made against the estate, as well.

Two other debts were found to be justified, namely £1,873 owed to the estate of Ward[348] & Hunter and £2,955 owed to the estate of Champe, Ward, & Hunter. The investigator wrote:

Falmouth, Oct. 26, 1822

John O'Rion was also granted liquor licenses in Stafford in 1830, 1832, and 1833.

[345] On Jan. 15, 1805 Mordecai placed a notice in the *Virginia Herald* that read:

> A MACHINE For separating COTTON from the seeds is just erected at the Allum S[prings] Mills, where seed cotton will be taken in & picked for the eighth part—Customers may calculate on dispatch as the Machine works 37 Saws. Cash will also be given for good Seed Cotton delivered in this place.
>
> Mordecai Barbour
>
> Freddg. Dec. 10th 1804

[346] Lionel (1682-1737:42) and Samuel were sons of Cornelius Lyde of Bristol, England. Lionel kept a store n Port Royal, Caroline County. He served as mayor of Bristol and married Elizabeth Moore (1716-1779).

[347] Samuel took his case to the American courts claiming that on Feb. 14, 1768 James Hunter became ndebted to the Lydes for £1,524.19.5 and £33.5.10 2/3 "for divers goods, wares and merchandize before hat time shipped, sold and delivered to them" (U. S. Circuit Court Records, "Lyde, svg ptr vs Hunter's xors, 1794"). No decision was given in the surviving court documents.

[348] Capt. Thomas Ward was a merchant from Liverpool and one of James Hunter's many business partners. The Virginia Colonial Records Project includes copies of papers pertaining to a claim submitted by Ward's administrators, William Dennison and wife Jenny and John Oddie and wife Ann of Great Britain. They claimed that Patrick Home and Adam Hunter, executors of James Hunter, held £500 Sterling equivalent to 2,200 that rightfully belonged to the estate of Thomas Ward. They presented a bond dated July 13, 1772

> These debts originated not in ordinary mercantile transactions but in the inhumane and impolite slave trade. Ward was captain of a ship in the service of John Backhouse[349] of Liverpool and about 1756 through the interest of his employer commenced a traffic in slaves from the coast of Guinea in connection with James Hunter the Elder of Stafford County near Fredericksburg and Col. John Champe[350] [c.1700-c.1763] of King George County...James Hunter the Elder at his death left a very considerable real and personal estate, but having at an immense expense erected a set of iron works at the falls of the Rappahannock, he was compelled to borrow large sums of money to assure the repayment of which he mortgaged the greater part of his slaves. One debt alone to James Mills[351] &

---

in which James Hunter had borrowed £500 from Ward. Patrick and Adam claimed that James had paid this during his lifetime. The Ward administrators produced two more bonds, both dated July 30, 1772, one for £1,000 equivalent to $4,400 and a third for £500 or $2,200. The court determined that James Hunter had not paid these debts and ordered Adam and Patrick to Ward's administrators £2,409.5.3 or $8,030.40 ("British Mercantile Claims," *The Virginia Genealogist*, vol. 28, pp. 222-224; VCRP, Loyalist Claims - A List of Claims, 1762-1798).

This suit continued in the U. S. Circuit Court. William Dennison, Ward's administrator, brought suit against Patrick Home for $2,222.21 due Thomas Ward from James Hunter's estate. Dennison claimed that on June 13, 1772 Hunter had signed a bond with Capt. Ward for £500, equal to $2,220. One month later Hunter signed another bond for £1,000, none of which he had paid. On Sept. 23, 1799 George Wheeler agreed to advance $2,222.21 to Patrick Home towards settling this claim. On July 3, 1798 Patrick Home and Robert Patton executed two bonds with David M. Randolph (1760-1830), Marshall of the Virginia District, to guarantee payment of $760.94 and $4,444.02 with interest for the benefit of Thomas Ward's heirs (U. S. Circuit Court Records, Ward's Admrs vs P. Home, 1800; U. S. Circuit Court Records Ward's Admr vs Hunter's exor, 1797).

The *Maryland Gazette* of Aug. 2, 1753 contains an advertisement from John Champe and Thomas Ward of King George County who advertised slaves for sale at Leedstown and Morton's Warehouse. The notice stated that Capt. Ralph Lowe had imported the slaves from Africa in the ship *Lintot.*

[349] These abstracts on slave imports list John Backhouse only one time. On July 30, 1752 he was recorded as having landed 124 slaves on the Rappahannock. These records are likely incomplete as neither James Hunter nor Thomas Ward were listed (Minchinton 151).

[350] John Champe resided in King George County and was a warden of Brunswick Parish which was formed from Hanover Parish in 1732. On May 21, 1761 the *Maryland Gazette* carried a notice from Champe and Hunter stating that Capt. Samuel Murdock, in the ship *Alice* had just arrived in the Rappahannock River from Angola with slaves. On June 11 they placed another notice in the same paper stating that Capt. Daniel Kerr, in the ship *Mary* had just arrived at Port Royal on the Rappahannock River with slaves for sale. They also announced that Capt. William Lowe in the ship *Bassa* from the Windward Coast, had arrived with slaves that would be sold at Hobb's Hole (Tappahannock) and Leedstown. Champe and Hunter placed yet another notice in the Sept. 17, 1761 issue of the *Maryland Gazette* in which they advised readers that they had slaves for sale in Fredericksburg that had just been imported by Capt. Cuthbert Davis from Africa on the ship *Peggy*.

In his reports compiled for the British Mercantile Claims, William Waller Hening wrote of John Champe, "He left but two children, his sons John and William Champe, both of whom died without issue before the peace. The widow of John Champe intermarried with Col. Lewis Willis of Spotsylvania, now Fredericksburg with whom she is now living and in whose hands the estate of her former husband has not been diminished...About 1775 a settlement of the long standing accounts between John Backhouse and these companies took place in Fredericksburg. This may account for the judgment carrying interest from 1775 only, although they had been running for many years before" ("British Mercantile Claims," *The Virginia Genealogist*, vol. 28, no. 3, July-September 1984, p. 224).

[351] James Mills (1718-1782) was a merchant in Tappahannock, Naylor's Hole in Richmond County and later, in Urbanna in Middlesex County. In 1743 he purchased the Ritchie property in Tappahannock with its brick dwelling and storehouse and eventually owned 8 lots in town. His store in Urbanna was erected c.1756 and is now the Urbanna Public Library. In 1752 he served as a justice for Essex County and in 176' became a vestryman of Christ Church Parish, Middlesex County. He married Elizabeth Beverley (1725

Co. of Urbanna exceeded £10,000 principal, which has been carrying interest from some period before the war. To provide for the payment of this debt a great proportion of the slaves were sold...A judgment has been obtained in the Federal courts against the executor of James Hunter for the full amount of these debts. The executor exhibited a statement of the assets in his hands which leaves a balance of some thousand pounds after paying these debts ("British Mercantile Claims," vol. 30, p. 144).

Rebecca Backhouse, widow and executrix of John Backhouse, brought several suits in the federal courts in an effort to obtain satisfy debts due by James Hunter to her husband's estate. One of these, filed in 1795, claimed that in April 1767 Champ Ward and Hunter became indebted to John Backhouse for £4063.3.9 "for divers goods, wares and merchandize before that time sold and delivered, and for divers sums of money before that time paid, laid out and expended by the said John Backhouse in his life time for the said Champ Ward and Hunter at their special instance and request" (U. S. Circuit Court Records, Backhouse's admx vs Hunter's exors, 1795). The court found that Hunter's estate owed Rebecca $16,000 plus costs, but Adam Hunter and Patrick Home failed to pay the judgment. This suit includes a letter from James Hunter to John Backhouse dated July 1, 1771. James explained to his partner that he was having difficulty collecting a bond due in Frederick County and couldn't pay Backhouse until that was settled. He asked John to be patient with him, but said if he sought "legal recourse...I will not defend myself, but instantly dispose of (even the most profitable part of) my Estate to avoid such Contest and ruinous Costs." James added that he was to receive "One thousand Sterling being sent from Jamaica to Philadelphia for my Acct. in part payment of my large claim on Mr. Campbell and I expect £500 more due the 1st Augt. next on said acct. The former I have submitted to Mr. Tabbs order which reduced his demand under £2000 currency." James offered to execute a deed of trust using his property as security. He added, "There were Freshes in this Country continuing almost uninterrupted since March, have hindered my Forge and Mills from working, and hurt me much in my dependence thereon this Season, but I have great reason to be thankfull in suffering so little damage while others whose property joined the River are totally ruined for the Waters rose to an incredible height."

Rebecca's suit also claimed that James Hunter was indebted to her husband for £2924.12.7 for wares and merchandise due from March 1776. Prior to his death, Abner Vernon had promised to pay this debt but had never done so. The court ordered the Marshall of the Virginia District to take goods and chattels of Hunter's estate valued at $8,238.45 plus costs. This suit was continued until long after the deaths of Adam Hunter and Patrick Home. On June 17, 1810 the Marshall was ordered to take from Adam's estate the $16,000 still due Rebecca Backhouse. This was demanded of then administrator Joseph Ennever "but none shewn and I do certify that I can find no goods and chattels of the said defendant at the time of his death in the hands of his admr whereof I can make the within mentioned debt and costs or any part thereof" (U. S. Circuit Court Records, Backhouse's admx vs Hunter's exors, 1795).

It appears that James' executors tried to settled the accounts due Backhouse. On July 5, 1786 Adam Hunter wrote to Thomas and John Backhouse, Jr., "I did flatter myself the Estate would have produced a much larger crop last year, but bad weather at the time of cuting [sic] reduced it much and the whole crop was but 23 Hhds and of which 16 are to your address as above and the other 7 go by same Ship to Messrs. Caddis for Accts of Captn. Wards Claim...Be assured I labour to reduce the large Balance of £7146.3.5 with all possible expedition...I have ordered suits in some very considerable sums and others I have got well secured by Bond...I have an offer made for the claims my Brother had in Jamaica and believe I shall accept it and secure the payment in this State" (U. S. District Court Records, Hunter's exors vs Backhouse's admx, 1797).

James Hunter, Jr.'s letter to his uncle Archibald Hunter of June 7, 1787 also made reference to the settling of the forge estate. He wrote that numerous suits and mortgages "[fell] on the property in times so distressing as the present, it will blow up everything. The ill natures would say everything which would add to a man's distresses and [illegible]...not to report that the estate cannot pay off the claims. The present

---

1795), the daughter of William and Elizabeth (Bland) Beverley of Essex County. She married secondly Thomas Griffin Peachy.

Hunter's debt to Mills seems to have been a result of their partnership. Mills was also involved in another partnership styled Mills, [Simon] Frazer and [Overton] Cosby. In 1834 Overton, originally from Urbanna, died in the lunatic hospital in Williamsburg, having been kept there since 1805.

management is hurtful for the sale of the 100 negroes has answered no material purpose, the amount being frittered away by the old managers of the Forge and others, converting the sale to their own payment, when every farthing should have been applied to rub out the old heavy interest amounts. I was astonished to see Backhouses claim amounting to 7000 and odd pounds sterling, which I imagined was only £2000. Adam had remitted them some [illegible] last year, but it was only like a drop in the ocean" (Hunter-Garnett Papers).

A summation of Hunter's estate was provided in one of the British Mercantile Claims reports. The agent wrote, "Although the estate of James Hunter the Elder was very considerable, it must be admitted that the claims against it were also very great. The close of the administration can alone determine whether the estate at the peace was sufficient to discharge the debts then due from it. If we accept the purchase of that part of the property made by Mr. Home and the annual expenses of Mr. Hunter which are stated to have been great and take into view the debts due from the estate at the peace, it is as solvent now as it was at that period. It is the general opinion of those most conversant with the affairs of Mr. Hunter's estate that it will not be able to pay all the debts contracted by him before his death. Should it be determined that a davistavit has been committed by the executors, their securities are very able to answer for the maladministration of their principals" ("British Mercantile Claims," vol. 28, p. 223).

From 1802 through 1803 most issues of the local newspapers contained at least one notice for a sale of some of Hunter's property as Patrick gradually sold off portions of his uncle's estate. On May 13, 1802 he sold 114 acres to Isaac Burton (Fredericksburg Deed Book D, page 381). This land was on the north side of Banks' Road (State Route 654), slightly west of Berea Church (Fredericksburg Court records, Deed Book D, page 381). On Feb. 18, 1803 Richard Young Wigginton, Carter Beverley (1774-1844), and Samuel Slaughter[352] (1774-1846) commissioners, announced the sale of land lying near Culpeper Courthouse, "late the property of James Hunter...to satisfy a debt due to the Estate of James Mills" to whom the property had been mortgaged (*Virginia Herald*, Feb. 18, 1803). On Feb. 22 John Minor advertised the sale of "the Mansion House, tract of Land & Appurtenances, Late the property of James Hunter, esq...Also, of a separate Tract of Land, Adjoining the above...generally admired for their natural advantages and for the taste with which they have been improved" (*Virginia Herald*, Feb. 22, 1803). On July 15, 1803 the executors again advertised the Marsh tract in Fauquier half of which was described as being "capable of being made excellent meadow" and the remainder "well calculated for Tobacco and small grains" (*Virginia Herald*, July 15, 1803).

On Feb. 17, 1802 Patrick sold the 95-acre forge along with three tracts of 320 ac. (Fredericksburg Court Records, Deed Book D, page 508),[353] 150 ac., 348 ac., and 281 ac. (Fredericksburg Court Records, Deed Book D, page 504)[354] to Robert Dunbar[355] (c.1745-1831). Included with the forge tract were Hunter's merchant and gristmills,[356] both being operated at the time by a miller named Robert Swan. In 1810 the

[352] Samuel Slaughter married first Frances Banks and secondly Virginia Stannard.
[353] Dunbar paid $960 for 320 acres adjoining Liberty Hall.
[354] Dunbar paid £281 for this tract described as beginning at the mouth of Rocky Pen Run where it empties into the Rappahannock, running southeast with the river, corner to Banks, and back to Rocky Pen Run.
[355] Robert was a Scottish merchant of Falmouth. He married in 1797 Elizabeth Gregory Thornton (c.1764-1851), the daughter of Frances Thornton (1737-1794) and Ann Thompson of Fall Hill, Spotsylvania County. His obituary appeared in the *Virginia Herald* of Dec. 14, 1831 and read, "In Falmouth, on Monday morning last, Robert Dunbar, Esq., aged 86 years—for many years a respectable merchant in that place."
[356] Hunter's Forge Mill was leased to a series of different men, among whom were William Cooch and John Hollingsworth. In 1806 Patrick Home petitioned the court to stop a marshall's sale of the merchant mill "and a dwelling house called the blacksmiths house & two acres of land adjoining the dwelling house called the coopers house garden and shop and half the tan house & half the coal house, late the property of the said James Hunter, deceased. As a part of this suit, Cooch and Hollingsworth summoned to court Cadwallader Evans, administrator of Amos Strettle. They claimed that on Aug. 1, 1801 they had rented the property from Patrick Home for a term of nine years. They further claimed to have expended a considerable sum "in putting the said mill in complete order which is now very nearly complete." Amos Strettle's estate had brought suit against Hunter's estate for about $1,000 and Evans hoped to force the sale of the valuable mill tract in order to satisfy this debt. Cooch and Hollingsworth won their injunction advertised that they would "give a liberal Price in CASH for good WHEAT, deliverable at the Mill" (*Virginia Herald*, July 9, 1802).

forge mill was described in a Mutual Assurance Society insurance policy as a two-story stone building 68 feet long by 36 feet wide, with a wooden roof. One of the larger mills in Falmouth, it was insured that year for $10,000 (Mutual Assurance Society Records, R5/V47/376).

Hunter's executors periodically advertised the sale of small groups of slaves. One such notice from 1803 read:

> To be Sold, on Thursday the 16th day of June next; at Rappahannock Forge, adjoining the town of Falmouth, and about a mile from hence, all the NEGROES, Belonging to the Estate of the late James Hunter, dec. Fifty or sixty in number, Men, Women, and Children; amongst whom are some valuable Tradesmen, such as Blacksmiths, Nailors, Forgemen, House Carpenters, and a ship Carpenter. There will also be sold, some household Furniture, Plantation utensils, a Waggon and team, some Horses and Cattle, and a quantity of blacksmith's Tools. The terms of sale are credit till the first of January next; Bond and approved security being given, to bear interest from the date if not punctually paid [sic] to Robert Patton, Administrator.
>
> Fredericksburg, May 11, 1803
> (*Virginia Herald*, May 24, 1803)

Robert Swan was a busy man but left little personal information in the public records. On Feb. 1, 1805 he purchased Brent's Mill in what is now Widewater. While there he joined in partnership with Daniel Carroll Brent (1759-1815) to operate the mill and drew an annual salary of $300 (Brent's Mill Ledgers, 1804-1806).

On June 1, 1808 Robert Dunbar made a new lease with Swan for several of the forge businesses. This was a five-year lease on buildings which included "the Manufacturing Mill, Corn Mill, Saw Mill, dwelling House now occupied by the said Swan, Kitchen, Meat House, Garden, Counting House, Cooper's Shop lately occupied by Frederick Pilcher, half the Tan House, the House lately occupied by Capt. Irvin as a School House and the low ground for the cultivation of small Grain and Clover, which was enclosed and worked by Cooch and Hollingsworth[357] supposed to contain about fifteen acres." The annual rent for this property was $900 "but in case of War between the United States and Great Britain and during such time as it shall exist, the rent to be at the rate of Six hundred instead of Nine hundred dollars." Swan agreed to annually saw for Dunbar "from fifteen to twenty thousand feet of plank and scantling at one shilling and six pence per hundred superficial feet." Swan further agreed not to "obstruct the free and certain passage of the water from the Forge race into the North fork of the River except when repairing the dam or cleaning out and repairing the race...that he will not use the corn kiln but let it be taken down."

Dunbar and Swan ended up in court as a result of a dispute regarding this lease. Shortly after taking over the property, Swan discovered that a pair of millstones needed to be replaced. He asked Dunbar to order a new set which the latter promised to do, but didn't. Swan eventually offered to order them himself if Dunbar would agree to pay for them, to which the latter agreed, but never produced the money. Severe flooding in 1812 left the millrace full of rocks and debris and Swan had it cleaned out at his own expense. As he considered both of these issues of maintenance for which Dunbar should have been responsible, the dispute went to court (Fredericksburg Court Records, Dunbar vs Swan).

For a number of years, Dunbar was one of the wealthiest men in Falmouth. By means of his marriage, he gained control of the Falmouth Bridge previously built by his father-in-law, Francis Thornton. This bridge crossed the river very near Payne's Cottage (also known as the sluice-gate cottage). Economic conditions in Europe were creating a tremendous demand for American flour. Dunbar, and many others like him, used easy credit to buy and build more mills to meet a seemingly insatiable need for flour exports.

During the early 19th century Cooch and Hollingsworth also operated a grist mill in present-day Old Mill Park on the Fredericksburg side of the Rappahannock River

[357] William Cooch and John Hollingsworth came to the Stafford/Fredericksburg area from Cecil County, Maryland. William was the son of Francis Lowen Cooch (died 1854) and Elizabeth Maris (1776-1840) of Philadelphia. He married Margaret Hollingwsorth, sister of John Hollingsworth of Cecil County. The Cooch family were long associated with milling, having built the first cotton mill in Cecil County c.1775 (*Cecil County Democrat*).

The situation, however, was too good to last and in December 1807 the flour business crashed.[358] Relations between Britain and France had so deteriorated by this time, that President Thomas Jefferson recommended that Congress impose a shipping embargo on goods bound for Europe. This poorly conceived plan devastated the nation's economy and, although it was repealed in March 1809, irreparable harm had been done to many businessmen, including Dunbar and John Strode in Culpeper.

In an effort to generate more income, Dunbar bought William Richards' Contest Mill and the Falmouth Canal that powered it. This was Dunbar's fourth mill in Falmouth and the purchase did little but drive him further into debt. In an effort to consolidate his largest debts, Robert and his wife Elizabeth mortgaged their property to settle their affairs (and a debt of $15,203.30) with the Bank of Virginia. Among these properties were "a tract of land in the county of Stafford on which the Forge, the Manufacturing Mills, the Gristmills, Saw Mill and many other improvements are erected; containing by estimation ninety five acres...together with the said Forge, Mills, improvements and all the rights of water and fixtures appurtaining to the same" (Stafford Deed Book AA, p. 79). Dunbar's holdings also included 514 acres adjoining the forge, the Contest Mill with its miller's house, granary, and 40 acres through which ran the canal from the Falmouth Mill Pond to Contest Mill. As a result of the economic collapse created by the shipping embargo, Robert was financially ruined. By Christmas 1822 his land and even his personal property had been auctioned to meet the demands of his creditors.

Sometime in 1802 court-appointed commissioners Robert Patton and Gen. John Minor[359] (1761-1816) assumed the management of James' estate. Assisting them was a long-time forge employee, Joseph Ennever. Joseph's father, also named Joseph, had worked for Charles Carter but stayed on at Stanstead after Charles and his family removed to Cleve. The court records of December 1802 state that Joseph was paid £385.10.3 for his services as a clerk of James Hunter's estate. These records noted that "The Estate of James Hunter has been represented to your Commissioner as being of that description which actually requires the aid of a Clerk of Superintendant [sic], in addition to all the Exertions of the administrator, which made it necessary for the administrator to retain, for that purpose, Joseph Ennever, who had been for many years employed by the Hunters as a Clerk, and was the only person to be found, who knew any thing

---

[358] Desperate warfare between England and France was revived during Jefferson's second administration. Both countries denied America's claim to neutral rights on the high seas and both seized American ships and cargoes. The British navy also claimed an alleged "right of search" of captured ships. In June 1807, within sight of Cape Henry, a new navy frigate named *Chesapeake* sailed through the Virginia narrows for a trial run. The British warship *Leopard* intercepted the *Chesapeake* and ordered the latter to prepare to receive a boarding party. The captain refused and the *Leopard* opened fire, killing three of *Chesapeake's* crew and wounding 18 others. The captain of the *Chesapeake* managed to return fire before surrendering. Virginians flared in anger and Patrick Henry suggested an invasion of Canada in retaliation. Jefferson decided, instead, to withhold goods from England and France, believing that American commerce was of such value that these two countries would recognize U. S. neutrality in order to keep it. In late 1807 Congress approved an act prohibiting any merchant ships, except those involved in local coastal trade, from leaving any American harbor. Deprived of expanding overseas markets for the tobacco, flour, and other products that had brought relative prosperity during the previous six years, Virginians joined with New Englanders who bitterly condemned the embargo.

Although Jefferson didn't believe that his embargo had been tried long enough for its impact to be fully appreciated by the combatants, he wasn't deaf to those who argued that the U. S. needed foreign trade more than the belligerents needed American trade. In March 1809 he joined Congress in repealing the embargo.

[359] Gen. John Minor was born at Topping Castle, Caroline County. He was the third child of Maj. John Minor (1735-1800) and Elizabeth Cosby. During the Revolution, John was present at the siege of Yorktown. He studied law under George Wythe and practiced law in Fredericksburg. During the War of 1812 John was stationed in and around the city of Norfolk. In 1812 Benjamin Henry Latrobe enlisted his help in promoting the Potomac Steamboat Company, this venture sponsored by Latrobe and Robert Fulton.

John married first in 1790 Mary Berkeley (died 1790) the daughter of Landon Carter Berkeley of Hanover County. He married secondly in 1793 Lucy Landon Carter (1776-1859), the daughter of Landon Carter (1751-1811) of Cleve and Mildred Washington Willis (died 1778). John resided at Hazel Hill on the opposite side of the Rappahannock River from Ferry Farm.

relative to the Business of the Estate." Joseph was paid £75 per annum to June 30, 1810 it "being the same salary which he had been allowed by Mr. Hunter, and the former administrator."

Patrick Home died, unmarried, on April 28, 1803 and John England (1755-1851), Joseph Fant[360] (1738-1812), and James Templeman inventoried his estate. This inventory consisted of 5 slaves, 1 mahogany bureau, musical instruments, "a fowling piece, a pair of pistols," books on religion, Scottish history, law, physiology, world history, travel, dictionaries, Greek grammar, poetry, farming, books written in Latin, French, and Greek, a pair of silver shoe buckles, a pair of silver knee buckles, a great coat, 2 pairs of breeches, drawers and waist coat, 2 pairs silk hose, 1 pair of worsted hose, 1 pair of boots, 2 waist coats, and 4 cattle. What became of the Rappahannock Forge records, which should at that time have been in his possession, is unknown. The *Virginia Herald* of May 24 1803 carried a notice announcing that Joseph Ennever had been appointed "to settle and collect the debts due on the Books of the late James Hunter, dec. and those of his late Executor, Patrick Home."[361] Patrick's slaves were sold on June 13, 1803.

Hannah Hunter, Patrick's aunt, brought several suits against her nephew's estate. One of these resulted in the sale of his real estate, most of which was in Tennessee. Alexander Vass, administrator of the estate, claimed that Patrick left little personal estate and only a limited quantity of real property. On Aug. 25, 1804 Vass auctioned 2,820 acres in Tennessee, 540 acres in Spotsylvania, and 96 acres in Stafford. Of the Tennessee property, one tract of 1,300 acres brought only $13.00 (Fredericksburg Court Records, Hunter's admx vs Patrick Homes' admr).[362]

In June 1802, about a year prior to Patrick's death, the Federal Court of Virginia assumed administration of Hunter's estate. This court ordered the sale of James Hunter's personal estate (see Appendix C) which was carried out at a three-day public auction. The following year the court ordered the sale of his remaining real estate. A notice was placed in the local newspaper:

> According to a Decree of the Federal Court Richmond, will be sold, on the Premises, on Thursday, the 15th of next December, All the remaining Lands, belonging to the Estate of James Hunter, deceased, in Stafford county, adjoining the Forge. The quantity, about 4500 acres, will be laid off in Lots pervious to the day of sale, and the whole sold on credit of twelve months, the purchasers giving bond with approved security, to bear interest from the date if not punctually paid. A Title will be made when the money is paid.
>
> The Commissioners
>
> (*Virginia Herald*, Nov. 11, 1803)

The court ordered the property surveyed by John Minor Herndon[363] (1768-1829) and named Robert Patton and John Minor commissioners to conduct the sale. The courts normally became involved in settling estates for one of two reasons. If the deceased was greatly in debt, the court assumed administration of the estate and sold the personal and real property to pay creditors. The courts would also take over an estate if there were no executors or heirs to handle the affairs. In the case of James Hunter, the federal court became involved in settling his estate because some of his creditors resided outside the United States. All three situations were applicable to Hunter's estate. Settlement of his estate outlived all his administrators. James Hunter's two younger cousins, William and James Hunter, were the only other close relatives living in America. William removed to Essex County where he died c.1787. Young James ended up in the Carolinas and died there in 1788. Thus, there was no alternative but for the courts to assume administration of the estate in order to finally settle it.

---

360 Joseph was the son of William and Catherine Fant of Stafford. He married first Sarah Penn(?) and secondly, in 1784, Hannah Lewis.

361 The Nov. 8, 1803 issue of the *Virginia Herald* carried a notice for the sale of the "land which formerly belonged to the late Henry Mitchell, and now to the Heirs of Patrick Home, dec." This land was described as being on the Rappahannock River about three miles above Fredericksburg. The tract contained 537 ½ acres was at that time occupied by Thomas Moffit.

362 The total earned from this sale was $3,602.67. This suit mentions Patrick's brothers James and John and his sister Helen, the wife of George Logan, Jr.

363 John Minor Herndon was the son of Joseph Herndon (1737-1810) and Mary Minor (1741-1822). He was a distinguished lawyer of Spotsylvania and, later, of Louisa. He died unmarried.

Although most of the deeds for this period are missing from the Stafford County Courthouse, a notation of the 1803 division and sale of the property appears in the list of alienations (changes) at the end of the 1804 land tax records as well as in the Fredericksburg Circuit Court records.

On Sept. 5 and Dec. 15, 1803 Hunter's land was sold at auction, total sales amounting to £5,309.12. Buyers and acreage were:

Robert Lewis[364] (1769-1829)—350 ac.
James Donavan—292 ac. (part of Baxter tract)
Joseph Ennever[368]—Stanstead house and 295 ac.
Samuel Marquess[370] (c.1780-c.1850)—291 ac.[371]
William Richards (1765-after 1803) --23 ac.
Elijah Marquess[373] (died 1815)—159 ac.
John Wallace[376] (1761-1829)—234 ac.
Capt. John Seddon (1780-1808)—286 ac.[377]
(part of the Baxter tract)
William Hore—151 ½ ac.
John Swetnam[365]—242 ac.[366]
Edward Hore's Estate[367]—223 ac.
Michael Wallace[369] (born 1753)--164 ac.
Alexander Vass (died 1814) --175 ac.[372]
Lot # 76 Falmouth
Meshack Massey[374]—156 ac.[375]
Thomas Collins—271 ac. (part Stanstead)
Robert H. Hooe (1748-1834)—133 ac.[378]
N[eri?] Swetnam[379]—228 ac.
Anthony Marquis[380] (1752-1821)—103 ac.

---

[364] Robert was a son of Fielding Lewis of Kenmore.

[365] This tract was not deeded until Oct. 8, 1807 and the deed remains on file in the Fredericksburg Circuit Court records, Deed Book F, page 140. The land was on the south side of Route 17, about ½ mile east of Berea Church.

[366] This property was deeded to John Swetnam on Oct. 7, 1812. The 242 acres adjoined Greenbank and John paid Hunter's estate £477.19 for it (Stafford Deed Book AA, p. 345).

[367] According to the 1803 land tax records, Anne and William Hore paid taxes on this parcel that year.

[368] In 1806 Joseph Enniver married Fanny Potts in in King George County.

[369] Maj. Michael Wallace, the son of Dr. Michael Wallace (1719-1767) of Ellerslie. Maj. Wallace married Lettice Smith Wishart (born c.1750).

[370] Samuel was the son of Anthony Marquess (1752-1821). In 1810 and 1820 Samuel was living in Spotsylvania County.

[371] On Mar. 14, 1811 Robert Patton and John Minor, commissioners appointed to sell Hunter's remaining real estate, deeded this 291 acres to Samuel (Stafford Deed Book AA, p. 208).

[372] On May 2, 1827 Robert Patton, surviving commissioner and William Marquess deeded this 175 acres to Isaac Burton of Stafford. According to the deed, Vass, "now deceased," purchased the property at the 1803 auction of the Hunter property. In 1805 he conveyed the tract to Elijah Marquess who sold it to his father Anthony Marquess. Anthony sold it in 1813 to his son William (Stafford Deed Book GG, p. 402).

[373] Elijah was the son of Anthony Marquess (1752-1821).

[374] Meshack married Agatha White.

[375] This property was deeded to Meshack on June 13, 1810. He had paid £280.12 for 156 acres that was described as beginning at the mouth of Rocky Pen Run and lying on the east side of the run (Stafford Deed Book AA, p. 110).

[376] John was the son of Dr. Michael Wallace (1719-1767) and Elizabeth Brown (born 1723) and lived at Liberty Hall, part of the Ellerslie tract. John married Elizabeth Hooe (c.1766-1850) and this 234 acres adjoined Liberty Hall.

[377] On Mar. 13, 1809 Robert Patton and John Minor deeded 385 acres "belonging to the estate of James Hunter, deceased" to John Seddon. He paid $481.25 for the property (Stafford Deed Book AA, p. 3).

[378] Adjoined Harwood Branch, the site of the Days Inn Motel.

[379] In 1803 Neri Sweatnam (1777-1862) of Culpeper County, but then a resident of King George, married Mildred Cross (1778-1860), also of Culpeper.

[380] This tract had actually been sold to Anthony by Patrick Home on Apr. 15, 1803, just two weeks before his death. Anthony paid £133 for land on the north side of "the main mountain road" on a branch of Fall Run (Fredericksburg Court Records, Deed Book D, page 471).

Anthony was probably the son of William Marquess (died c.1790) for whom he was named executor. Anthony married Elizabeth Winlock. During the first half of this century, his gravestone was located ½ mile above the old Berea Post Office at the home of Mrs. W. A. Bullock, though she remembered as a child seeing it in place on a nearby farm. The stone reads, "In Memory of Anthony Marquess—Born

Isaac Burton—114 ½ ac. Robert Dunbar—150 ac. (Potomac Run)
Anthony Marquis—50 ac.[381]
Fielding Alexander[382] (died 1847)—235 ac. (part of Baxter tract)
Thomas and John Strode purchased 1,100 acres "on the great marsh in Fauquier" for £1,650[383]

Not included on this list was the 200 acres at Accokeek, which also appears to have been sold at this time. A later deed for this tract dated June __, 1826 reveals that Francis Foushee had conveyed it to his son, Dr. William Foushee (1749-1824), on Apr. 1, 1804 (Stafford Deed Book GG, page 82 and Deed Book GG, page 186). Another tract not listed above was comprised of 104 acres purchased by Lawrence Sanford who paid Hunter's estate £130 for the property. This land was described as situated primarily on the west side of Rocky Pen Run in the corner formed by Licking and Rocky Pen runs (Stafford Deed Book AA, p. 206).

---

the 10th July 1752—Departed this life Decr—15th 1821, Aged 69 years—5 months and 5 days—Remember man as you pass by—As you are now so once was I—As I am so you must be—Prepare for death & follow me." Mrs. Bullock died many years ago and the farm changed hands. The author visited the old Bullock farm but was unable to locate the stone.

[381] On Jan. 11, 1805 Joseph Ennever placed an advertisement in the *Virginia Herald* announcing his intention of selling 185 acres adjoining the lands of John Wallace and Anthony Marquis. The tract was described as being "formerly a part of the Forge Woodcutting, but at this time [is] covered with a thriving young growth of timber." This is of interest because from it may be assumed that Hunter cut wood not only to the north of the forge, but to the west as well.

[382] In 1804 Fielding was paid £46.6 "it being the amount of his claim as an Overseer of Hunter." From 1801-1803, Fielding was employed as overseer of Stanstead and Rocky Pen Plantation. Based upon the land tax records, Fielding seems to have resided on the 235 acre Baxter tract. By 1817 the farm included buildings assessed at a modest $150. In 1842 Fielding's heir, Thomas T. Alexander (c.1816-1850) conveyed the property to John M. O'Bannon (1800-1870) of Falmouth. Two years later, John conveyed the property to John McDermot of Washington who sold it in 1845 to William Bloxton (born c.1804). The Bloxton family held the property until well into the 20th century. This tract is on the northeast side of Abel Reservoir between Kellogg Mill Road (State Route 651) and Mountain View Road (State Route 627).

[383] The deed for this conveyance is contained in the Fauquier County Circuit Court Records, Chancery Papers, Dec. 17, 1803—1803-018. Hunter's Marsh Tract was in the southeastern part of Fauquier not far from that county's border with Stafford. In a file of unrecorded deeds in Fauquier County is a deed dated Dec. 17, 1803 in which John and Thomas Strode purchased this tract for $5 per acre. It was described as 1,100 acres bounding on Great Marsh Run and the lands of Nicholas Wycoff, dec., Mrs. John Quisenberry, Robert Beverley, William Skinner, and Thomas Allen. While not deeded until December of that year, the Strodes had actually purchased it on Sept. 5. John and Thomas executed a deed of trust with Robert Patton, the $5,500 to be paid on or before Sept. 5, 1804.

John Strode later brought suit against Robert Patton, one of the commissioners responsible for selling Hunter's real estate in 1803, claiming that he had discovered the tract did not contain 1,100 acres. He explained his reasons for purchasing the property:

> Your Orator John holding (on account) a considerable Claim for himself and for Abner Vernon decd. for whom he is administrator and having long and for many years laid out of those just Claims advised his son Thomas your other Complainant with a view of obtaining a part of his long with held Claims to attend at the said Sale and become a purchaser and he should have one undivided equal share." They bought the land, but after taking possession, cultivating it and building improvements, they had it surveyed and discovered that 420 acres of the tract earlier granted to Reuben Wright who sold it to James Hunter, actually belonged to someone else, leaving John and Thomas with only 624 acres. They then discovered that the Allen tract was actually 50 acres smaller than had been previously assumed (U. S. Circuit Court Records, Strode vs Patton and als, 1832).

Another parcel of the Hunter land not included on the alienation list was 231 acres purchased by Charles Withers (Stafford Deed Book AA, p. 296). In this instrument Charles Withers sold the 231 acres to James Templeman for $1,000. Although he had purchased the property from commissioners Robert Patton and John Minor, the deed was not deeded until Oct. 12, 1807. According to the meets and bounds given in the deed, this land began at a fork of Banks' Road and Mountain Road, corner to Joseph Fant, Anthony Marquis, and Isaac Burton.

The commissioners also had to deal with a number of long-standing debts against the Hunter estate. On Mar. 8, 1805 Robert Patton, Commissioner "placed in the hands of Gen. John Minor, for collection, sundry bonds and notes, amounting in principal to the sum of £4600.8.1 plus interest totaling £5051.2.10." Two years later, William Fitzhugh[384] of Chatham instituted a suit against Joseph Ennever, administrator of Adam Hunter. This was a long-standing complaint that had involved some tobacco sold by Fitzhugh to Hunters & Taliaferro. According to William Fitzhugh, the store owed him £130 plus interest since Jan. 1, 1774. In 1793 William and Adam had agreed to put the dispute in the hands of arbitrators and to abide by their decision. The arbitrators found in favor of Fitzhugh, but Adam never paid the debt (Fredericksburg Court Records, Fitzhugh vs Hunter's Admr). Based upon copies of ledgers included in the Fredericksburg records, Hunters & Taliaferro was a dry goods store that sold such items as shoes, combs, cotton, "hard mettel Tankard[s]," candle boxes, pearl barley, tea, butter, rum, sugar, molasses, window glass, playing cards, broad cloth, Irish linen, silk, thread, bobbins, buttons, osnabrig, buckram, broad axes, foot adz, grindstones, nails, saws, hinges, scythes, files, salt peter, tea kettles, dutch ovens, frying pans, candle sticks, tea cups and saucers, wool cards, twine, rope, locks, paper, paint, lead, veregrease,[385] linseed oil, sweet oyl,[386] soap, hats, flint tumblers, pencils, and augers.

Another case involving Hunter's estate was heard in the Superior Court of Chancery in Richmond on Feb. 14, 1810. In the case of Robert Dunbar adr of Alexander Vass (died 1814), dec. vs. Robert Patton, the plaintiff complained of a debt of £51.15.0 owed to Vass by the estate of James Hunter "for inoculating 69 slaves and carrying them through the Small Pox." This was "objected to by the Representatives of Hunter." Dunbar also claimed that the Vass estate was due £60 "for attending Hunter's slaves at Rocky Pen Plantation [which was] objected to by Hunter's Representatives." According to the suit, on May 20, 1793 Dr. Vass (of Culpeper) and Patrick Home had entered into a written agreement whereby Vass was "engaged to attend, prescribe for, and give medication to all the Negroes belonging to the Estate of James Hunter dec, at Rappahannock Forge, Stanstead Plantation and its vicinity, also all the Negroes at the Mansion House Plantation adjoining and to treat them agreeable to the best of his skill and ability, for the sum of thirty six pounds per annum." The case says that by affidavit of Thomas Withers in 1796 Patrick "settled or established a Plantation called Rocky Pen, and fixed a number of Hunter's Negroes there." Vass provided medical care for those slaves for eight years. Joseph Ennever also swore to Patrick Home's establishment of Rocky Pen Plantation, saying, "the settlement of the Quarter was made by a part of the slaves from the Mansion House Quarter, which was at that time broke up, and the remainder of the slaves from that place were sent to the Estate's quarters in Culpeper and at the marsh in Fauquier, where the doctor was not to attend in the agreement" (Fredericksburg Court Records, Vass vs Patton Admr).

By the time commissioners Minor and Patton received the Hunter estate, Patrick Home had sold much of James' personal property and some of the real estate; yet there remained many outstanding debts, some of which had outlived Hunter's creditors. Only a handful of official records pertaining to the final settlement have been found, making it difficult to reconstruct the final disposition. From a variety of local

---

[384] William Fitzhugh (1741-1809) was born at Eagle's Nest in King George County and was the only son of Col. Henry Fitzhugh (1706-1742) and Lucy Carter (1715-1763). From 1772-1775 William was a burgess from King George. He was a member of the Constitutional Conventions of 1775 and 1776, member of the Continental Congress 1779-1780, served in the Virginia House of Delegates 1780-1781 and 1787-1788. William was also a state senator 1781-1785. He married Anne Randolph (1747-1805), the daughter of Peter Randolph and Lucy Bolling of Chatsworth, Henrico County. William Fitzhugh built Chatham where he lived until 1799 when he removed to Ravensworth in Fairfax County. While a resident of Stafford he was known as a breeder and racer of fine Thoroughbred horses.

[385] Probably verdigris, a disacetate of copper. In an impure state it was used as a green pigment. In a pure state it was a remedy for dizziness.

[386] Sweet oil was a medicine used to treat earaches.

sources such as newspapers and court records, it is possible to track the forge property after it left the estate. On June 1, 1814 Robert Dunbar and his wife sold the 95-acre forge tract and an adjoining 328 acres to Thomas Williamson[387] (1777-1846) and Miles King, Jr.[388] (1786-1849) of Norfolk (Dunbar, Robert to Thomas Williamson). This included the mill, mill seat, and houses for which Williamson and King paid $3,000. Dunbar applied this to his mortgage with the Bank of Virginia (Fredericksburg Court records Bank of Virginia vs. Dunbar). On Aug. 22, 1815 Williamson paid Dunbar $181 for an additional 11 acres "contiguous and binding upon the Forge Tract lately sold and conveyed to Messrs. Williamson and King" (Fredericksburg Court records Bank of Virginia vs. Dunbar). This land was on the east side of the forge. Williamson agreed not to build a bridge or road from any part of the land across the north branch of the river to Vicaris' Island. This second deed was lost from the Stafford County records, but a copy remains in the Virginia Historical Society among Robert Dunbar's papers.

In 1822 George Newton[389] of Norfolk paid the land taxes on the 353-acre forge tract. A margin notation said that the land had come "from Williamson." George was listed in the tax books as owning the property in fee simple. There is no surviving deed for a conveyance between Williamson and Newton either in Stafford or Fredericksburg and, based upon later records, it would appear that he had been a part owner in the property from the time Williamson and King had purchased the tract from Dunbar. From 1823-1835 George and William Newton paid the taxes on the 353 acres plus an adjoining 36 ½ acres. From 1818-1835 the building assessment remained stable at $2,000. This was a substantial assessment and it seems reasonable to assume that some industry continued at Rappahannock during this time.

At this time the forge merchant mill was rented out to Messrs. Brooke and Ficklen who placed a notice in the local paper:

> Forge Mills
>
> The subscribers having rented the Forge Property for a term of years, bet leave to inform the public, that they will continue the Grinding of Corn, and will assure every dispatch to those favoring them with their custom, as well as having the work executed in the best manner. They will supply retailers of meal, at all times, on good terms.
>
> Brooke & Ficklen,
>
> Who Have for Sale,
>
> A general assortment of Staple Goods, & Groceries, with Family Flour, of superior quality.
>
> Cash given for Wheat.
> Falmouth, Oct. 26.
> (*Virginia Herald*, Oct. 26, 1822)

The fate of John Strode's massive armory remains a mystery. It was included on the 1809 Adjutant General's report, but this may be a result of clerks simply re-copying the previous year's data (year after year). Its inclusion on that report does not necessarily mean that the structure was still standing at that time. The building's presence on the Adjutant General's reports implies that it had been paid for with public funds and it would, therefore, have been included on state inventories from the time of its completion. Few of the Adjutant General's reports compiled prior to the War Between the States survive,

---

387 Thomas was the son of Thomas Williamson II of Norfolk and the grandson of Thomas Williamson (born 1708). Thomas married first in Richmond in 1800 Elizabeth Galt (died 1807). He married secondly in 1809 Anne McC. McWalke. Until his death he was cashier of the Virginia Bank in Norfolk and was mayor of that city in 1829.

388 Miles was the son of Miles King, Sr. (1747-1814) and his second wife Martha Kirby (1765-1849) of Hampton and, later, Norfolk. Miles, Sr. was four times mayor of Norfolk and was an officer and surgeon's mate during the Revolution. Miles, Jr. graduated from William and Mary College, was a member of the House of Delegates in 1815 and other years, and mayor of Norfolk in 1833. He was also navy agent for Norfolk and captain of the Norfolk Light Artillery Blues. He married first in 1805 Rebecca Calvert (1788-1827) and secondly in 1833 Mary L. Fisher.

389 Probably George Newton (1786-1835), the son of Col. Thomas Newton (1742-1807) of Norfolk. George married Courtney Tucker, the daughter of Daniel Norton of Norfolk.

making it impossible to track the building on those documents. According to Gordon Barlow, numerous armories were built using public funds and in these situations the state held title to the property. While some of these titles have survived, a title for Rappahannock Forge is not among them. Mr. Barlow notes that by about 1800 the state, no longer in need of these buildings, simply abandoned them or allowed the titles to revert to the former owners. By that time, guns were being manufactured in centralized armories that were more cost effective and efficient to operate. It is possible that Strode's armory was still standing in 1814 when Williamson and King purchased the forge tract from Robert Dunbar. The land tax records don't provide a separate column for building assessments until 1817. At that time, the buildings on the forge tract were valued at $3,000. A year later, that had decreased to $2,000. Thomas Williamson and Miles King seem to have purchased the property for speculation and may have leased it to someone who operated the mills or had some other business there. A search of the Fredericksburg newspapers from 1817-1850 failed to reveal any advertisements submitted by Williamson, King, or Newton.

Exactly when Rappahannock ceased making iron is unknown, though the furnace probably continued in use for many years after the Revolution. Iron products were an essential to everyone in the community and the closing of one facility would likely have spawned the building of another. That may have happened in the case of Rappahannock Forge and the Fredericksburg area. In August 1830 Daniel Davis of Baltimore placed a notice in the local paper:

> I shall put in blast on my lot, situated on the east side of the Tobacco Warehouse, in the Town of Fredericksburg, Va. a FURNACE, for casting Iron and Brass in all diversified forms which may be demanded.
>
> Those who may be disposed to patronize the Furnace, will please send on their patterns. If patterns should be required of me; persons so requiring, must address their descriptions to me in Baltimore, where I shall procure them upon the lowest terms, and remit them to Fredericksburg. Castings will be done with great neatness and dispatch, and upon low terms.
>
> (*Fredericksburg Political Arena*, Aug. 6, 1830)

By Jan. 14, 1831 Davis' furnace had been taken over by James Peyton and John Clark[390] (1804-1882), millwright and Baptist minister. They offered to make "Mill Gearing and all kinds of Machinery [and] Agricultural Implements." Clark, who had already been engaged in the making of wheat machines added a postscript to the notice stating that that business would be "carried on as usual, or probably more extensively" (*Fredericksburg Political Arena*, Jan. 14, 1831). Unfortunately, the 1830s to 1850s are nearly bereft of records pertaining to Rappahannock Forge, Robert Dunbar, or Daniel Davis. With the exception of a handful of newspaper notices and court records, there is not enough extant material to make any definitive statements regarding the forge or the furnace in Fredericksburg. It may or may not be a coincidence that Davis' forge opened during a period of financial collapse at Rappahannock. The relationship, or lack thereof, between the two facilities is simply not clear.

The economic depression of the early 1830s impacted businesses and individuals. For years business at Rappahannock had been gradually waning as the facility aged and the overall economy stalled. Robert Dunbar overspeculated in the flour and milling business and lost all he had. The extent of activities at Rappahannock during Dunbar's tenure is unknown but was probably limited as his bankruptcy followed closely behind his acquisition of the property. If the lack of newspaper advertisements is any indication, Newton and King seem to have done little with the property. Miles King bankrupted and his estate was sold, the forge tract being purchased and held briefly by his son.

The Stafford land records include a deed of Sept. 24, 1835 that reveals some of the history of the forge during this period. William B. Lamb, sergeant of the Borough of Norfolk, sold King's property which "was taken in execution by the said William B. Lamb sergeant as aforesaid on the 29th day of July one thousand eight hundred & thirty five by virtue of two capiases ad satisfaciendum issued from the Clerk's office of the circuit superior Court of Law & Chancery for the Borough to satisfy two judgments rendered in

[390] John Clark became minister at White Oak Baptist Church in 1830 and preached there for most of the remainder of his life. He also preached at Chopawamsic Baptist Church in the north end of the county and when the 1850 census was taken, resided near that church. His son, John W. Clark (born c. 1834) was a miller.

the said Court at the June term next preceeding the one at the suit of Thomas Williamson for the benefit of the said President Directors and Company of the Bank of Virginia against the said Miles King and the other at the suit of the President Directors and Company of the Bank of Virginia against the said Miles King and John Hodges, and the said Miles King being so in Custody was discharged thereupon in pursuance of the several acts of the General Assembly of Virginia concerning execution and for the relief of insolvent debtors, he having subscribed and delivered in a schedule of his estate, which said execution, with the schedule aforesaid, were duly returned to the clerk's office aforesaid." Lamb sold Miles' interest in his real and personal estate, the Bank of Virginia purchasing that interest for a mere $15.

George Newton died in 1835 and his estate was settled by court-appointed commissioner Tazewell Taylor (1810-1875). The following year Taylor conveyed the forge tract to John Moncure (1793-1876), Joseph Burwell Ficklin, Sr. (1800-1874), and Walker Peyton Conway (1805-1884). This was described as being "the same which the said Miles King acquired title to by deed from Robert Dunbar and Elizabeth his wife, dated the 1st day of June 1814, to Thomas Williamson and Miles King—and by deed from the said Thomas Williamson and Ann his wife to the said Miles King dated the 1st day of December 1819...and which said premises became vested in the Sergeant of Norfolk Borough, upon the said Miles King taking the oath of insolvency and then the said Sergeant sold, and all the interest and estate of the said Miles King being therein conveyed to the President, Directors, and Company of the Bank of Virginia, by his deed dated the 24th day of Sept. 1835" (Stafford Deed Book LL, p. 96). On Aug. 19, 1836, just prior to the sale, Tazewell Taylor placed an advertisement in the *Fredericksburg Political Arena* that read:

> TO MANUFACTURERS!--VALUABLE WATER POWER AND REAL ESTATE—FOR SALE.
>
> By virtue of a Decree of the Circuit Superior Court of Law and Chancery, for the Borough of Norfolk, pronounced on the 24th day of June, 1836, in a suit depending in the said Court, in which Williams[son] and Wife were plaintiffs, and George Newton's administrix and heirs were defendants—the undersigned, acting as Commissioner, will expose to sale at Public Auction, to the highest bidder, for Cash, on Friday the 28th day of October next, before the Tavern door of Major James Corbin, in the town of Falmouth, opposite to Fredericksburg, Va., at 12 o'clock,
>
> *One undivided half of the long established Mills called the* FORGE MILLS, situated about one mile above the Town of Falmouth, on the Rappahannock River, and of 73 acres of land, attached to the Mills. The other half of these Mills belongs to Miles King, Esq. of Norfolk, who will no doubt sell his interest on advantageous terms—or, if not, a partition of the Land, and right of water, would be decreed by a Court of equity as a matter of right.
>
> The great and valuable rights of water that this property possesses, (commanding at low water almost the whole of the Rappahannock River,) render it an object worthy of the attention of Capitalists who may be disposed to enter into any description of manufacture. At one period, very extensive Iron-Works were conducted at this place, and afterwards a large Flour-Mill; but, from long neglect, the Mills and Buildings have been suffered to decay. There being no Cotton Factory in this neighborhood, no situation in Virginia is better adapted for a profitable investment of this sort. The Land is of ordinary quality.
>
> At the same time and place, I will sell a farm of 350 acres, immediately adjoining the Mills.
>
> J. B. Ficklen, Esq. of Falmouth, will shew the Mills and Farm to persons inclined to purchase, and also give any information that may be desired. Refer also to R. B. Maury, Esq.[391] of Fredericksburg.

---

[91] Richard Brooke Maury (1794-1840) was the son of Fontaine Maury (1761-1824) and Elizabeth Brooke died 1800) of Spotsylvania. Richard served as President James Monroe's private secretary, was the first lerk of the Navy Department, and served in the War of 1812. In 1824 he married Lucy P. Hunton of .ancaster County. He married secondly Ellen Magruder (1798-1879) of Upper Marlborough, Maryland.

Tazewell Taylor, Comm'r.
Norfolk, August 15, 1836.

According to a margin notation in the 1836 land tax records, John Moncure immediately sold his interest in the property to Ficklin. At this point, the building assessment dropped from $2,000 to $0, suggesting that there had been a fire.

On Apr. 24, 1837 the Bank of Virginia deeded to Newton Calvert King[392] of Norfolk all right, title, and interest to the real, personal and mixed estate of Miles King for which Newton paid the bank $4,000. The court awarded the following judgments to the Bank of Virginia:

--one against Miles King and John Hodges for $3,700 and another for $3,412

--one in favor of Thomas Williamson suing for the benefit of the Bank of Virginia as plaintiffs for $5,911.39, "all of which judgments so renewed are still wholy unsatisfied by either of the defendants" (Stafford Deed Book LL, p. 96)

Two days later, Miles King and his wife, Mary L., and Newton Calvert King and his wife, Lucy deeded to Walker P. Conway all right, title, and interest in one half of the mills, mill seat and houses, with an adjoining 73 acres.

Ficklen and Conway saw economic potential in a cotton mill and published the following notice:

> NOTICE. The undersigned owners of the Water Rights on the Rappahannock, about half mile above the town of Falmouth, known by the name of the Forge Mills, hereby inform the public, that Books of Subscription for stock of a Cotton Factory, to be established there, will be opened at the Compting room of Joseph B. Ficklen, in the town of Falmouth, and at the Store of Jeremiah Carter, in the town of Fredericksburg, on Saturday, the 15th day of July (instant).
>
> It is unnecessary in the vicinity to speak of the peculiar fitness of the location, or the sufficiency of water for the establishment of the proposed Factory—they are both known to the public. The Forge dam, receives the water of the River, which conduct to Falmouth on the one hand, and to the mills on the south side of the river on the other, and at dry seasons commands nearly the whole water of the River. It is confidently believed, that in the driest seasons of the year, there would be more water than would be required for a Factory on a large scale—the fall is good—and nothing is wanting but capital and enterprize to ensure the benefits which result from their judicious employment.
>
> The undersigned are willing to receive a fair annual rent for the use of the water and ground, for the erection of the necessary buildings—or they will receive a moderate amount in stock—as the company when organized under the act of Incorporation, obtained at the last session of the Legislature, may elect.
>
> Jos. B. Ficklen
> W. P. Conway

(*Fredericksburg Political Arena*, July 7, 1837)

Ficklen and Conway dissolved their partnership later that year, Walker selling his 2/3 undivided interest to Ficklen for $1,500 (Stafford Deed Book LL, p. 123).

According to the land tax records, by 1840 Ficklen had built his cotton mill which was assessed a $2,363.53. This became known as the Falmouth Cotton Factory, employed some 50 people, and ran 1,800 spindles (Johnson, "The Falmouth Canal" 33). Based upon the 1850 census data, approximately $40,000 was invested in the cotton factory by that time, and that year it produced $30,000 worth of cotton yarn Despite the 1850 production record, for some unknown reason, this seems to have been an unsuccessful venture. The mill disappeared around 1857 as in that year the building assessment dropped from a previous $2,500 to $0 (Stafford Land Tax Records, 1858). This sudden change in the building assessment may indicate a disaster such as a fire or flood and seems to mark the end of any significant activities a

[392] Newton Calvert King (c.1811-1859) was the son of Miles King, Jr. (1786-1849) and his first wif Rebecca Calvert (1788-1827).

Rappahannock Forge. There do not appear to have been any buildings on the tract from that year through at least 1882 when Joseph's estate paid taxes on the 391-acre "Forge above Falmouth." The economic depression that followed the physical decimation of the War Between the States eliminated all hopes for a revival at Rappahannock. Trees eventually displaced the scrubby undergrowth that typified Stafford after the war and the great industrial site was largely forgotten.

Labor at Rappahannock Forge

The Rappahannock Forge industrial site was utilized from c.1759-c.1857 with the height of activity occurring between 1770 and 1784. An operation as extensive as Rappahannock employed a great many people, black and white servants, unskilled laborers and highly skilled artisans, masons, carpenters, wheelwrights, millwrights, farmers, coopers, millers, textile workers, teamsters, blacksmiths, cooks, clerks, and many more. The few acres that actually made up this industrial complex fairly teamed with people, each fulfilling duties critical to the business. The number of employees engaged at Rappahannock is staggering unto itself. In 1783, the earliest year for which there are personal property tax lists, Hunter paid taxes on 260 slaves, 43 cattle, and 81 horses. Not included in these numbers were the skilled craftsmen and apprentices who were engaged in his various mills and shops. Conservatively, these must have numbered 100 or more. Nor does this figure include wives and children who would have increased the size of the forge family several times again. It seems well within the realm of reason to propose that Hunter had from between 800 to 1,000 people living in the immediate vicinity of the forge. In several newspaper articles Hunter offered prospective employees half-acre lots on which they could build homes and, for a reasonable rent, live forever. Most seem to have accepted his offer; consequently their names are not listed in the land tax records. Further, these families rarely owned cattle, horses, or wheeled vehicles, prerequisites for being included on the personal property tax records. The result is that Hunter's workers constituted an invisible population that left little trace of their presence in Stafford County.

Hunter provided his workers with everything they needed within the self-contained village at the forge. When his workers died, they were probably buried somewhere on Stanstead. Furnaces and other industrial sites were noted for frequent and profound injuries and deaths caused by dangerous equipment with rapidly moving parts. There were no safety regulations in place and common sense and luck often defined the line between life and death. With so many people working at Rappahannock, it is a given that there would have been a cemetery nearby on the tract. Hunter, however, would have borne no obligation to pay the expense of an engraved headstone and the deceased's relatives probably could not have done so. A simple wooden cross or fieldstone marker would likely have been used to mark a final resting place and these have long ago rotted away or been bulldozed. Even the highly skilled gunsmiths were transients, rarely remaining in any one factory for more than three or four years. They came from all parts of the colonies and simply moved on when they tired of a particular armory or heard of a better offer somewhere else. Only a handful of gunsmiths employed at Rappahannock are known by name.

Each name, then, becomes precious and each constitutes a bit of a mystery as we try to piece together who worked at Rappahannock. In 1781 many of Hunter's workers were drafted into the militia, one of whom was John Mulberry (c.1753-1838). Hunter petitioned the Council asking that John be returned to the forge, describing him as "very serviceable to him in the Gunfactory Business." On June 24, 1777 the Council ordered that Mulberry be discharged "(if he be willing) from the service of this State, upon the said Hunters returning the Bounty and enlisting Money" (McIlwaine, vol. 1, pp. 438-439). For whatever reason, John was not released. He served three years with Capt. Walter Bowles and later received a pension for his military service. John was listed in the Stafford County personal property tax lists from 1785-1790, but was not included on the land tax records. During those years, he paid taxes on 1 or 2 horses and 10 or 11 cattle. In some years he owned 1 or 2 slaves and in some years he had none. He may have lived in one of the forge houses, which would account for his absence from the deeds and land tax rolls. What is obvious, however, is that John was not a common laborer. He had far too much personal property and could rightly be described as middle class. On May 24, 1779 he signed the petition to move the Stafford courthouse to its present location (Virginia Legislative Petitions, 954-P). John Mulberry married widow Elizabeth Spiller. He and his brother Jacob, and father, John, Sr.[393] (died c.1797) migrated from Stafford to Kentucky before 1790. Both John, Sr. and John, Jr. died testate in Scott County, Kentucky (Scott County Will Book A, p. 49; Jan. 22, 1798; Will Book E, p. 69, Sept. 26, 1829).[394]

A branch of the Mulberry family, sometimes spelled Mayberry, Maybury, or Marbury, settled in Pennsylvania and had a long association with iron working in that colony. The Mulberrys were also employed as ironworkers at the Principio and Kingsbury furnaces in Maryland, though the name was listed in the company ledgers as "Maybury" or "Mayberry." On Feb. 27, 1725 Thomas Maybury worked as a finer and hammerman at Principio. From 1740-1777 Francis Maybury worked at Kingsbury as a founder and hammerman, both highly skilled occupations. In 1745 Richard Maybury was employed at North East Forge in Maryland. On May 10, 1742 William Vestal, John Tradan, Richard Stevenson, and Daniel Burnet contracted with Thomas Mayberry to build a "Bloomery for making Barr iron, upon the present Plantation of William Vestal lying upon Shunandore" (Wayland, Hopewell 166). In 1772 and 1773 Joseph Mayberry was employed by the Principio Furnace in an unspecified position.

In 1757 a John Mayberry operated a charcoal-fired blast furnace and forges in Breconshire, England (Gross xiii). One is left to suppose that John Mulberry of Rappahannock Forge was a highly skilled craftsman, perhaps a founder, who was essential to the operation of the forge. Based upon what is known of James Hunter's Scottish nature, it seems highly unlikely that he would have considered purchasing this man's discharge if he could have replaced him.

John's brother, Jacob Mulberry (died c.1822), was listed in the Stafford County personal property tax lists of 1783 and 1785 and, based upon the names above and below his on this list, it appears that he lived very near the forge. Jacob was later named in the Central Kentucky Militia Lists of 1786 as serving with Capt. Saunder's Company. He was included on the Fayette County, Kentucky tax lists of 1789, on the Woodford County (from which Scott County was formed) lists in 1790, and in Scott County in 1800. He owned at least one and possibly several saw mills on Eagle Creek in Scott County.[395]

The 1787 personal property tax records for Adam Hunter list tithables Abner Vernon, Benjamin Bussell, Joseph and Robert Lavender, and John Rogers. Abner Vernon has been discussed at length. Benjamin Bussell appears on the 1790 tax lists with 1 slave and 4 horses. Joseph and Robert Lavender were also listed on the tax records of 1788 and 1789. John Rogers first appeared in the personal property tax records of 1783 with 7 slaves, 2 horses, and 5 cattle. He continued in the tax records until 1790. John Rogers may well have been a blacksmith or manager of the forge prior to Thomas West's taking over in 1793. In the tax lists of 1790 his entry was written, "Rogers John forge." Because these men owned enough personal property to make them typical middle class, they may have held management positions as opposed to being craftsmen. These same names are also listed on the petition of May 24, 1779 to move the Stafford Courthouse to its present location. As luck would have it, the carrier of the petition made his last stop seeking signatures at Rappahannock Forge. Many of the last 58 names on this petition are known to have been involved with forge operations and the remaining ones may well have also worked there. These 58 names include:

Jesse Harlan (born c.1743)[396]	William Burgess, Jr.
Israel Robinson	Micajah Hughes
Frederick Klette[397]	William Kirk[398]

---

[393] John Mulberry, Sr. was listed on the 1783 Stafford personal property tax records as Johannes Maltsbury which is the German spelling of Mulberry.

[394] In his will John mentioned his wife Mary (Mallory) Mulberry and her children. John had no children of his own. He also mentioned his brothers and sisters, Elizabeth Trotter, James Mulberry, Polly Neale, Nancy Pigg, William Mulberry, Betsy Mallory, Lucinda Pack, Catherine Fields, and Joel Mulberry (died c.1866).

[395] Jacob's will is on file in Scott County (Will Book C, pp. 374 and 390 and Will Book F, p. 187).

[396] Jesse was a Quaker miller, probably brought to Stafford by John Strode c.1776. He was born in Bradford Township, Chester County, Pennsylvania. In 1764 he married Sarah Harlan (born 1748) in an Episcopal ceremony at Holy Trinity (New Swedes) Church, Wilmington, Delaware. Sarah was the daughter of Joseph Harlan and Edith Pyle. After leaving Rappahannock Forge, Jesse and Sarah removed to Maryland and later to Albemarle County, Virginia where they died and were buried.

[397] Along with John Strode and Joseph Purkin, Frederick Klette was a key element in the success of Rappahannock Forge. He was considered one of the finest lock makers of his day, following a tradition se

Francis Asman
John English[400]
Edward Singleton
John Ferney
Edward Wells
Isaac Rose[404] (c.1750-1822)
Joseph Lavinder[405]
James Thompson[406]
John Lavinder[407] (died c.1847)
John Stanley[399]
William Allen[401]
Samuel Robert Brooke[402]
Edward Ferney
George Hood[403]
James Saunders
Josiah Greenwood (died c.1782)
Robert Lavinder
William Haner[408]

---

by his father and grandfather. In 1760 Frederick also worked as a gunsmith in the Stevensburg Armory near Culpeper. He was a German from Philadelphia.

398 The Kirks were Quakers from Chester County Pennsylvania and seem to have been included in the first group of Pennsylvania Quakers who came to the forge c.1760. A William Kirk appeared on the 1768 Stafford quit rents though a margin notation stated, "Says he pays in Fauquier." William Kirk was the son of Jeremiah Kirk, Jr. (1758-1819) and Ann Monroe (born c.1760), the daughter of George Monroe. Ann was a second cousin of President James Monroe. William was the grandson of Jeremiah Kirk, Sr. (died c.1792) and Ann Thomas of King George. Jeremiah, Jr. was born in King George but moved to Stafford on 120 acres willed to him by his father. This land was near Holloway's Mill in the north end of the county. William eventually moved to Fairfax County.

399 John Stanley was a Quaker from Hanover County, Pennsylvania.

400 John English was a Quaker from Burlington County, New Jersey.

401 The Allens were Quakers from Chester County, Pennsylvania and seem to have been among the first group of Quakers who came to the forge c.1760. In 1768 William Allen's executors paid quit rents on 905 acres in Stafford County.

402 From 1762-1767 Samuel Robert Brooke was a member and minister of the Henrico Monthly Meeting. In 1767 he was dismissed for "disorderly and inconsistent conduct." He was from Hanover County, Virginia.

403 Possibly the father of George Hood (c.1766-1835) who submitted an application for a Revolutionary War pension. In his deposition, George stated that he was the son of George Hood and had been born in Virginia though his father was originally from Pennsylvania. In 1794 he married Catherine Mullen (born c.1777) in the Swedish Church of Pennsylvania. She was the daughter of Daniel Mullen. George, Jr. was a ship cooper and served as a private in the Virginia line.

George Hood submitted a pension application for service during the Revolution, Claim W8939, National Archives and Record Administration, Washington, DC.

404 Isaac Rose of Stafford County was on the Virginia Pension roll and served as a private in the Virginia line. He was granted a $96 annual allowance beginning on Mar. 6, 1819.

405 Joseph, John, and Robert Lavinder were all listed in the 1789 Stafford County personal property tax records. The Lavinders seem to have settled around Franklin County, Virginia. A number of them were wheelwrights or involved in iron manufacture. A William Lavinder was paid wages according to a ledger recording forge accounts on file in Fredericksburg. A Thomas Lavinder, born in the late 18th century, was proprietor of an iron works in Roanoke. A letter included in the Virginia Colonial Records Project microfilm at the Library of Virginia may provide a clue as to when and why the Lavinders came to Virginia. On Aug. 18, 1764 John Semple, manager of Occoquan Iron Works, wrote to James Russell in Maryland. He stated that John Lavender, founder at one of the Principio works in that colony, had agreed to come to Occoquan and bring his four sons where they were to work for five years. A later letter stated that they arrived at Occoquan in the spring of 1765 (VCRP, Court of Session Unextracted Processes, June 1756-March 1779).

406 James Thompson was a Quaker from New Castle County, Delaware. He listed in the Stafford County personal property tax records of 1787 and 1788 with 6 horses and 14 cattle.

407 The will of John Lavinder, Sr. was recorded in Franklin County, Virginia on June 7, 1847 (signed July 10, 1846). In it he mentions his wife, Katharine and children Thornton, Chilton, John, William, Joseph, James, and Jesse Lavinder and daughters Anna (Lavinder) Campbell, Mary (Lavinder) Turner, Frances (Lavinder) Turner, and Emily Lavinder.

John Conyers[409] (c.1754-1819)	William Woodside[410] (c.1756-after 1833)
Benjamin Griffith	Robert Smith[411]
John Shepherd[412]	Archibald Rollow[413] (1755-1829)
Amos Thorp	Daniel Northup[414]
John Reids	William Palmer
Harris Winlock	Abel Griffeth
John Pollitt[415]	Henry Day[416]
William Kayley	John Brown[417]
John Banks (1756-1784)	Henry Banks (1761-1836)
George Follis[418] (1727-1798)	Thomas Turnham, Jr. [419] (1749-1830)
Joseph West	John Grat[420]
Isaac Green[421]	Isaac Holloway[422] (born 1735)
Asa Holloway[423] (born 1744)	John Holloway[424] (born 1732)

[408] William Haner paid taxes in 1785 on 5 slaves, 2 horses, and 8 cattle. He lived in the south end of the county. There was another William Haner listed that same year who had 3 horses and 5 cattle and also lived in the south end. One of these was the son of Feneral Harmon Haner (died 1794) who signed the 1779 petition to move the courthouse to the center of the county. Feneral's son William Married Mary Hardin and removed to Scott County, Kentucky.

[409] John Conyers was a millwright who later built a mill for William Richards, just downstream from the forge.

[410] William served 23 months as a private in the North Carolina line in a regiment commanded by Col. Davidson. On Mar. 4, 1831 he was granted a pension of $76.66. At the time of his pension application William lived in Iredell County, North Carolina (Revolutionary War Pension Applications).

[411] Robert Smith was a Quaker from Hunterdon County, New Jersey who came to Stafford via Loudoun County, Virginia. In 1776 he signed the petition to change the Stafford/King George County boundary.

[412] John Shepard was a Quaker from Chester County, Pennsylvania.

[413] Archibald married Ann "Nancy" Jett (1757-1830), the daughter of Francis Jett (c.1735-1791) and Barsheba Porch (c.1738-1817). He is believed to have been a Scottish immigrant. Archibald was buried about five miles east of Falmouth on the south side of White Oak Road (State Route 218). His will, on file in Fredericksburg, mentions Robert Finnall who married daughter Polly Rollow and daughter Mary who married William Cox (Fredericksburg Court Records, Will Book C, p. 93).

[414] Possibly Daniel Northup (1738-1811) who was born in Washington County, Ohio. He married in 1777 Ann Hampton, widow of Robert Collins of Virginia.

[415] Possibly John Pollard who paid taxes in 1782 on 220 acres in the south end of the county.

[416] Henry Day appears in the 1783 personal property tax records. He was living in the south end of the county and had one slave.

[417] John Brown was listed on the 1782 land tax records as owning 600 acres in the south end of the county.

[418] Around 1776 George Follis, a Quaker, purchased 712 acres of the Carter estate. George was either a millwright or a miller and built a mill on Deep Run. By 1780 he was a member of the Hopewell Meeting in Frederick County, Virginia.

[419] Thomas' Revolutionary War pension application revealed that he enlisted in the fall of 1775 and served in Capt. William Taliaferro's company, Col. Woodford's second Virginia regiment. He was at the Battle of Norfolk and at several other skirmishes. After a year of service, he was discharged, but re-enlisted in Fredericksburg in the winter of 1776. He served a short time in Capt. Walter Vowles' company, Col. Stevens Virginia regiment. He was transferred to Capt. Churchill Jones' company, Col. George Baylor's regiment of Light Dragoons. He saw action at the battles of Brandywine, Germantown, Monmouth, Savannah, Cowpens, and was present at the capture of Cornwallis. He was discharged on Oct. 25, 1781. By 1823 he had moved to Spencer County, Indiana where he died. Thomas married first Nancy ___ and secondly Eve Liston (Revolutionary War Pension Applications).

[420] This may be the Quaker John Grigg who purchased 281 acres from Charles Carter's administrators.

[421] Isaac was a Quaker. Around 1776 he bought 136 acres from Charles Carter's administrators and, in 1781, added another 125 acres from the old Carter holdings. In 1781 he witnessed the wedding of John Green of Stafford and Ruth Holloway held at Stafford.

[422] Around 1781 Isaac Holloway, a Quaker, purchased 126 acres of the old Carter estate.

William Branson[425] (c.1714-before 1783)
George Shinn[426] (1740-1782)
Daniel Antram[427] (1721-1819)
Robert Painter[428] (c.1738-before 1814)
William Mullen[429] (died c.1791)
Abner Vernon (died 1792)
James Allen[430]

William Allason's[431] business ledgers on file in the Library of Virginia contain the names of three "forgemen," David Hanson, John Macrae, and John Hunt. Allason was a Scottish merchant who kept a store in Falmouth during the mid-18th century. As there was no other forge in the area, there is a strong possibility that these three men also worked at Rappahannock.

Jarrett Burton (c.1759-1833) was employed in Stafford County from 1777-1781 as an armorer and gunsmith (Whisker, Arms Makers of Virginia, p. 25). According to Jarrett's pension application, he served in Stafford County under Capt. William Garrard (c.1715-c.1786) and Maj. Henry Fitzhugh (1723-1783), serving principally on the Potomac River watching British ships. This continued for about three months, then he was discharged. In the fall of 1777 Jarrett was called up to work at the Fredericksburg armory under Col. Fielding Lewis and Charles Dick. William Grady was superintendent there and Jarrett worked at that armory for four years. At the time he applied for his war pension in 1833, he was a resident of Mason County, Kentucky (Revolutionary War Pension Applications). The Burton family is still well represented in the southern end of Stafford County.

Two other forge workers were Thomas Harwood (1769-1845) and John England (1755-1851). According to records in the Fredericksburg Circuit Court, Thomas Harwood was employed at

---

423 Asa was the son of George Holloway (born c.1709) and Ruth Wood, Quaker residents of Springfield, New Jersey. Asa married Abigail Wright.

424 John was the son of George Holloway (born c.1709) and Ruth Wood. He married Margaret Buck. On Nov. 13, 1792 he was appointed overseer of the road "from Poplar Road to the Old Furnace."

425 Around 1781 William Branson, a Quaker, purchased 135 acres of the old Carter estate. The Bransons came to Virginia from Springfield Township, Burlington County, New Jersey. William was the son of Thomas Branson, Sr. (died c.1744) and Elizabeth Day. Thomas, Sr. owned property in the Shenandoah Valley near White Post, now in Warren County. He devised this land to his sons Thomas, Jr. and Jacob. William married Elizabeth Osmond (c.1737-1778) and lived for some years in Stafford County. From him are descended the Bransons of Frederick County, Virginia.

426 George Shinn was the son of Francis Shinn (1706-1789) and Elizabeth Atkinson (died 1783) of Burlington, New Jersey. A Quaker, he was either a millwright or miller. Around 1776 he purchased 242 acres on Deep Run from the Carter estate. In 1771 he and his family transferred from the Burlington Meeting to the Hopewell Meeting near Winchester. Between 1771 and 1776 he served as a committee member for the Hopewell Meeting.

On Dec. 22, 1783 Charles Carter and his wife Eliza and Robert Painter and his wife Eunice sold to Rachel Shinn, widow of George Shinn, a 13-acre tract with a grist mill. Fielding Lewis had sold the land in 1776 to Carter, Painter, and Shinn and they had built the mill upon it. (Stafford Deed Book Z, p. 158) Rachel paid taxes on the mill for many years thereafter.

427 Daniel Antram (1721-1819) was a Quaker born in Springfield Township, Burlington County, New Jersey. In 1758 he married Susanna Weaver in Burlington. They moved to Stafford in 1769 at which time he, John Antram, and Joseph Wright requested permission from the Hopewell Meeting near Winchester to establish the Stafford Meeting. Around 1776 Daniel purchased 107 acres of the Carter estate. Daniel later moved to Fayette County, North Carolina.

428 Around 1776 Robert Painter, a Quaker, purchased 105 acres of the Carter estate. The Painter family came from the border between England and Wales and had settled in Philadelphia by 1705. Robert was born in Burlington, New Jersey and married Eunice Osmond (born 1742).

429 Around 1776 William Mullen, a Quaker, purchased 185 acres of the Carter estate.

430 Around 1776 James Allen, a Quaker, purchased 260 acres from Charles Carter's administrators. This came to be known as Poplar Grove and it was probably upon this plantation that the Quaker meeting house was built.

431 William Allason (1731-1800) was the son of Zacharius Allason and Isbel Hall of Glasgow, Scotland. He arrived in Virginia in 1757 and, shortly thereafter, opened a mercantile business in Falmouth.

Rappahannock as a millwright. Thomas' father, John Harwood,[432] also probably a forge employee purchased part of the Stanstead tract upon which he resided. Thomas inherited this property after his mother's death. The house was located in the vicinity of the Days Inn Motel on Warrenton Road (U. S Route 17). John England was born in Bucks County, Pennsylvania and died in Richmond, Virginia. He was employed at Rappahannock as a maker of gunlocks. One of John's locks survives, it being engraved with his initials and mounted on the Wertenberger pistol (Swayze 24). In 1782 John married in Stafford Ann "Nancy" Musselman (c.1760-1823), the daughter of Christian Musselman (c.1730-after 1801) and his wife Elizabeth. Among their children was a son named Abner Vernon England (1797-1838). On Sept. 4 1834 John submitted a pension claim stating that in 1780,

> while engaged as an artesan at Hunter Iron Works in the County of Stafford and State of Virginia, he was drafted according to Law to serve as a soldier for Eighteen months in the Revolutionary War and was immediately marched by Captain Mason Pilcher who was under the Command of Col. William Gerrard, who was the Col. Commandant of the Militia of Stafford, to the place of Rondevous, in Fredericksburg under the Command of General George Weedon and ordered to return to the place from whence he had been taken and where arms were made and repaired for the public use—stating to him that he was one of the best workmen whose services could be obtained, that arms were much wanted rendering his country more benefit in that capacity than be remaining in the Army & from that declared a furlough, with orders to be at all times ready in case of necessity to march at forty eight hours notice. The declared in obedience to his orders returned to the Iron Works and continued there the eighteen months for which he had been drafted faithfully employed in making and repairing guns for the army and always ready and willing to the call of his country, whenever they might chuse to change the nature of the service (Revolutionary War Pension Applications).

John continued by stating that he had "received no compensation for his work during the term for which he was drafted, considering himself as much a soldier as if in the army." In his deposition John also stated that he was bound an apprentice in Bucks County before coming to Stafford County and Rappahannock Forge. Claiming to have lost his discharge papers, John named Robert Howson Hooe (1748-1834), Samuel Gordon (1759-1843) Basil Gordon (1768-1847), Nathaniel Greaves (c.1769-after 1850), Joseph Ennever (died c.1849), Thomas Harwood (1769-1845), and George Banks (1779-1837) as men who could testify as to the validity of his statements. His pension request was denied because the law required that he render at least six months service in a military capacity and the alleged service was not considered as such. While in Stafford, John and his family resided at Seine Pocket Farm located on the Rappahannock River at the end of Riverside Parkway (County Road 1445). By 1848 he had removed Richmond, Virginia, though his farm remained in the England family for many years.

There were other Quakers who purchased land from Charles Carter's administrators though they cannot be proven to have any connection to the forge. These included:

Joshua Brown[433]
Gabriel Jones[434]
John Gregg
Edward Godley
James Primm

Two or three members of the Reveley family were also involved with Rappahannock. Thomas Reveley[435] (c.1720-1785) was born in near Whitehaven, England and was apprenticed there to an iron

---

[432] Because of the many John and Thomas Harwoods in Virginia and Maryland during this period, it is difficult to determine the background of this branch of the Harwood family.

[433] Joshua was a Quaker from Little Britain, Pennsylvania.

[434] According to the Stafford land tax records, Gabriel bought "Carter's Mill Seat."

[435] Thomas Reveley's obituary stated that he "drowned in Rappahannock opposite Falmouth foarding [sic] the river. His horse fell and he was lost." (Hudgins, n.p.)

master. Prior to coming to Virginia, he married Elizabeth Nicholson (c.1742-1814) in Cumberland, England. According to family history, Thomas worked at Rappahannock, though no forge records survive to document this (Hudgins, n.p.). The Reveley family Bible notes that Elizabeth died while visiting Rappahannock Forge. Thomas Reveley died in Stafford leaving several children, among whom were two sons, John (1741-1802) and William[436] (1743-1788). William purchased a tract of land on the northeast side of Greenbank. Named Woodend after the family home in England, the property remained in the Reveley family until 1812 when it was sold to George Banks (1779-1837). Again according to family history, both John and William were employed at the forge. William lived at Woodend, a farm on the east side of Greenbank. This tract was first owned by Enoch Innis, then James Threlkeld, and thirdly by Edward Moore. Sometime between 1767 and 1783 Moore conveyed it to William Reveley. The deed for this tract described it as consisting of 144 acres on the western fork of Gravelly Run and bounded on one side by Banks Road (Stafford Deed Book AA, p. 302). John later became involved with John Ballendine in the Westham Foundry and went bankrupt.

There are no surviving court documents in Stafford that make reference to the Woodend tract. Again depending upon Reveley family history, in 1807 Elizabeth (Nicholson) Reveley sold Woodend to George Banks (1779-1837) for £100, though she retained a lifetime lease on the property. After Elizabeth's death, Woodend became part of the large Greenbank estate. Today, part of Woodend is occupied by Berea Plantation subdivision.

It is even more difficult to determine the identities and numbers of slaves working in and around Rappahannock Forge. Other than Hunter's inventory in which his slaves were listed by name, the only other reference to a slave by name yet found by the author pertains to a runaway named Daniel. On May 17, 1799 John Ferneyhough[437] (died 1815), a coachmaker in Fredericksburg, placed the following notice in the *Virginia Herald*:

> Thirty Dollars Reward, Ran away from the subscriber on Wednesday last, a negro man named DANIEL, by trade a Blacksmith, well acquainted with making keys and repairing locks; also a good hand in the Coach Smith's Business. He is about five feet six inches high, stout made, the top of his head bald, full faced, rough visage, very black, and speaks fairly, loves liquor, and is about fifty years of age. He was born on North River, Gloucester county, Virginia, and was first the property of John Page esq. who sold him to Mr. James Hunter, at Rappahannock Forge, with whom he lived many years, from whose estate I purchased him. He has a wife and children at William Fitzhugh's esq. of Chatham, opposite Fredericksburg—I cannot give a particular description of his clothes as he has several suits, one of which is gray, as also a blue cloth coat, and drab colored breeches and jacket, almost new. Any person who will secure him in any gaol in this state, shall receive Twenty Dollars Reward, and if taken out of the State, Thirty Dollars and all reasonable charges paid by John Ferneyhough. Virginia, Sligo, near Fredericksburg, May 4.
>
> N.B. It is likely he may have obtained a pass, changed his name, and endeavor to impose himself on the public as a freeman. All masters of vessels and others, are forewarned from carrying him out of this State.

---

[436] Around 1783 William married Anne Towles (1757-1823), the widow of Thomas Carter.

[437] John Ferneyhough came to America from England in 1788 and settled first in Albemarle County. Shortly thereafter he moved to Fredericksburg. The local newspapers contain numerous advertisements for his coach making shop and in 1792 he was elected to the town council. He was also a member of the Fredericksburg Masonic Lodge. John and several succeeding generations of his family resided at Sligo on the outskirts of Fredericksburg.

The forge property was occupied by Union troops during the War Between the States. How many, if any of Hunter's buildings were still standing at that time is unknown. Any structures on the site were destroyed or dismantled and utilized as bridge and road building material, all but erasing the last traces of Hunter's complex. Over time, the trees grew up along the riverbank and hillside and "the greatest iron works that is upon the Continent" was largely forgotten. In 1942 the site was scavenged for scrap iron, destined for use during World War II.

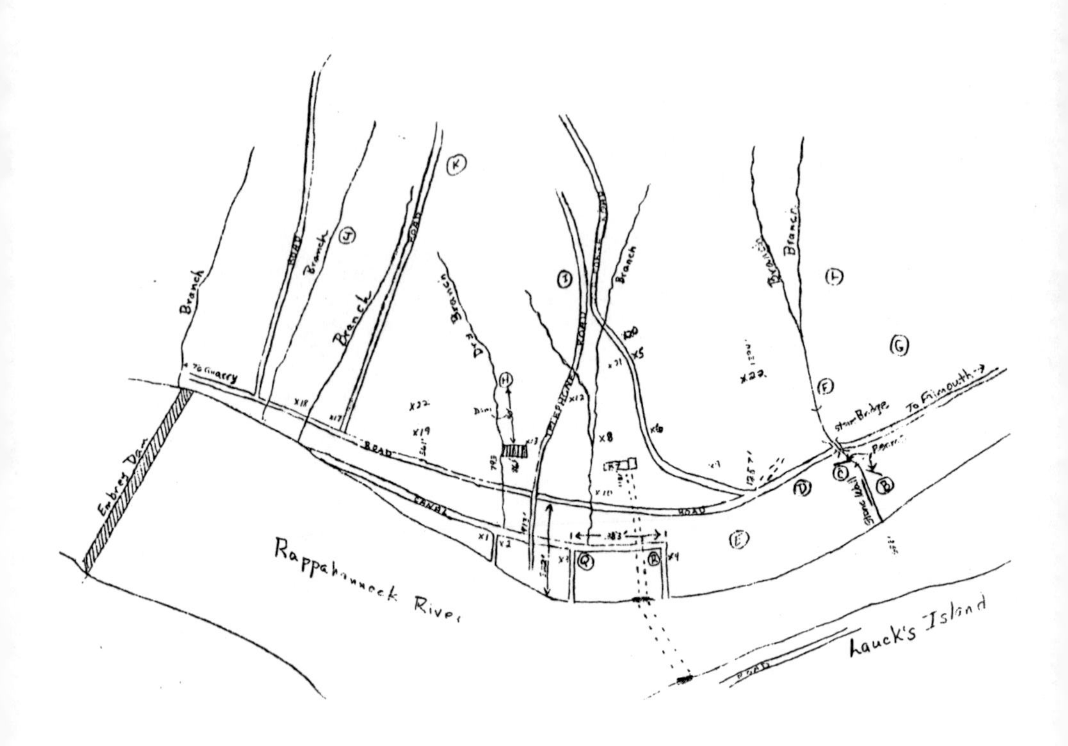

Fig. 2-9
Physical remains on the Rappahannock Forge site.

Site Map Legend

B—Slitting Mill—foundation—berm
C—Stone foundation—berm—appears to have had basement
D—Stone foundation—approximately 20' north to south—bits of dross, a few shell fragments
E—Stone foundation—40' x 84'—coal house?--held together with cement made from ground shells and river sand
F—Stone foundation—20' x 27'
G—Stone foundation
H—Stone foundation—20' x 40'—appears to have had a brick chimney on northeast corner
I—Stone foundation—16' x 20'
J—Stone rubble—unknown
K—Blaisdell home, site now beneath driveway had brick floor and fireplace with seven pigs laid in the bottom.
L—Portion of a stone foundation, two finely-worked stone blocks with decorative molding
Q—Stone foundation
R—Stone foundation

X1—Possible site—no visible foundation
X2—Stone foundation
X3—Stone foundation—found slate (roof)—roughly 30' x 30'—piece of iron fireback found on east side of this foundation—also 2 pieces dross and one small piece of mineral coal
X4—Stone foundation
X5—Stone foundation—located on the north side of the road and canal and well up the hill—approximately 25' north to south and 47.5' east to west—cut sandstone with groove running length of the block—several cut sandstone and hard rock blocks. The west side of this building appears to have been constructed of stone which has collapsed down the hillside along the road
X6—Stone rubble—finely cut sandstone block
X7—Furnace—many oyster shells, cinders, charcoal, sand casting pit, dross, found spade and long narrow nail(?)
X8, X9, X10, X11—Stone foundations
X12—on top of hill—cut hard stone—pieces of brick—approx. 20' east to west, 24' north to south—found pieces of oyster shell cement
X13—Stone foundation—original furnace(?)—approx. 30' north to south by 40' east to west—broken pieces of brick—directly below and to the immediate west of this foundation, the soil contains a layer of coal dust, some cinders, charcoal, broken bricks, and a good quantity of dross. The hill was cut away to make room for the structure. The road below this site is littered with pieces of dross. The small branch to the immediate west of this foundation shows evidence of having been dammed. The hill behind this structure has two terraces above the level of the foundation. An oyster shell was found on the edge of the upper-most terrace. Foundation H is 50' northwest of X13 and on top of the uppermost terrace.
X14—Elaborately cut sandstone and bricks—perhaps an office or house(?)—brick building on stone foundation
X17—Stone foundation—bricks on east side—possibly chimney
X18, X19, X20—Stone foundations
X21—Large dressed sandstone blocks and a few bricks
X22—Well-cut stone blocks

Letters refer to sites located on the 1971 survey by the dept. of the Interior.
Number designations beginning with "X" are sites found during visits in the winters of 1999-2001.

Appendix A

Warrants and Orders

Warrants/Payments

Date	Amount	Description
May 4, 1776	£ 132.3.0	"…for intrenching Tools for the use of Col. Mercers Regiment"[438]
	£5.0.0	"…for a Gun furnished the 3rd Regiment."[439]
June 7, 1776	£150	25 muskets, complete[440]
	£160	Paid Hunter "per order to Richard Graham for 533 1/3 Dollars" (£160)[441]
	£250	"…on account to purchase Arms, Iron, and intrenching Tools"[442]
June 17, 1776	£84.0.0	14 muskets and bayonets[443]
	£10.3.9	"…for Ferriage"[444]
Aug. 8, 1776	£361.5.0	60 muskets, 2 pair bullet molds, and 3 chests[445]
	£103.5.0	Iron supplied James Anderson, public armorer[446]
	£50.0.0	For gunpowder used to prove guns at Rappahannock
	£62.17.2	For a balance due and for arms, ferriages, bar iron, cannon balls, and flour furnished for the navy
	£37.19.4 ½	Three 6-pound cannon
	£253.11.0	Spades and shovels "furnished for the use of the Publick Salt Works"[447]
Aug. 14, 1776	£114.16.0	Unspecified number of cannon balls[448]
Aug. 24, 1776	£149.5.6	Unspecified number of cannon balls[449]
Oct. 17, 1776	£700.0.0[450]	
Nov. 5, 1776	£700.0.0[451]	
	£116.8.6	"for Sundries"[452]
Nov. 6, 1776	£1,300.0.0[453]	
Nov. 11, 1776	£1,347.12.1 ½	"…being the amount of his Account for Arms, Camp Utensils, Waggonage and other Necessaries, as settled by the Commissioners."[454]
Nov. 22, 1776	£175.15.0	25 muskets and bayonets and for "a Chest for Package also for Bar Iron, Steel and Waggonage."[455]

---

[438] Journals of the Council of the State of Virginia, vol. 2, p. 496.
[439] Journals of the Council of the State of Virginia, vol. 2, p. 496.
[440] Journals of the Council of the State of Virginia, vol. 1, p. 7.
[441] Virginia Committee of Safety, Account Ledgers, 1775-1776, Miscellaneous Reel #301.
[442] Virginia Committee of Safety, Account Ledgers, 1775-1776, Miscellaneous Reel #301.
[443] Virginia Committee of Safety, Account Ledgers, 1775-1776, Miscellaneous Reel #301.
[444] Journals of the Council of the State of Virginia, vol. 1, p. 25.
[445] Journals of the Council of the State of Virginia, vol. 1, p. 113.
[446] Journals of the Council of the State of Virginia, vol. 1, p. 113.
[447] Journals of the Council of the State of Virginia, vol. 1, pp. 114-115.
[448] Virginia Navy Board, Ledger, July 10, 1776 - July 10, 1777, Miscellaneous Reel #302.
[449] Virginia Navy Board, Ledger, July 10, 1776 - July 10, 1777, Miscellaneous Reel #302.
[450] Virginia Navy Board, Ledger, July 10, 1776 - July 10, 1777, Miscellaneous Reel #302.
[451] Virginia Navy Board, Ledger, July 10, 1776 - July 10, 1777, Miscellaneous Reel #302.
[452] Virginia Navy Board, Ledger, July 10, 1776 - July 10, 1777, Miscellaneous Reel #302.
[453] Virginia Navy Board, Ledger, July 10, 1776 - July 10, 1777, Miscellaneous Reel #302.
[454] Journals of the Council of the State of Virginia, vol. 1, p. 235.
[455] Journals of the Council of the State of Virginia, vol. 1, p. 248.

	£165.15.0	25 muskets and bayonets, bar iron, steel, and waggonage[456]
Dec. 1776	£1,856.9.7[457]	
Dec. 18, 1776	£956.5.0	"...being the amount of fifteen hundred Bushells of Salt purchased by the Governor and Council."[458]
	£122.5.0	Warrant issued to William Fitzhugh for the use of James Hunter for muskets and waggonage[459]
Dec. 19, 1777	£300.0.0	"...on account of the Gun factory."[460]
July 15, 1778	$417 35/90	"...a warrant issued on John Gibson...to enable them to pay James and Adam Hunter."[461]
Aug. 20, 1778	$3,000	"...to pay the wages due the seamen on board the brigantine Morris, now under their care"[462]
Nov. 13, 1778	$7,735	"to pay a bill drawn on "the Committee of Commerce] by Messrs. James and Adam Hunter...in favour of Mr. Amos Strettle"[463]
Jan. 20, 1779	$460	Bill from James and Adam Hunter to Amos Strettle[464]
Nov. 11, 1779	£5096.10.0	"for arms &c furnished cavalry."[465]
Nov. 16, 1779	£458.10.0	"for wire for the use of the State Shop."[466]
Nov. 16, 1779	£16,990.0.0	"for accoutrements &c for [cavalry]."[467]
Nov. 16, 1779	£16,094.6.0	"for anchors and other necessaries purchased from him by the commissioners of the navy for the use thereof."[468]
Feb. 4, 1780	£3,100.0.0	"for payment of two anchors purchased of him for the Brigantine Jefferson."[4
June 29, 1780	$123.077 45/90	"in part paymt of ye drafts of Congress in his hands"
1780?	£50,000[470]	
Nov. 12, 1782	13,097 lbs. tob.	Paid James and Adam Hunter[471]
Nov. 23, 1782	£10,365	Paid James and Adam Hunter for waggonage[472]
Dec. 22, 1782	£110,734	"James Hunter rec. in full of this date"[473]
Dec. 22, 1781	£54,571.0.2/3	"for Tobo. advanced you, being so much purchased by them on Acct of the S
July 28, 1782	£51,225.15.6[475]	
June 27, 1786	£15,000	For bar iron and nail rod omitted[476]
Unknown	£130,000[477]	

[456] Journals of the Council of the State of Virginia, vol. 1, p. 248.
[457] Virginia Navy Board, Minute Books, 1777-1778, Miscellaneous Reel #302.
[458] Journals of the Council of the State of Virginia, vol. 1, p. 288.
[459] Journals of the Council of the State of Virginia, vol. 1, p. 289.
[460] Journals of the Council of the State of Virginia, vol. 2, p. 49.
[461] American Memory, Journals of the Continental Congress.
[462] Stewart, Robert A. History of Virginia's Navy of the Revolution. Richmond, VA: Mitchell an Hotchkiss Pub., 1933.
[463] American Memory, Journals of the Continental Congress.
[464] American Memory, Journals of the Continental Congress.
[465] Virginia Board of War, Letterbook, and Index, 1780-1782, Miscellaneous Reel #264.
[466] Virginia Board of War, Letterbook, and Index, 1780-1782, Miscellaneous Reel #264.
[467] Virginia Board of War, Letterbook, and Index, 1780-1782, Miscellaneous Reel #264.
[468] Virginia Board of War, Letterbook, and Index, 1780-1782, Miscellaneous Reel #264.
[469] Virginia Board of War, Letterbook, and Index, 1780-1782, Miscellaneous Reel #264.
[470] Felder, Paula. Fielding Lewis and the Washington Family, p. 283.
[471] Commercial Agent Records, Day Book, July 18, 1782 - June 27, 1786, Miscellaneous Reel #634.
[472] Commercial Agent Records, Day Book, July 18, 1782 - June 27, 1786, Miscellaneous Reel #634.
[473] Commercial Agent Records, Day Book, Mar. 17, 1781 - July 17, 1782, Miscellaneous Reel #634.
[474] Commercial Agent Records, Day Book, Mar. 17, 1781 - July 17, 1782, Miscellaneous Reel #634.
[475] Commissary of Stores, Ledger, 1781-1785, Miscellaneous Reel #636.
[476] Commercial Agent Records, Day Book, July 18, 1782 - June 27, 1786, Miscellaneous Reel #634.
[477] Felder, Paula. Fielding Lewis and the Washington Family, p. 283.

Unknown	$123,077[478]	
Unknown	£727.7.0	"sundry Military Stores &c furnished Col. Baylors Regiment of Light Dragoons"[479]
Unknown	£10,161.12	From Col. Baylor[480]

James Hunter's account with the Navy Board:
November 5, 1776

To Warrant on the Treausry	1700.0.0
To Isaac Zane for Cannon Ball &c as per Acct	1856.9.7
To Fielding Lewis for Cannon Ball &c as per Acct.	147.7.0
To Warrant on the Treasury	1300.0.0
To James Hunter Jr. for sundried Credited Contra this mistake	116.8.6
	£4122.3.9
	3770.19.3

Credits to James Hunter's Account

July 24, 1776—By Alexander Dick for sundries	116. 8.6
Aug. 6—By Richard Taylor for sundries as per acct.	37.19.9
By Ditto for Ditto as per acct	9.10.1
By William Frazer for cannon balls &c as per Acct.	114.16.0
By Richard Adams for Cannon Balls as per Acct.	72.10.0
	351.4.4
Ballance to new Account per Contra	3770.19.3
	£4122.3.7[481]

Orders

Date	Amount	Description
Mar. 27, 1776		Committee of Safety agreed to "take of Mr. James Hunter so many Falling axes, spades, shovels and mattocks as will be necessary to supply 4 regiments, and as many Picks as will be necessary for the third Regiment."[482]
Apr. 24, 1776		"Anchors of any size you have ready for the Navy" along with 4 dozen saw files and 200 lbs. of steel.[483]
June 6, 1776		"...for as many good muskets, with bayonets, sheaths, and steel ramrods as he can manufacture within twelve months from this time" for £6 per stand.[484]
June 15, 1776		Hunter to furnish the commissioners of the Fredericksburg Manufactory with spades and shovels[485]
July 12, 1776		52 muskets and bayonets[486]
July 18, 1776		"Iron of different sizes and Anchors"[487]

478 Journals of the Council of the State of Virginia, vol. 1, p. 263.
479 Virginia Board of War, Journal, 1777-1780, Miscellaneous Reel #302.
480 Virginia Board of War, Journal, 1777-1780, Miscellaneous Reel #302.
481 Virginia Navy Board, Ledger, July 10, 1776 - July 10, 1777, Miscellaneous Reel #302.
482 Journals of the Council of the State of Virginia, vol. 2, p. 469.
483 Virginia Navy Board, General Correspondence, 1776-1787, Miscellaneous Reel 462.
484 Virginia Committee of Safety, Account Ledgers, 1775-1776, Miscellaneous Reel #301.
485 Journals of the Council of the State of Virginia, vol. 1, p. 24.
486 Journals of the Council of the State of Virginia, vol. 1, p. 67-68.

Aug. 1, 1776		About 20 tons of iron and 2 anchors weighing 1,000 and 900 lbs. respectively[488]
Aug. 10, 1776		"...for such arms as they may have already finished for the public."[489]
Aug. 14, 1776		"...four Ton of Iron for the purpose of Building the Rowe Gallie on Mattapony River lately undertaken by Caleb Herbert."[490]
Oct. 4, 1776		A salt pan for Richard Parker's salt works in Northumberland County, " pan to contain about one thousand Gallons."[491]
Oct. 17, 1776		"...three Ton of Iron for the Purpose of Building the two Gallies on the Eastern Shore."[492]
Dec. 19, 1776		Anchors of various weights—2 of 1,200 lbs. each, 2 of 1,000 lbs. each, 8 of 800 lbs. each, 8 of 600 lbs. each, 8 of 500 lbs. each, 6 of 350 lbs. each.[493]
Feb. 1777		"...so many arms as may be sufficient to arm a regiment [680]."[494]
Mar. 13, 1777		80 stands of arms or as many more as can be procured" for Col. Stephen of the 10th Virginia Regiment.[495]
Apr. 20, 1777		"...Anchors, Cutlases and Pike as Capt. [Eleazer] Callender may be in want of for the use of the *Dragon* Galley now building at Fredericksburg[496]
Apr. 24, 1777		"Anchors of any size you have ready for the Navy," 1 dozen crosscut and 4 dozen pitsaw files, and 200 pounds steel[497]
June 3, 1777		Board of War ordered William Aylett[498] to "send a Vessel to Mr. James Hunter at Fredericksburg for four Tons of such sized Bars as the said [James] Anderson may want."[499]
June 11, 1777		6 dozen pit saw files, 2 dozen crosscut saw files, 50 pounds of steel[500]
June 24, 1777		"...all the Muskets compleatly fitted which [Hunter] can make within twelve Months from this Time" for £8 each "providing they shall be as well filled [sic] & finished as those formerly purchased by this Board of the said James Hunter."[501]
July 18, 1777		"Iron of different sizes and anchors"[502]
Aug. 1, 1777		"Twenty Tons of Iron, 2 anchors, one of 1000 lbs and the other of 900 pound
Aug. 6, 1777		4 tons iron for the use of the Navy to be delivered to William Frazier[504]
Dec. 19, 1777		2 anchors of 500 pounds each[505]

[487] Virginia Navy Board, Letterbook, July 15, 1776 - Oct. 9, 1777, Miscellaneous Reel #302.
[488] Virginia Navy Board, Journal, 1776-1779, Miscellaneous Reel #301.
[489] Journal of the Virginia Council of Safety from Morgan, William J. Naval Documents of the American Revolution, vol. 6, p. 142.
[490] Virginia Navy Board, Journal, 1776-1779, Miscellaneous Reel #301.
[491] Journals of the Council of the State of Virginia, vol. 1, p. 185.
[492] Virginia Navy Board, Journal, 1776-1779, Miscellaneous Reel #301.
[493] Virginia Navy Board, Letter Book, July 15, 1776 - Oct. 9, 1777, Miscellaneous Reel #302.
[494] Journals of the Council of the State of Virginia, vol. 1, p. 332.
[495] American Memory, Journals of the Continental Congress, Mar. 13, 1777.
[496] Virginia Navy Board, Letter Book, July 15, 1776 - Oct. 9, 1777, Miscellaneous Reel #302.
[497] Virginia Navy Board, Ledger, July 10, 1776 - July 10, 1777, Miscellaneous Reel #302.
[498] William Aylett (died 1829) was a delegate from King William County. During the Revolution, h served as Commissary General and was one of George Washington's aides-de-camp. He lived at Fairfiel in King William.
[499] Journals of the Council of the State of Virginia, vol. 1, p. 424.
[500] Virginia Navy Board, Ledger, July 10, 1776 - July 10, 1777, Miscellaneous Reel #302.
[501] Journals of the Council of the State of Virginia, vol. 1, p. 440.
[502] Virginia Navy Board, Ledger, July 10, 1776 - July 10, 1777, Miscellaneous Reel #302.
[503] Virginia Navy Board, Ledger, July 10, 1776 - July 10, 1777, Miscellaneous Reel #302.
[504] Virginia Navy Board, Minute Book, 1777-1778, Miscellaneous Reel #302.
[505] Virginia Navy Board, Minute Book, 1777-1778, Miscellaneous Reel #302.

Sept. 1777-June 1779		32 bridle bits, 15 swords and scabbards, 16 pairs of pistols, 1 bell, 1 sword blade, and the "hilting and polishing &c of an Of[ficer's] Sword."[506]
May 1, 1778		Unspecified number of arms, swords, and pistols[507]
Summer 1780		1,000 camp kettles[508]
Feb. 12, 1781		Nail rods and 1,000 camp kettles[509]
March 30, 1781		500 horsemen's swords[510]
Mar. 26, 1781		800 spades, 400 common axes, 200 broad and grubbing hoes, 100 pick axes, 300 fascine knives and small hatchets, 6 crosscut saws, carpenters' tools and "nails of the larger sort."[511]
Summer 1781?		1,000 horsemen's swords[512]
May 14, 1782	£100,000	"Sundries"[513]
July 28, 1782	£15,000	Slit iron, bar iron, and nail rod[514]
Undated	In excess of £11,000	See pp. 111-112 [515]

[506] Account ledger, Col. George Baylor's account with James Hunter, Sept. 27, 1777 to June 19, 1779, Alderman Library, University of Virginia, Charlottesville, Virginia.
[507] Fitzpatrick, John C., ed. The Writings of George Washington from the Original Manuscript Sources, 1745-1799, vol. 11, Washington, DC: U. S. Government Printing Office, 1944, pp. 338-339.
[508] Boyd, Julian P., ed. The Papers of Thomas Jefferson, vol. 4, pp. 666-667, Princeton, NJ: Princeton University Press, 1952.
[509] War Department Collection of Revolutionary War Records, Entry 9—Photographic Copies of State Records, c.1775-1783, Record Group 93, Box 42, File #035548, National Archives and Records Administration, Washington, DC. Boyd, Papers of Thomas Jefferson, vol. 4, p. 551.
[510] Boyd, Julian P., ed. The Papers of Thomas Jefferson, vol. 5, letter from Thomas Jefferson to Maj. Richard Call, Mar. 30, 1781, p. 284.
[511] War Department Collection of Revolutionary War Records, Entry 9—Photographic Copies of State Records, c.1775-1783, Record Group 93, Box 42, File #035676, National Archives and Records Administration, Washington, DC.
[512] Palmer, William P. ed. Calendar of Virginia State Papers, vol. 2, p. 618, NY: Kraus Reprint Corp., 1968.
[513] Commissary of Stores, Ledger, 1781-1785, Miscellaneous Reel #636.
[514] Commissary of Stores, Ledger, 1781-1785, Miscellaneous Reel #636.
[515] Papers of the Continental Congress, Board of Treasury, Expenditures for Old Accounts, October 1785, M247, r154, v2, p. 393, National Archives, Washington, DC.

Appendix B

The Will of James Hunter[516]

In the Name of God Amen. I James Hunter of the County of Stafford do make publish & declare this to be my last & only Will hereby revoking all others. I give and devise all my Estate real & personal to my brother Adam Hunter, to him & his Heirs forever subject however to following devises—My debts to be paid first. I give to cousin James Hunter of Virginia One Thousand pounds Virginia Currency upon condition he does not bring any suit against my Estate on [account] of my fathers Estate, or on account of my brother, William Hunters Estate. I give to cousin William Hunter's children (The oldest excepted who in not to have a part.) the like sum of One thousand pounds—upon condition that their fathers representative or them selves do not bring any suit against my Estate on account of their Grand fathers Estate. It is my wench "Aggy" & her children be freed from all service after my decease; and that my brother Adam allot her a piece of Land out of the Culpeper Tract, sufficient for her & her children & two female slaves to work on for the maintanance of the sd "Aggy" & her children during her life, the female slaves are to be such as do not bring Children, & in case, one or both die during "Aggys" life, one or more to be supplied from my Estate, my brother Adam is also to provide for the sd "Aggy" as soon as may be housed on the sd land, where she shall choose, fit for her use & to her liking & to furnish cattle & Hogs fit for the support of her and the place. And I give the sd "Aggy" my old chair (A) add to clause at .A. a Mare called hers—with two colts & a horse called Billys pad, with two Horses fit for work one to be the one I bought lately. Also all the furniture, now in the house she sleeps in & corn & pork for the ensuing year her & childrens support, and corn sufficient for feeding the two slaves & horses, allotted her but all to return to my estate after the death of the sd "Aggy". It is my desire that my two nephews in Scotland to Wit, only son of my deceased brother John, & the son of my sister in Scotland, now with my brother at Dunkirk, may be sent for to assist in the adjustment of my affairs. And I give & devise to them respectfully one fourth part of my Estate, that shall remain after my death, legacies & other demands be discharged, lastly I constitute & appoint my said brother Adam, & Mr. Abner Vernon Executors of this my will with full power to dispose of any part of my Estate for payment of debts and legacies. And in case my nephews before mentioned or either of them shall come to Virginia & choose to act, I desire that they may be joined in the executorship in witness whereof I have hereunto set my hand & Seal the eighteenth day of November 1784.

James Hunter

Signed Sealed & published by the aforementioned James Hunter as & for his last Will in our presence which we attest by subscribing the same as witnesses in his presence and at his desire, James Mercer Lawrence Brooke, W. J. Stone.

[516] Stafford Deed Book S, p. 197. This will was recorded in Stafford on Apr. 11, 1785. At some point during the 19th century, the book containing this will was re-copied. It appears that in the process some words were left out as some sentences seem incomplete.

Appendix C

Inventory of James Hunter
April 1785

Anthony McKitterick,[517] Daniel Payne[518] (1728-1796), William Taylor, and David Briggs completed an inventory of James Hunter's personal estate in Stafford in April 1785. Other appraisers listed and valued his property in Culpeper and Fauquier counties. This was an extensive estate totaling £12,735.16.4 and filling five pages in the deed/will book. It is of interest because it was listed room by room and building by building, providing a broader picture of the extent of Hunter's operations at Rappahannock and Stanstead.

Appraisement of the Estate of James Hunter Esq., decd to us produced agreeable to an order of the Court of Stafford dated April 1785.

In Stafford
At the Mansion House, Furniture in the Chamber

Viz—1 Desk Walnut £6, 2 Oval Tables £3, 1 Iron Chest £15, 6 Walnut Chairs £3, 1 Old dressing glass 10/, 6 large China Cups & 6 Saucers 30/, 3 three quart China Bowls 45/, 1 two quart do[519] 12/, 2 Quart do 12/, 2 brass pots 4/, 2 Bun Cans 5/, 1 pr. And Irons 12/, 1 Sett Back Gammon Tables 10/.

Furniture in the Hall

2 Oval Tables 72/, 11 Walnut Chairs 115/, 1 Pair large Looking Glasses 10£, 2 3 qt. Bowls 30/, 1 two qt do 12/, 2 one qt do 6/, 12 Beer Glasses 12/, 2 large tumblers 3/, 1 Dozen China Cups & 1 Dozen Saucers 24/, 1 Tea Pott 3/, 1 Doz Wine Glasses 6/, 1 Glass Salt Cellar 6/, 1 pickle Stand 3/, 1 Bottle stand 2/6, 12 Dish Matts 5/, 1 Pair And Irons 12/.

---

[517] Anthony McKitterick was a merchant in Fredericksburg and Falmouth. In 1782 he paid taxes on 2 lots in Falmouth. His name was listed on Stafford's personal property tax rolls of 1783-1790. During those years, he averaged 9 to 11 slaves, 8 to 12 horses, 4 or 5 cattle, and a two-wheeled vehicle called a "riding chair."

[518] Daniel Payne was born at Red House, then King George and now Westmoreland County, the son of John Payne (c.1693-1750). At an early age he assisted his father in a mercantile business in Leedstown. He later moved to Dumfries where he was a merchant. During this time, he served as a banker for about 350 residents of the Dumfries area. In 1764 he served as sheriff of Prince William and was a justice for that county in 1768.

In 1771, with a starting capital of £5,000, Charles Yates (c.1728-1809) of Fredericksburg, Daniel Payne of Dumfries, and Edward Moore of Falmouth formed a partnership of Payne, Moore, and Company, merchants of Falmouth. Their store was in the large brick building now known as Lightner's Store. Daniel was the principal manager of the firm, which, prior to the Revolution, carried on a lucrative foreign trade. At the outset of the war, Moore removed from the area as an epidemic of smallpox swept the Falmouth region. Payne remained and took over sole management of the firm's affairs. Yates was occupied with his own business in Fredericksburg. At the close of the war, the partnership was dissolved and Daniel continued in business on his own.

In 1777 Daniel Payne, Arthur Morson (1734-1798), James Buchanan (c.1737-c.1780), William Love, and Edward Moore were named trustees for Falmouth, Daniel serving as treasurer.

Daniel never married and died at his residence above one of his Falmouth warehouses. He was buried in the family cemetery at Cedar Hill, three miles from Leedstown.

[519] In these early inventories, "do" is used as an abbreviation for "ditto."

Mr. A. Hunter's Room

1 Small Chest of drawers £3, 1 do. Oval Table 20/, 5 Walnut Chairs 55/, 1 Bed bedstead & furniture £12, 1 pr. Broken And Irons 6/, 1 Warming pan 6/, 1 Towel 2/6.

Little Room next the Counting House

1 Chest of drawers £4, 1 Bed furniture &ct £12, 1 small Table 5/, 2 Walnut Chairs 20/, 15 old pewter dishes 50/, 40 do. plates & 40 do. plates 40/, 12 Queens China Dishes 40/, 72 do. plates & 11 small 28/, 1 Tea box 5/, 1 Soop Spoon 36/, 12 Tables do 144/, 10 do spoons 30/ 4 Candle sticks and 5 V. Sunfford 50/, 2 doz. Knives & 2 do forks with 2 boxes 20/, 4 Small Chafin dishes 10/, 60 Tumblers unpacked 30/, 2 Bun Cans 5/, 1 Coffee Pot 5/, 5 Stone potts 6/3, 14 Wine glasses 7/4, 1 Large bottle 3/, 2 Punch Cans 5/, 3 Small Decanters 10/, 2 Mugs 4/, 9 China Cups 10 do. Saucers & Tea Pott and Board 16/, 1 dozen Phiols 1/6, 6 p. Tongs & 3 Shovels 40/, 1 pr. And Irons 6/, 14 ½ pr. Sheets at 15/ £10.17.6, 5 Counterpanes £6, 5 pr. blankets £3.10, 6 damask Table Cloths £7.10, 2 do. 40/, 2 ditto 12/, 10 Virginia made ditto £9, 3 Small do. £1.10, 10 do. Towels 25/, 1 ditto 2/6, 2 Twilights 5/, ½ doz bro. lin, ditto 5/, 17 Pillow Cases 36/, 2 Beds with Bedsteads & 2 Counterpanes £16, 3 old Walnut Chairs 26;k 1 Small pine Table 5/,1 Wash bason 1/3, 1 Pair And Irons 16/.

Left Room up stairs

1 Bed bedstead & furniture £10, 3 Walnut Chairs 30/, 1 Small pine table, 1 Corner to ditto, and 1 Towel 7/6, 1 Pair And Irons 10/.

Counting House

1 Bed and furniture at Mitchells £8, 1 Bed Bedstead & furniture £8, 1 Writing table & two stools 50/, 1 Walnut Table 15/, 1 Liberty Stove[520] and Irons, Tongs, & Shovel £6, 1 pr. old hand bellows 3/9, 1 Jugg 3/, 1 pr. Money scales, weights, Box &c 10/, 1 wash bason & stool 1/6, 3 old chairs 7/6, and shovel 15/, 1 pr steelyards 15/, 6 Cast Iron Kettles 28/, Bed & furniture 120/, 1 Case with pint bottles 12/6, 1 pr Copper Scales, and weights 48/, 1 pr Money do & do 6/, 2 Bottles sweet Oil 1/6, 22 Gun locks, @ 5/, 2 pr Boot legs @ 17/, Old scales 1/, 46 Gallons Chain oil[521] at 3/, 165 B_____ Buckets at 2, Carpenters & wheelright tools valued at £15, 1 Sett Coopers ditto, ditto £1.10.

Kitchen

2 Roasting spits, Seives, Ladles, Skinners, Flesh Forks, Cullender, Grid Iron, frying pan, & dripping pan, Tongs, Shovels, Pott Hooks, Griddle, Trivett, and Tea Kettle, £3, 1 pair Large and Irons and 4 Pott Racks £1.10, 7 Potts, 4 Ovens, 5 Iron Skilletts, and 1 Chafin dish £4, 2 Tinn Coffee Potts 3/9, 1 old Copper Kettle 48/, Tubs & Pails 12/, 2 Marble Mortars &c 18/, 4 Iron basons 6/1, 1 Brass Mortar &c 20/.

In the Store

8 Buckskins 80/, 1 Doz Candle Moulds 24/, 1 Bundle T___ 20/, 1 Stock lock, Key & staple 10/, 1 Pad lock and Key 3/, 7 Matts for dishes 9/, 12 drawing Knives 24/, 4 doz. Grinders 5/, 19 small files 8/6, 1 bundle shoe makers tools 6/, 1 Marble slab &c for grinding paint 20/, 1 Piece Ozreal(?) 9s contg. 108 Ells at 1/3 £6.15, 1 Piece ditto 187 ½ £4.7, 7 White Washing b____ 7/, p old money scales 1/6, 6 hammers 12/, 8 Grits 10/, 2 pair stirrup leathers 4/6, 44 pair Negro shoes @ 4/, 13 large Juggs 45/6, 1 Large Bag of Corks 100/, 3 do Cases wh 10 Bottle each 180/, 1 pr Copper scales and weights 36/, 2 Garden Polls 10/, 8 x Cut saws 96/, 1 small -d- Oven 5/, 10 do Kettles 36/, 264 Soldiers Coats at 12/, 172 pr do overhauls at 6/, 183 ditto Jacketts at 6/, 1 Iron Tea Kettle, spider and griddle 10/, 5 Scythes 35/, 3 Small Cases and 15 Bottles 24/, 2 Negro Shirts 6/, 2 do Jackets & Breeches 20/, part of Chest of____ damaged 40/, Quantity of Iron 60 lbs. & staples, Tools, Smiths Tongs, and vices, Sledges &c Damaged valued at £3.10, 1 Cask Sheet Tin 8/, 1 ditto German Steel 60/, 1 pare large Sheers 1/3, 1 Funnel 9/, 12# Saine brads 4/, 208 hats £20, 4 ½ yards Blue Cloth £6.15, 1 pr. Botts 40/, 3 pieces black princes Stuff or Everlasting, about 110 Ells at 6—£33, 1 ditto check Linnen 82 yards at 2/, 6 ½ yards bro Cloth at 7/, 3 do stripes at 20/, 1 remnant of coloured silk 7 yds at 5 ½, 4 pare shoes at 10/, 4 Packs &c paper pins, 20/, 16 Pare White Cotton

---

[520] This was the brand name of a wood stove.

[521] To prevent rusting, iron chains were dipped in a barrel of oil.

stockings £4, 74 Flemish Ells sheeting @ 1.10 ½ £6.18.9, 12 doz. large & 9 do small Buttons 12 ½, 5 pcs. Black Linnen @ 25/, 32 Ivory & 6 horse combs 40/, 1 doz razors 7/6, 1 ½ white thread 18/, 6 yds Shalloon @ 3/6, 8 ½ yds coarse white linnen at 2/, 6 ½ yds coarse white linnen at 2/2, 3# bro thread 10/, 3 pcs narrow Tape 3/, 1 M. Pins 2/, 4 bundles Twine 3/8, 5 Ink stands 7/6, 3 Carpenters Rules 4/6.

Store Loft

9 Sides Sole Leather @ 8/, 10 sides upper ditto at 10/, 55 Small Kettles, Potts, and dutch Ovens @ 3/, £8.5, 56 Slender bars of S___ about 5# each, at 2/6 £7, Old Irons from scow about 300# at 2 £2.10.

76 Negroes Viz. at Mansion House and lower Plantation

Old Jonathan & Judith his wife £30, Sam £20, Kate £15, Pallis and young child £75, Grace £75, Maria £10, Robin 70£, Nelly 30£, Unity, Gu_____, Billy, & Molly 120£, Carter 70£, Betsy 60£, Robin and Peggy 70£, Maria, David, Delphi, and Fanny 120£, Betty, Sally, Philis, Lucy, Winney, Harry & Billy £220, Grace, Sarah, Jimmy & Ben 115£, York 10£, Rodgerd £211, Betty, Cato, Flora, & Dick 85£, Cupid 70£, Tanner(?), Lucy, Lyddia, Billy & Eve £150, Aaron £80, Tina & Cato £65, Sam & Lucy, Elijah & Page £150, Phil & Lucy £120, Tom £50, Fender(?) & Fanny £100, Charlotte, Sally, Charlotte, Live(?), Betsy, and Peter £235, Tim £70, Judith £60, Judith & Molly £70, N___ £45, Kate 45£, Billy £45, Molly £50, Ned & Harry £70, Cuffy £100, Mason, Ben £80, Gerrard £70.

Stock Viz—

1 Bay Mare £5, 1 Grey do £5, 1 Sorrel Horse £3, 2 horses £20, 1 Bay ditto £5, 1 ditto £3, 1 ditto 16£, 1 Old white horse 15£, 1 Bay do 14£, 4 Cows with Calves 14£, 9 Cows 27£, 10 Steers £35, 15 yearlings 19£, 26 old sheep at 9/, 14 Lambs @ do £18, 3 Sows & 10 Pigs £4, 20 Shoats 5£, 1 Boar 20/.

7 axes 35/, 8 broad hoes 24/, 7 Hilling do 14/, 3 Grubbing do 10/, 1 adze 5, parcel old plows &c 20/, 4 pitch forks 18, 5 Scythes & Cradles 25/, 4 half share plows and Coulters at 15/, 1 Barshare do 20/, 1 Dutch p__ 40/,2 pair wedges 8/, 5 pair chains 25/, 5 pair harness 12/6, 2 ox chains 18/, 1 Grind stone 2/6, 1 Old wagon £8, 1 Cart £4, 1 old Pott 4 ½.

At Rappa. Forge 66 Negroes Viz—

Cupid £100, Molly 20£, Titus £100, Ambrose 100£, Kirby Dick 120£, Mochas, Patrice, Jack, Dick, and Charity 320 £, Little Dick 75£, Chick 90£, Sampson 80£, Charles 50£, Liverpool 80£, Tom 80£, Matt 75£, Aaron 75£, Nea 80£, Cheshire 80£, Cain 100£, Peter 100£, M. Jack 100£, N. Billy 100£, Geo 80£, Isaac 80£, Solomon 80£, Caesar 90£, Tony 70£, London 70£, Jim 75£, Tam__ Scipio 50£, w. Jos 100£, Pope 80£, John 70£, S. Jack 70£, Armstead 80£, Jos P. 75£, C. George 70£, Lewis 75£, s. Peter 100£, Daniel 90£, Johnny 75£, Scipio 90£, Daniel 90£, Hector 65£, Ben(?) 70£, Cou__ 90£, Anthony 75£, Sancho 50£, Jack 45£, True blue 60£, Mercules 60£, Solomon 45£, Danekl 65£, Jos P. 90£, Dick 80£, Bob 45£, Neptune 70£, Sam 50£, John G. 36£, Tony 45£, Edward and Saunders 50£, Dangerfield 100£, Jos 100£.

Anchor Shop

4 pair large bellows £20, 1 pr com# do# pr nailers £6, 2 Large Cast anvils about 1000 # at 12/, 2 wrought do 345# £7.10, 2 Bed(?) Irons 30/, 2 Vices 45/, 5 large & 5 small sledges 70/, 3 Setts Hammers & 11 pr tongs part wore 45/, 34 finished anchors & 1 grapling W.G. 546 __ @ 7^{d} £150.7.11, 1000 Bushels pitt Coal w 1/3 £62.10.1, 1 Ton Iron & unfinished work £30.

Smiths Shop &c

18 Pair Smiths bellows £40, 2 __ vices & 1 pr Stillyards[522] 10/, Sett Hammers & Tongs at 15/, 2 Beak(?) Irons 30/, 8 wrot anvils at 100/, 4 Cast do at 12/, 2 Setts horse shoeing tools 15/, 1 Ladle mould 5/, 1 Cast Mandrill 12/, 1 wrot ditto 25/, 2 pr __ Shaves(?) 25/, 167 weeding Hoes at 3/6, 338 hilling do at 3/, 98 grubbing do at 3/, 45 Mattocks at 6/, 101 Broad Axes at 6/, 80 Poll do at 5/, 11 adzes at 4/, 22 Small B__s

[522] More commonly spelled "steelyard." This was a heavy steel or iron arm that could be hung from a ceiling beam or other support. Heavy objects were hung from the short end and counterweights hung on the long arm were used to create, essentially, a balance beam for weighing heavy objects such as iron.

at 2/9, 10 pr Chains at 9/, 16 Curry Combs at ½, 1 Fr__ 3/, 2 Dutch plows 13/6, 5 pr polished and irons @ 18/, 23" spikes at 10d[523], 4 Barshare plows 63# 40/, 44 Hammers 22/, 14 Spades at 5/, 7 do half finished at 2/6, 1 Drawing Knife 1/6, 1 Ladle 2/, 20 wedges 10/, 4 Lock Chains 48/, 2 pr breast Trigpins &c ¾, 1 Pitman Iron &c 4/, 160# brands(?) &c @ 6s, 186 Ton & nails @ 5/, 38 part finished work @ 5/, 51# Slit Iron at 4 ½/, 2 Millspindles &c 187# at 9d, 49 __ Bar Iron at 4d, 14# German Steel at ___, 95 Musket barrels @ 6/, 700 old Iron at 4d.

Counting House &c

32 Horsemans Swords at 5/, 1 Large Rifle 36, 21 pr Common Stirrups @1/, 135 Curb bitts 2/, 116 Bridle Bitts @ 1/, 19# Nails @ 9d, 1 Muskett with Bayonet 45/, 15 ½# Tallow 7 ½, 340 Sole Leather @ 1/, 2 Kip skins, 1 hog do 30/, 16 sides Negro upper leather & do harness @ 7, 2 writing tables and stools 45/, 1 small Table 5/, Johnsons Dictionary 2 volumes 18/, 3 old chairs 3/9, table 5/, Johnsons Dictionary 2 Volumes 18/, 3 old chairs 3/9, 2 Jugs 8/, 2 Brass Candlesticks 5/, 1 pr Andirons and Tongs

Tannery

2 Fleshing Knives, 1 Currying do., 1 Steel 15/, 179 Hides in Tan @ 10/--£89,10.17, 17 Kip skins @ 7/6, 5 Calf do @ 5/, 7 Sheep do at 3/, 2 dog skins @ 2/, 6 horse hides 18/.

Iron House

5.9.2 C 2 grs. o.m. Bar Iron at £33 169£.2S.6D., 8b.3g.0h Anchors at 20/, £8.15, 16 Tons 15 __ Pig Iron a £12 £201, 6b, 1 qrs. old Iron hoops, &c @ 4& £11.13.4-C- 2-2g-11u ditto at 4 £4.1, C.14—2 qr Sl___ Iron at 33/4 £24.3.4, 1 pair Scales and sett weights £20, 80# Rolled Iron 40/, 12 Sheet @ 9, 5 Setts Wagon Boxes 12(?), 1 New ox wagon valued £20, 250 Loads Coal at 18, 350 cords wood at 2 £260, Ben's Team with four horses, waggon & Harness £80, Dick's ditto £100, Billie's do £70, Hecktors do £80, Three spare horses £40, Three Coal waggon bodies £6, two pair of new wheels 72/, One old Log carriage[524] £60.

Gristmill

Forty nine bbl Flour packed at 30/, 110 do 17/, 1300 Seconds @ 10/, 19 4# 2 Ship Stuff @ 6/, 11 bus. of Bran 6/10 ½, 35 ditto Shorts @ one Third, 52 dito Corn @ 2/, 4 ½ Screenings 2/9, 24 do wheat @ 5/6, 3 ½ do Rye @ 2/6, 19 empty flour bbls @ 2/3, 1 pr Scales, weights &c., 12 bbs, 1 pr french burr grind Stones, new, £35.

Houses &c

Five beds with furniture &c, £24, 15 axes £3.15/, 2 dishes, Six plates, 1 pine Table, 6 spoons & 2 basins 20/, five Cows, four calves & bull £27, 3 pine tables 15/, 18 chairs 36/, 1 arm chair 4/, 2 pr Andirons, 1 pr tongs & shovel, 30/6, 1 Tureen, 4 queens China dishes 13/6, 26 dito plates 6/, five Stone Pots 15/, 17 Pewter plates & three dishes 30/, 1 Coffee pot 4/, 1 pine chest 5/, 1 Trunk 15/, 6 China Cups & six saucers 9/,9 queen China dito & nine do 4/, 3 dito Tea Pots 3/6, 2 milk Pots 3/9, 1 Tin Tea Kettle & one iron dito 5/6, 2 large dishes 5/2, 2 Tea Canisters 2/6, 1 2 qt bowl queen China 2/, 1 Sugar box 12/, 1 qt do 1/, 2 pt do 1/, 1 pint do China 1/6, 1 Decanter & ½ doz wine glasses 6/, ½ doz queen China Basins 7/6, 1 large Kettle 21/, 2 do pots 20/, 1 middle sized pots five shillings, 1 small do 3/9, 1 Chaffen dish 3/, 1 dutch Oven 7/6, 1 frying pan 3/, 3 pot racks & four pr hooks 27/, 1 pr broken and irons & shovel 7/, 1 grid iron 5/, _erbs Pails &c 18/.

"Stanstead Plantation"

Robin & Daphney £100, Aggie & her child Daphney £80, Bob £75, Clara £60, Alexander £50, Christiana £40, Mary £25, Esther £30, Patty £50, & her children Viz. Nero £40, Yorick £35, Dangerfield £30, Billy £15, Sucky £70, Serger £40, Jenny £15, Jenny £ 20, Rhoda & child £70, Isom £35 Jenny £65, Eve £20, Tina £60, Willis £15, Rose £35, Jerry £35, Tillian £15, Old Frank £20, Amy £35, Dover £36, 4 work horses £40, 1 yearling cold £3, 7 do. £24.10/-, 1 Steer £3.10/-, 1 Bull £4.10/-, 3 cows £30.10/-, 2 Heiffers 2 years old 100/-, 2 yearlings 40/-, 2 sows @ 25 shillings, 1 Boar 25/-, 34

---

[523] In these old inventories, use of the lower case letter "d" means pence.

[524] Possibly a wheeled device used for hauling logs from the woods.

shoats @ five shillings, 8 oxes 32/-, 6 grubbing hoes 22/-, 8 hilling do 28/-, 9 weeding do 26/-, 2 pr wedges 10/-, 1 dutch plow & coalter 8/-, five pr chains 35/-, four half fluke & coalters 30/-, four pr Hames 10/-, 1 Drawing knife 2/-, 1 grindstone 7/6, 1 two winged plow six shillings, 1 ox Waggon £15, 2 Ox chains 15/-, 1 Harrow 8/6, 2 pitch forks 2/6, 1 curry Comb 1/-

Aka. [Accokeek] Furnace

Godfrey £100, Milly £70, Aaron £30, Sarah £20, Priscilla 15£, Salt £60, Kate 40£, Cyrus £75, M__ron £35, Peggy 60£, 3 old horses £21, 3 Cows 9£, 1 Heifer 50/-, 1 yearling 18/-, 3 calves 37/6, 1 Bull 25/-, 4 Oxes at 65/-, 1Steer 60/-, 2 Sows at 20/-, 13 Shoats at 12/-, 15 Pigs at 2/-, 5 Axes at 4/-, 3 Wedges 7/6, 4 Grubbing hoes @ 3/9, 6 Weeding hoes at 3/-, 5 hilling do @ 3/-, 5 half shares @ 7/6, 1 Barshare &c 20/-, 1 Ox Cart & chains £5, 2 Pick axes 6/-, 1 Grind Stone 2/6

Forge K.[itchen]

1 Skillet 10/-, 1 Iron spice mortar 4/-, 3 Candlesticks, Snuffer & Stand 12/, 1 Chocolate Pot 1/6, 1 Metal Mortar 10/-

Rodger £70 Amount £12735.16.4

Anthony M. Kitterick
Daniel Payne
William Taylor
David Briggs

Stafford Sct.
November Court 1785

This inventory & appraisment of the estate of James Hunter Esquire decd, in Stafford County, being returned by his Executors, was ordered to be recorded & is so.

Test.
Thos. G.S. Tyler C. St.

In Culpepper

Appraisment of the Estate of James Hunter Esq decd. to us produced as agreeable to an order of the Court of Stafford dated April 1785.

Negroes viz.

George £25, Cully £25, Joe 40, Sam 90£, Harry 75£, Edmond 75£, Peter 50£, James 65£, Dick 60£, Hannah 60£, Effy 85£, Judy 65£, Milly 90£, Phebey 65£, Joan 80£, Aggy 85£, Pegg 70£, Daphney 85£, Rachel 85£, Molly 70£, Nan 70£, Nan 60£, Robin 30£, Cheshire 40£, Griffith & child Frank 75£, Kate 50£, Daniel £25, Dianna, Jack, and Molly 130£, Alice & child Jack 105£, Hannah, Pallis, & Will 130£, F___ 50£, John 30£, Winney 25£, V___s and Flora 105£, Alice 35£, Sou__ 50£, Lucy and child Sam 60£, Nelly 75£, Tiller 75£, Bob 40£, Phebe, Emanuel & Esther 150£, Judy 20£, Adam 40£, Will 45£, Susie 45£, Nancy and Betty 55£, Harry 50£, Joe 60£, Charles 70£, Harry 40£, Ben 90£, Perry 50£, Patty 80£, Fanny & child Betty 105£, Jenny & child Sylva 105£, Dianna 40£, Cooper Billy 90£, Hezekiah 70£, Mary 75£, Baylor £90, 1 old Bay Mare 6£, 1 Black do & colt 12£, 1 do horse 10£, 1 Bay Colt 5£, 1 ditto 7£, 1 Black Colt 10£, 1 Bay do £7.10, 1 Bay Colt 9£, 1 do horse 12£, 1 do horse 12£, 1 do Mare 9£, 1 dark do mare 15£, 1 do horse 15£, 1 Cht sorrel horse 10£, 1 do 18£, 1 black do 9 £, 1 Bay Mare 5£, 1 Do 12£, 1 Road do 8£, 1 Black do & colt 8£, 1 Colt 2£, 1 Colt Harry 8£, 1 Roan Colt 3£, 1 Bay horse 15£, 1 do mare and young Colt 13£, 1 Three year Colt 8£, 1 yearling colt 4£, 64 head of Cattle @ 35/--£112, 2 Work Oxen 10£, 18 do at 40£, 19 sheep at 7—10 ditto 88/-29 Hogs @ 20/-, 27 Shoats @ 6/-, 3 Sows & pigs @ 26/8, 45 hogs at 17, 26 Pigs at 3/-, 11 Small do @ 1/6, 4 Sows 100/-, 22 Shoats @6/-, A parcel Carpenter & Coopers tools 3£.10, 33 Hoes 2/6, 15 Axes @ 4/-, 4 Wedges 8/-, 1 Harrow 12/-, 4 old Plows 4/6, 6 pr Chains @ 5/-, 6 C___ @ 1/-, 7 Singletrees 1/3-, 1 Cutting box 3/-, 2 Hoes 6/-, 1 old wagon wheel 20.-, 1 Barshare Plow 18/-, 1 old wagon 5£, 8 hoes @ 3/-, 2 axes 5/-, 23 Hoes 3/-, 1 pr wedges 5/-, 2 pr hard mill stones 40/-, 5 Empty Casks 5/-, 23 Hoes 3/-, 1 pr wedges 5/-, 4 Axes 4/-, 1 Spade 5/-, 13 old plows @ 5/-, 4 do fixed @ 10/-, 1 Barshare do 18/-, 2 Coalters &c 6/-, 5 pair hames & 5 pr Chains 45/-, 1 old Cart 4£, 1 do wagon 9£, 2 Grind stones 6/-, 4 Scythes & 4 Cradles @ 7/6, 2 Grass Scythes @ 5/-, 1 old hand saw 2/6.

Amt. in Culpepper £4542.17.3

Stafford Sct. November Court 1785

Robt. Coleman[525]
Gabriel Long[526]
William Green

This Inventory & Appraisment of the Estate of James Hunter Esq dec in Culpepper County being returned ____ by his Executors, was ordered to be recorded and is so.

Test.
Tho. G.S. Tyler C.S.C.

In Fauquier

Appraisment of the Estate of James Hunter Esq. Decd. to us produced, made agreeable to an order of ye Court of Stafford dated April 1785.

27 Negroes Viz.

Mucker Val 30£, Jacob £45, Will £90, Marcus £90, Will 90£, Ben 75£, Boy Marcus 65£, Ned 50£ Alex £60, Phillis & her child Isaac £55, Cibby 60£, Jenny 70£, Judy 30£, Sarah 70£, Dorcas £70, Milly & child Jenny £100, Betty £30, Aggy £50, Daniel £40, Charity 55£, Milly 40£, Aaron £27, Nancy £40 Salt £40, Emanuel £40, 1 young bay mare 40£, 1 do 8£, 1 Roan Mare £12, 1 Bay horse 10£, 1 Bay yearling Colt £6, 1 old gray horse 30/-, 6 work oxen £21, 6 cows £18, 6 cows with calves £21, 12 young Cattle 18£, 34 Sheep @ 5/--£8.10, 2 large hogs @ 17/--£17.17, 32 Pigs @ 2/6—£4, 24 Shoats @ 6/--£7.4 2 Hides 8.-, 1 Cutting box 4/-, 1 old wagon £10, 1 Cart 4£, 1 do wheat Fan £3, 7 half share plows 52/6 5 pair chains & hames 55/-, 4 Dutch plows &c 32/-, 7 Weeding hoes 21/-, 7 Hilling do 21/-, 7 axes £1.15, 7 Grubing hoes 21/-, 2 old grindstones 5/-, 2 Ox Chains 15/-, 5 wedges 7/-, 2 chisels, 2 augurs, 1 Old hand saw & pr old steelyards 12/-

Amount in Fauquier £1606.9.6
Wm. Grant
Lunfield Sharpe
William Woodside

Stafford Sct. November Court 1785

This Inventory and Appraisment of the Estate of James Hunter Esq decd in Fauquier County being thus returned by his Executors, was ordered to be recorded and is so.

Teste
Tho. G. S. Tyler C. S. C.

---

[525] Robert Coleman (died c.1793) died intestate in Culpeper County. According to the British Mercantile Claims, Robert "was far advanced in age and had for many years before broken up house keeping and lived alternately with his children. Although he did not possess much property he was generally considered an honest man."

[526] Gabriel Long (1751-1827) married first Elizabeth Stubblefield and secondly Ann "Nancy" Slaughter.

Appendix D

Will of Adam Hunter
May 16, 1798

In the Name of God amen I Adam Hunter of the County of Stafford and State of Virginia, being of a sound and disposing mind and memory, and intending to dispose of what Estate and Interest I now have, Do make, publish, and declare this as my last Will and testament , hereby revoking all others.

Imprimis, I give and devise all the Estate I have both real and personal unto my much esteemed Friends Mr. James Somerville and Mr. Henry Mitchell of the town of Fredericksburg Merchants to them their Heirs and assigns forever, upon Trust, nevertheless to maintain and support the uses of Trust herein after mentioned and none other, and confiding in the well known Honour and Integrity of the said James Somerville and Henry Mitchell It is my Will that they have the most ample power over my Estate that any Trustees would have, and desire that the most liberal conduct may be observed between my Brother Archibald Hunter, my intended Heir, in the settlement of all accounts relating to my Estate, with a genteel and friendly allowance for their trouble in manageing my Estate during the Continuance of their Trust.

Secondly, The Trust I mean are as follows, namely, To the use of my said Brother Mr. Archibald Hunter, now residing at Dunkirk in the Kingdom of France, and to his heirs and assigns forever, in fee simple, and it is my Will, that so soon as my said Brother, or any Devisee of his, shall become a Citizen of the United American States, and be thereby capable of accepting of a grant of real Estate in America, that the said James Somerville and Henry Mitchell release and convey until the said Archibald Hunter, in Fee Simple, all the Estate I have now in America or else where; of if my said Brother shall depart this Life, before he shall have become a Citizen as aforementioned, then I desire that the said James Somerville and Henry Mitchell release and convey to the Child or Children of the said Archibald Hunter, according to their relationship to my Brother, and the discretion of the said James Somerville and Henry Mitchell. But should my Brother die Testate, then the Conveyance to be made to his Devisee or Devisees, according to the Devises in his will, and the better to secure the use of my Estate to my said Brother his Children and Devisees. It is my Will that in Case the aforementioned Devisee of the use of my Estate to the sd. Archibald Hunter in Fee Simple cannot be supported according to the law of the American United States, then I declare them void, and it is my will that the said James Somerville and Henry Mitchell, have and hold the fee simple Interests in my said Estate in Fee upon Trust as follows—To the use of the said Archibald Hunter his Executors, Administrators, Assigns for and during, and to the full end and term of none hundred and ninety nine years, Remainder to the use of my right heir, or such other Heir as may then be a citizen of America.

Also it is my Will, that the Devise in my late Brother James Hunter's Will in favour of his two nephews James Hunter and Patrick Home be carried into due and full Execution, and that the said James Somerville and Henry Mitchell be aiding in perfecting their Title to such Estates as my said Brother James has devised them, according to the true intent and meaning of such Devises, and in Case the Law shall determine that the Estates so devised the said James Hunter and Patrick Home by my said Brother James, does not vest in them for want of capacity as Citizens of America, but is vested in me, under the general Devise of my said Brother James's Will then I desire that moiety so devised to the said James Hunter and Patrick Home by the said James Hunter and so vested in me unto the said James Somerville and Henry Mitchell in Fee Simple upon Trust to, and for the use of the said James Hunter and Patrick Home respectively, as fully and amply, and under the same Conditions, Proviso's and Limitations as are herein before expressed and declared in favour of my Brother Archibald, as to my own proper Estate and desire, that the said James Somerville and Henry Mitchell may Confirm the Interests of the said James Hunter and Patrick Home, therein as is directed to be done to my Brother Archibald.

Lastly I hereby charge my Estate with the payment of all just debts due from my Brother James Hunter deceased or myself, and as the same is fully sufficient to pay all demands against it, I direct by Executors herein after named, shall not be obliged to give Security upon undertaking the execution of this my last Will and I appoint my said Brother Archibald Hunter, and my said Friends James Somerville and Henry Mitchell my Executors. In Witness whereof I have signed sealed, and published this as my last Will and Testament—the whole being wrote with my own hands, this Seventh day of June, in the year of our Lord, one Thousand Seven hundred and Eighty Seven.

Sealed published & declared to be the
last Will & Testament of Adam Hunter
by the said Adam in our presence, &
attested by us & by subscribing our
names as witnesses thereto, in his
presence & at his desire.

H. McAusland[527]
James Robb[528]
Andw. Glassell[529]

At a District Court held at Fredericksburg May the 16th 1798 The Last will and Testament of Adam Hunter decd. was proved by the Oath of James Robb one of the Witnesses thereto, and the Executors named in the said Will being dead, on the motion of Joseph Ennever who made Oath and together with David Briggs Nicholas Payne and Patrick Home his Securities entered into and acknowledged their Bond in the penalty of Five Thousand Pounds conditional as the Law directs Certificate is granted him for obtaining Letters of Administration on the said Decedents Estate with his will annexed in due form.

---

[527] Humphrey McAusland was a Scots merchant of Fredericksburg.

[528] James Robb was a merchant and resident of Orange County. He was in business with Robert Bogle, Sr., Robert Bogle, Jr. and William Scott in the mercantile firm of Bogle, Scott and Company, merchants of London.

[529] Andrew Glassell was a merchant from Galloway, Scotland. The son of Robert Glassell and Mary Kellon, he came to Virginia in 1738. His brother John established himself in Fredericksburg and Andrew settled in what is now Madison County, Virginia on a plantation he called Torthowold.

Appendix E

Inventory of the Estate of Adam Hunter
May 31st, 1798

We David Briggs, Nicholas Payne, and Patrick Home, Agreeable to an Order of the District Court of Fredericksburg dated May 16th, 1798 met at the late dwelling house of Adam Hunter Esqr. decd. & thereon the 31st of the same month appraised the Estate of the said Decedent according to Law.

An Old Bay horse	12. 0. 0
A Park Phaeton	15. 0. 0
1 Bureau	5. 0. 0
1 Chest of drawers	3. 0. 0
1 portable Mahogany Desk	.12. 0
1 Black Walnut bedstead	.5. 0
1 pr. of Silver shoe buckles	1. 0. 0
1 pr. do. Knee do.	.7. 6
1 pr of Spectacles	.1. 6
Case of Mathematical instruments, Pocket microscope	1. 0. 0
A stone stock buckle	.12. 0
Do. breast do.	.7. 6
2 Snuff Boxes	.10. 0
1 Pair of horsemans Pistols Silver mounted	5. 0. 0
A Silver Segar tube	.2. 0
Masonic medal	.1. 6
1 Pr. of Gold Sleeve buttons	2. 2. 0
2 Seals and a watch key	1. 10. 0
1 Gun	2. 0. 0
1 Trunk covered with leather	.10. 0
Annual Register 16 vol 8 v.	3. 4. 0
Anecdotes of the Emperor Joseph the 2nd	.1. 0
Atkinside on the pleasure of Imagination	.2. 0
Andersons History of Commerce 2 vol Flo.	2. 0. 0
Anecdotes of Samuel Johnson	.1. 6
American Atlas	2. 0. 0
Bible 41o	.10. 0
Bible 12 mo.	.2. 0
Biographical Dictionary 11 vol 8 v.	3. 6. 0
Beaths Essays 41o	.6. 0
Bloody Register	.3. 0
Boyers French Dictionary	.4. 0
Boswell Tour	.5. 0
Bozzy & Piozzy by P. Pindar	.1. 6
British Antiquities	.2. 6
Beauties of Fox North & Burke	.1. 6
Bells Travells 2 vol 8v	.8. 0
Baileys Dictionary	.9. 0
Byrons Narations 8vs	.3. 0
Churchills Lemons 8vs	.5. 0
Castellies Bibles 4 vol 2 ms	1. 10. 0
Curate of Craman 2 vol. 8vs	.6. 0

Cooks Voyages 2 vol 4. to with 1 vol of F plates	5. 0. 0
Dalrymples Essays 41o	.2. 6
Dalrymples Travells 41o	.4. 0
Devils on Sticks 12 ms	.2. 0
Dialogues of the Dead 8vs	.6. 0
Darts Antiquities of Canterbury F	.12. 0
Dictionary of Husbandry 2 vol 8vs	.5. 0
Epigoniad 8vs	3. 3. 0
Excursions from Paris to Fontaine bleau 8v	.2. 0
Essays on Crimes 8vs	.5. 0
Every man his own Broker 8vs	.1. 6
Forbes Works 2 vol 12 mo	.5. 0
Fieldings do 6 vol 4 to	1. 10. 0
Footes plays 2 vol	.5. 0
Fragments of Ancient poetry	.1. 6
Folly of Reason	.3
Fugitive pieces in verse & prose	.1. 6
Grays Debates of the house of Commons 10 vol 8 ms	2. 10. 0
Gordons History of the American War 4 vol 8 vs	1. 4. 0
Gutheries Geographical Grammar 2 vol 8 vs	.15. 0
Gentlemens Magazine from AD '83 to 87	.18. 0
Harveys Tracts 12 ms	.2. 0
History of England Smollet 15 vol 8vs	3. 15. 0
Do. of the Revolution of Sweeden by Sheridan 8vs	.4. 0
Do. of Sterlingshire by Nimnas 8vs	.5. 0
Do. of Philip the 2nd King of Spain 3 vol 8vs	.15. 0
Do. of Ayder Ali Khan 2 vol 12 ms	.5. 0
Hutchersons Moral Philosophy	.3. 0
Journey to the Highlands of Scotland 12 ms	.1. 6
Kalms Travells in N. America 2 vol 8vs	.10. 0
Keils Philosophy 8vs	.4. 0
London Magazine 2 vol for AD '44 & 45	.8. 0
Leightons Sermons 8 vs	.5. 0
Lexiphanes 12 mo	.2. 6
Love of fame 8 vs	.2. 0
Life of Gardner 12 mo	.2. 0
Mulsos Cattistus 12 mo	.2. 6
Memoirs of Mrs. Pilkington 8 vs	.1. 6
Mariners Compass 8 vs	.1. 0
Monthly reviews from AD '83 to 87	.18. 0
Margot Lee Raadueze 12 mo	.6
Naval History	.5. 0
Hew Dispensatory 8 vs	.5. 0
Observations upon the prophecies of Daniel by Sir Isaac Newton 41o	.5. 0
Philips Poems 12 mo	.2. 0
Pollitical Essays 4 to	.10. 0
Plays (a Collection of 16)	.8. 0
Posttewayts Dictionary F	1. 0. 0
Psalms of David	.5. 0
Parliamentary History 26 vol 8 vs	6. 10. 0
Do. Registers 30 do.	7. 10. 0
Peerage of Scotland 8 vs	.5. 0
Pamphlets	.10. 0
Quincys Dispensatory 8 vs	.4. 0

Rosseau Eloisa 4 vol 12 mo	.10. 0
Review of the history of the man after Gods own heart 8vs	.3. 0
Rational Recreation 4 vol 8 vs	.12. 0
Robertsons history of Scotland 2 vol 4 to	.15. 0
Roderick Random 2 vol 8 vs	.4. 0
Scots Magazine 2 vol 8 vs for AD '77 & 78	.8. 0
Spectator 8 vol 8 vs	1. 4. 0
Sheffield on Commerce 8 vs	.3. 0
Stackhouse's history of the Bible 2 vol F	3. 0. 0
Savarys Letters on Egypt 2 vol 8 vs	.12. 0
Samson Oratorio	.1. 0
Smollets plays	.3. 0
Salust Works 12 mo	.1. 6
Siege of Gibraltar 4 to with plates	.10. 0
Town & Country Magazine 1 vol 8 vs	.5. 0
Tristram Shandy 9 vol 12 mo	.18. 0
Telemachus 2 vol 8 vs	.4. 0
Tour through Great Britain	.7. 6
Titsot on Health 8 vs	.5. 0
Thicknesse Journey 2 vol 8 vs	.8. 0
Treatise on Bees 8 vs	.3. 0
Town Ecilogues by Erskine	.1. 0
Tillotsons Worls 12 vol 8 vs	3. 0. 0
Vie Du Marchall Duc De Villars 41 mo	.8. 0
Williamses Northern Government 2 vol 4 to	.10. 0
Welsted on the conduct of Providence	.2. 6
Yoricks Sermons 6 vol 12 mo	.15. 0
Aristotles Compendium of the art of logic	.1. 6
Ceasars Commentarys	.2. 0
Cicero's on the nature of the Gods	.1. 6
Do. works compleat 4 vol 41o	1. 4. 0
Coles Dictionary	.2. 0
Grotius upon the truth of the Christian Religion	.1. 6
History of Alexander the great	.1. 6
Horace	.2. 0
Livy 2 vol	.4. 0
Newtons institutes 2 vol	.5. 0
Salust	.2. 0
Terence	.1. 6
Ovids Metamorphasis with notes	.2. 0
Latin Grammar	.5. 0
Euripedes [Greek]	.1. 0
Grammar [Greek]	.1. 0
Testament [Greek]	.3. 0

David Briggs
Nichos. Payne
Pat. Home

At a District Court held at Fredericksburg October the 16th, 1798

This Inventory and appraisement of the Esta. of Adam Hunter decd. was returned and ordered to be recorded.

Teste

J. Chew[530] Ck. D. C.

a Copy Teste
R. Hening Ck. D. C.

[530] John Chew (1753-1806) was the son of Robert Chew (c.1730-1778) of Spotsylvania County. Around 1778 John married Elizabeth Smith (1759-1806). He was clerk of the court for Spotsylvania County from 1787-1802 and also served as clerk of the District Court of Fredericksburg.

Bibliography

Abbot, W. W., ed. The Papers of George Washington, Colonial Series. Charlottesville, VA: University Press of Virginia, 1988.

Abbot, W. W., ed. The Papers of George Washington, Revolutionary War Series. Charlottesville, VA: University Press of Virginia, 1988.

Abercrombie, Janice L. and Slatten, Richard. Virginia Revolutionary Publick Claims. Athens, GA: Iberian Publishing Co., 1992.

Account ledger, Edward Voss with John Strode, 1790-1791, Maryland Archives, MSA SC 2728-B9-F2.

Account ledger, Col. George Baylor's account with James Hunter, Sept. 27, 1777 to June 19, 1779, Alderman Library, University of Virginia.

Allason, William, records, 1722-1847, reel nos. 1359-1382, Library of Virginia, Richmond, Virginia.

American Memory. http://lcweb2loc.gov:

Journal of the Continental Congress, Mar. 13, 1777, p. 173, order for 80 stands of arms.

Journal of the Continental Congress, Aug. 20, 1778, p. 823, James and Adam Hunter to pay semen on brig *Morris*.

Journal of the Continental Congress, Nov. 13, 1778, p. 1130, warrant to Amos Strettel for $7735 45/90.

Thomas Jefferson Papers, Series 1, General Correspondence, 1651-1827, Letter, John Strode to Thomas Jefferson, Feb. 26, 1801, Image #1099.

Thomas Jefferson Papers, Series 1, General Correspondence, 1651-1827, Letter, John Strode to Thomas Jefferson, June 26, 1803, Image #683.

Thomas Jefferson Papers, Series 1, General Correspondence, 1651-1827, Letter, Jefferson to Strode, Mar. 11, 1805, Image #795.

Thomas Jefferson Papers, Series 1, General Correspondence, 1651-1827, Letter, John Strode to Thomas Jefferson, Mar. 25, 1805, Image #42.

Thomas Jefferson Papers, Series 1, General Correspondence, 1651-1827, Letter, Jefferson to Strode, Dec. 15, 1805, Image #1302.

Thomas Jefferson Papers, Series 1, General Correspondence, 1651-1827, Letter, John Strode to Thomas Jefferson, Jan. 7, 1806, Image #187.

Thomas Jefferson Papers, Series 1, General Correspondence, 1651-1827, Letter, John Strode to Jefferson, Jan. 12, 1806, Image #242.

Thomas Jefferson Papers, Series 1, General Correspondence, 1651-1827, Letter, John Strode to Henry Dearborn, Jan. 18, 1806, #319.

Thomas Jefferson Papers, Series 1, General Correspondence, 1651-1827, Letter, John Strode to Jefferson, Jan. 20, 1806, Image #317.

Thomas Jefferson Papers, Series 1, General Correspondence, 1651-1827, Letter, John Strode to Thomas Jefferson, Apr. 10, 1806, Image #1019.

Thomas Jefferson Papers, Series 1, General Correspondence, 1651-1827, Letter, John Strode to Jefferson, Apr. 18, 1808, Image #380

Anderson, Patricia A. "Fielder Gantt (c.1730-1807) of Fielderia Manor," *Western Maryland Genealogy*, vol. 15, no. 1, (January 1999), pp. 3-15.

Balch, Thomas, ed., Papers Relating Chiefly to the Maryland Line During the Revolution. Philadelphia, PA: T. K. and P. G. Collins, Printers, 1857.

Banks, Henry. The Vindication of John Banks of Virginia. Frankfort, KY, 1826.

Barry, Joseph. The Strange Story of Harper's Ferry. Martinsburg, WV: Thompson Bros., 1903.

Bathe, Greville and Dorothy. Oliver Evans: a Chronicle of Early American Engineering. NY: Arno Press, 1972.

Bining, Arthur C. British Regulation of the Colonial Iron Industry. Philadelphia: University of Pennsylvania Press, 1933.

Bining, Arthur C. Pennsylvania Iron Manufacture in the Eighteenth Century. Harrisburg, PA: Pennsylvania Historical Commission, 1938.

Bining, Arthur C. Pennsylvania's Iron and Steel Industry. Gettysburg, PA: The Pennsylvania Historical

Association, 1954.
Bockstruck, Lloyd D. Virginia's Colonial Soldiers. Baltimore, MD: Genealogical Publishing Co., 1988.
Bowen's Virginia Sentinel, Sept. 15, 1794, George Wheeler's advertisement for a hatter
Boyd, Julian P., ed. The Papers of Thomas Jefferson, Princeton, NJ: Princeton University Press, 1952.
Branchi, E. C., translator and editor. "Memoirs of the Life and Voyages of Doctor Philip Mazzei." *William and Mary College Quarterly Historical Magazine*, Second Series, vol. 9, no. 3, July 1929, pp. 161-174.
Brent's Mill Ledgers, 1804-1806, Alderman Library, University of Virginia, Charlottesville, VA.
"British Mercantile Claims." *The Virginia Genealogist*, vol. 28, no. 3, July-September 1984, pp. 222-224.
Brown, Stuart E. Virginia Baron: the Story of Thomas 6th Lord Fairfax. Berryville, VA: Chesapeake Book Company, 1965.
Bruce, Kathleen. "The Manufacture of Ordinance in Virginia During the American Revolution," *Army Ordinance*, The Army Ordinance Association, vol. VII, no. 39, pp. 187-193, (part 1), November/December 1926; vol. VII, no 41, pp. 385-391, (part 2), March/April 1927.
Callcott, Margaret L., ed. Mistress of Riverside: the Plantation Letters of Rosalie Stier Calvert, 1795-1821. Baltimore, MD: The Johns Hopkins University Press, 1991.
Caroline County Order Book, 1778-1781, p. 178, June 10, 1779, suit between Alexander Hanewinkel and Robert Gilchrist.
Carroll, Charles of Carrollton, Papers:
Item #5415, deed, June 28, 1792, Manuscript Department, Maryland Historical Society.
Item #5416, deed, June 28, 1792, Manuscript Department, Maryland Historical Society.
Carter, Edward C., ed. The Virginia Journals of Benjamin Henry Latrobe, 1795-1798. New Haven, CT: Yale University Press, 1977.
Catanzariti, John, ed. The Papers of Thomas Jefferson, Princeton, NJ: Princeton University Press, 1990.
Cecil County Democrat, Sept. 8, 1849, first cotton mill in Cecil County.
Chandler, J. A. C., ed. and Swem, E. G., ed. "Revolutionary Manufactures (petition of Alexander Hanewinkel)." *William and Mary College Quarterly Magazine*, vol. 1, (1921), p. 207-208.
Channing, Marion L. The Textile Tools of Colonial Homes. Marion, Mass: Channing Books, 1971.
Chapel, Charles. Gun Collector's Handbook of Values. NY: Coward, McCann and Geoghegan, 1983.
Church, Randolph W., ed. Virginia Legislative Petitions, 6 May 1776 - 21 June 1782. Richmond, VA: Library of Virginia, 1984.
Claghorn, Charles E. Naval Officer of the American Revolution: a Concise Biographical Dictionary. Metuchen, NJ: Scarecrow Press, Inc., 1988.
Clark, Charles B. The Eastern Shore of Maryland and Virginia. New York: Lewis Historical Publishing Co., 1950.
Clark, Victor S. History of Manufactures in the United States, volume 2, 1607-1860. NY: McGraw-Hill Book Company, Inc., 1929.
Clark, William B. Report on the Iron Ores of Maryland. Baltimore: The Johns Hopkins Press, 1911.
Coakely, R. Walter. "The Two James Hunters of Fredericksburg: Patriots Among the Virginia Scotch Merchants." *Virginia Magazine of History and Biography*, vol. 56, no. 1, Jan., 1948, pp. 3-21.
Coldham, Peter W. American Loyalist Claims, vol. 1, Washington, DC: National Genealogical Society, 1980.
Coleman, Elizabeth D. "Guns for Independence," *Virginia Cavalcade*, Winter 1963-1964, pp. 40-47.
Continental Congress, Papers, National Archives and Records Administration, Washington, DC.:
Board of Treasury, Expenditures for Old Accounts, October 1785, M247, r154,v2,p.393,
Board of Treasury, Return of Provisions, Sept. 30, 1777, M247, r95, i78, v9, p. 31
"The Corbin Family." Genealogies of Virginia Families from *The Virginia Magazine of History and Biography*, vol. 2, Baltimore, MD: Genealogical Publishing Co., 1981, pp. 303-370.
Cousins, Willard C. "Josiah Meriam, Patriot Gunsmith, Concord, Massachusetts: 1726-1809." *The Gun Report*, November 1973, pp. 16-20.
Cromwell, Giles. The Virginia Manufactory of Arms. Charlottesville, VA: University Press of Virginia, 1975.
Crozier, William A. Virginia County Records, Vol. 1: Spotsylvania County, 1721-1800. Baltimore, MD: Genealogical Publishing Co., 1978.

Culpeper Court Records:

DB B-146—June 26, 1754—deed—William Russell to James Hunter
DB B-306—Oct. 16, 1754—deed—George Hume to James Hunter
DB E-415—June 17, 1767—mortgage—Benjamin Rowe to James Hunter and John Glassell
DB F-59—Oct. 18, 1769—deed—George Hume to John Strode
DB F-433—Dec. 16, 1771—deed—James Hunter to John Strode
DB H-61—May 12, 1774—lease—Stephen Fisher to John Glassell, James Hunter, and Andrew Johnston
DB H-294—Feb. 6, 1775—lease—John Strode to Zephaniah Nooe
DB H-286—Apr. 26, 1776—lease—Zephaniah Nooe to John Strode
DB H-353—Nov. 22, 1776—deed—Obadiah Pettit to John Strode
DB H-627—Oct. 20, 1778—deed—Edward Voss to John Strode
DB K-59—Oct. 23, 1779—deed—Cadwallader Slaughter to John Strode
DB M-289—Aug. 11, 1780—deed—Bowles Armistead to John Strode
DB K-218—Mar. 10, 1781—deed—John Strode to Abner Vernon
DB L-110—Mar. 8, 1782—deed—William Lightfoot to John Strode
DB N-142—Jan. 2, 1786—deed John Strode to Francis Hume
DB N-297—July 8, 1786—deed—John Strode to Abner Vernon
DB Q-494—May 7, 1791—acknowledgment—Executors of James Hunter and Overton Cosby
DB R-506—Feb. 7, 1791—deed—Joseph Sanford to John Strode
DB S-91—May 3, 1794—deed—Edward Voss to John Strode
DB S-183—Feb. 1, 1795—agreement—Baylor Banks and John Strode
DB S-249—June 1, 1795—bill of sale—Armistead Green to John Strode
DB S-247—Aug. 1, 1795—deed—Armistead Marquess to John Strode
DB S-321--Dec. 10, 1795—deed—Robert Brooke Voss to John Strode
DB S-328—Dec. 15, 1795—deed—John Strode to Samuel Ball Green
DB T-124—Mar. 3, 1797—power of attorney—Heirs of Abner Vernon to John Strode
DB T-450—Oct. 31, 1797—mortgage—John Strode to Jonah Thompson and Richard Veitch
DB U-73—Sept. 17, 1798—bond—John Strode, French Strother, Philip Slaughter, Robert Latham, Jr., and Mordecai Barbour to Gov. James Wood
DB X-9—May 18, 1802—agreement—John Strode and William Ball
DB CC-163—Feb. 17, 1808—mortgage—John Strode and Thomas Strode to James Green, Jr.
DB CC-287—Feb. 26, 1808—deed of trust—Mordecai Barbour to John Strode
DB DD-127—Aug. 11, 1808—marriage contract—John Strode to Thomas Strode
DB DD-162—Apr. 12, 1809—deed—John Strode to Thomas Strode
WB E-344—Apr. 17, 1809—will of George Wheeler
DB DD-296—July 15, 1809—deed—John Strode to Gabriel Gray
DB DD-216—July 20, 1809—mortgage—John Strode to Mary Stuart
DB DD-158—Sept. 3, 1809—trust—John Strode to John W. Green
DB DD-210—Sept. 23, 1809—mortgage—John Strode to Bank of Virginia
DB CC-333—Nov. 30. 1809—mortgage—Thomas Strode and wife to George Pickett
DB FF-119—Dec. 25, 1809—deed—Thomas Strode to William Richards
DB EE-168—July 9, 1810—deed—John Strode to Reubin P. Michael
DB EE-247—Nov. 19, 1810—trust—Thomas Strode to William Ward
DB EE-323—June 20, 1811—deed—John Strode to Thomas Spilman
DB EE-321—Aug. 13, 1811—deed—Thomas Spilman to Thomas Strode
DB CC-44—Aug. 31, 1811—trust—John Strode to John Shelton
DB FF-117—Aug. 31, 1811—mortgage—John Strode to John Shelton
DB EE-534—Nov. 11, 1811—mortgage—Thomas Strode and wife to John White
DB FF-278—May 1, 1812—deed—Thomas Strode and wife to John Alcock
DB FF-327—Dec. 6, 1812—trust—Thomas Strode to Reynolds Chapman
DB GG-13—Aug. 14, 1813—trust—Thomas Strode to George F. Strother
DB GG-14—Aug. 14, 1813—bill of sale—Thomas Strode to James Richards
DB HH-203—Aug. 14, 1816—deed—Thomas Strode and wife to William Richards
DB LL-14—Aug. 16, 1819—deed—John S. Marshall, commissioner to Philip Lightfoot

DB LL-15—Aug. 16, 1819—deed—John S. Marshall, commissioner to John W. Green
DB LL-123—Nov. 15, 1819—deed—John S. Marshall to Thomas Goodwin
DB NN-228—Feb. 28, 1822—deed—Thomas Strode to William Broadus
DB WW-381—Apr. 2, 1830—trust—Harriet Strode to Thomas Seddon

"Culpeper County, Virginia, 1800 Tax List." *The Virginia Genealogist*, vol. 16, no. 4 (1972), pp. 277-280.

Culpeper Historical Society. Historic Culpeper. Culpeper, Virginia, 1974.

Darling, Anthony D. "An American Rifled Wall Gun." *National Muzzle Loading Rifle Association*, October 1971, pp. 2-5.

Darter, Oscar H. Colonial Fredericksburg and its Neighborhood in Perspective. NY: Twayne Publishers, 1957.

Diderot, Denis. Recueil de Planches, sur les Sciences, les Arts Liberaux, et les Arts Mechaniques. Paris, 1763.

"Dixon," Genealogies of Virginia Families from the William and Mary College Quarterly Historical Magazine, vol. II, Baltimore, MD: Genealogical Publishing Co., 1982.

Dobson, David. Scots on the Chesapeake, 1607-1830. Baltimore, MD: Genealogical Publishing Co., 1992.

Dunbar, Robert to Thomas Williamson, Deed, Virginia Historical Society, Richmond, Virginia.

Encyclopedia Britannica: or a Dictionary of Arts, Science, and Miscellaneous Literature; Constructed on a Plan, by Which the Different Sciences and Arts are Digested into the Form of Distinct Treatises or Systems. Edinburgh, Scotland: A. Bell and C. MacFarquhar, 1797.

Executive Communications, Correspondence, 1774-1920, Accession #36912, Library of Virginia, Richmond, VA
Letter, James Mercer to Gov. Thomas Jefferson, Apr. 14, 1781.
Letter, John Strode to Patrick Henry, Feb. 19, 1777.
Memorial of James Hunter to George Wythe, May 31, 1777.
Report of Commissioners Appointed to Locate Land for Iron Works, Nov. 15, 1777.
Letter, James Hunter to Thomas Jefferson, Jan. 25, 1781.
Letter, Fielding Lewis to Col. George Brooke, Feb. 9, 1781.
Letter, James Hunter to Thomas Jefferson, Feb. 25, 1781.
Letter, Mann Page to Thomas Jefferson, Oct. 16, 1781.
Letter, James Hunter to William Davies, Apr. 23, 1782.

"Extracts from the Carroll Papers," letter from Charles Carroll, Sr. to Charles Carroll, Jr., Mar. 26, 1772, *Maryland Historical Magazine*, vol. 14, p. 141.

Fall, Ralph E. Hidden Village: Port Royal, Virginia, 1744-1981. Self-published, 1982.

Fauquier County Circuit Court Records:
Chancery Papers, Hunter, James vs Thomas Chinn—1774-005
Mill Records, Adam Hunter and Abner Vernon vs Rev. Rodham Kenner—1788-001
Chancery Papers, deed between John Minor and Robert Patton, Commissioners to John and Thomas Strode, Dec. 17, 1803—1803-018
Unrecorded Deeds, Deed of trust between John and Thomas Strode to Robert Patton, Dec. 17, 1803—1803-005
Mill Records, Petition of John Strode, Oct. 14, 1806—1807-001

Fauquier County Deeds and Wills:
DB 2-264—Apr. 21, 1765—lease—John Allen and wife Mary to James Hunter
DB 2-266—Apr. 22, 1765—deed—John Allen and wife Mary to James Hunter
DB 8-167—Sept. 22, 1784—deed—Reubin Wright and wife Mary to James Hunter
DB 13-123—Mar. 26, 1796—agreement—Patrick Home to Nicholas Wykoff

Fauquier County Mill Records, petition of John Strode, Oct. 14, 1806, 1807-001.

Felder, Paula S. Fielding Lewis and the Washington Family: a Chronicle of 18th Century Fredericksburg. Fredericksburg, VA: The American History Co., 1998.

Felder, Paula S. Forgotten Companions: the First Settlers of Spotsylvania County and Fredericksburgh Town. Fredericksburg, VA: American History Co., 2000.

Felder, Paula S. "Guns for the Revolution," *The Free Lance-Star*, Jan. 22, 2000.

Fitzpatrick, John C., ed. The Writings of George Washington from the Original Manuscript Sources,

1745-1799. Washington, DC: U. S. Government Printing Office, 1944, pp. 338-339.
Flayderman, Norm. Flayderman's Guide to Antique American Firearms. Northfield, IL: DBI Books, 2000.
Fleet, Beverley. "The Will of Richard Corbin." Virginia Colonial Abstracts, vol. 2, Baltimore, MD: Genealogical Publishing Co., 1988, pp. 160-162.
Franklin County, Kentucky Court Records:
WB 6-226—July 7, 1847—will of John Lavinder, Sr.
Frederick County, Maryland Court Records:
DB J-262—Mar. 30, 1764—deed—Adam Ramsburg to Fielder Gantt
DB J-677—June 18, 1764—deed—Robert Lamar to Fielder Gantt
DB K-174—Jan. 10, 1765—agreement—James Hunter to Fielder Gantt
DB K-145—June 15, 1765—deed—Thomas Taylor to Fielder Gantt and James Hunter
DB K-1461—Oct. 21, 1767—deed—Fielder Gantt to James Hunter
DB K-1165—Oct. 21, 1767—deed of trust—Fielder Gantt to James Hunter
DB K-1176—Oct. 21, 1767—lease—Fielder Gantt to James Hunter
DB 10-137—July 13, 1791—deed—Adam Hunter and Abner Vernon to Col. Baker Johnson
DB 10-139—July 13, 1791—deed—Adam Hunter and Abner Vernon to Baker Johnson
DB 10-714—June 28, 1792—deed—Adam Hunter, Abner Vernon, and Patrick Home to Baker Johnson
DB 11-80—June 28, 1792—deed—Adam Hunter, Abner Vernon, and Patrick Home to Baker Johnson
DB 11-245—June 28, 1792—deed—Adam Hunter, Abner Vernon, and Patrick Home to Baker Johnson
DB 11-247—June 28, 1792—deed—Adam Hunter, Abner Vernon, and Patrick Home to Baker Johnson
DB 11-249—June 28, 1792—deed—Adam Hunter, Abner Vernon, and Patrick Home to Baker Johnson
DB 11-252—June 28, 1792—deed—Adam Hunter, Abner Vernon, and Patrick Home to Baker Johnson
Land Record Book, vol. WR 15, p. 253—June 8, 1792—deed—James Hunter's executors to Baker Johnson
Frederick-Town Herald, Nov. 14, 1807, obituary of Fielder Gantt
Fredericksburg Courier, July 21, 1800, advertisement for the sale of George Wheeler's tobacco, wheat, and corn farm
Fredericksburg Court Records:
DB A-16—July 3, 1789—deed of trust—John Strode to John Head, Jr. and William Sansome
DB A-262—Dec. 28, 1790—deed—Thomas Strode to Abner Vernon
DB C-307—June 20, 1800—deed—Patrick Home to Mary Sullivan
DB D-292—July 28, 1801—deed—Patrick Home to Dr. Alexander Vass
DB D-381—May 13, 1802—deed—Patrick Home to Isaac Burton
DB D-504—Nov. 26, 1802—deed—Patrick Home to Robert Dunbar
DB D-508—Nov. 26, 1802—deed—Patrick Home to Robert Dunbar
DB D-471—Apr. 15, 1803—deed—Patrick Home to Anthony Marquiss
DB E-513—Oct. 9, 1806—deed of trust—John Strode to James Ross
WB A-267—Sept. 26, 1801—Will of William Glassell
WB A-3, p. 94—May 16, 1798—will of Adam Hunter
WB A-3, p. 334—May 26, 1827—will of William Richards (of Culpeper)
WB C-93—Sept. 10, 1829—will of Archibald Rollow
Bank of Virginia vs Dunbar and Strode—SL/L/1808/576-59
SL/L/1808/576-61
SL/L/1808/576-62
SL/L/1808/576-64
SL/L/1808/576-65
SL/L/1808/576-68
SL/L/1808/576-71

SL/L/1808/576-78
SL/L/1808/576-81
SL/L/1809-576-60
SL/L/1809-576-66
SL/L/1809-576-69
SL/L/1809-576-70
SL/L/1809-576-72
SL/L/1809-576-80

Home Exor of Hunter vs Richards CR-DC-V/389-46/1794
Bet. Richards and Hunter's Exors CR-DC-V/388-96/1794
Home et al Exors and Devisees of James Hunter dec. vs William Richards CR-DC-V/ 390-27/1796
Hore vs Strode DC-L/387-79/1791 (signature of John Strode)
Justices vs Hunter &c DC-L/561-21/1797
Hunter's Exors vs Richards CR-DC-L/392-76/1797
Lawson vs Tayloe CR-SC-H/171-5/1786
Hunter's Exor vs Vernon's Admr CR-DC-V/560-95/1799
Strode vs Home &c CR-DC-L/563-104/1801
Patrick Home ads William Richards CR-DC-V/562-39/1801
Patrick Home vs William Richards CR-DC-V/562-40/1801
Hunter's Exors vs Patrick Home CR-SC-H/124-11/1804?
Hunter's Admx vs Patrick Home's Admr CR-SC-H/87-11/1804
Vass vs Patton Admr SC-H/284-159/1804
Home vs Home's Admr CR-SC-H/87-11/1804
Hunter vs Home CR-SC-H/124/11/1804?
Fitzhugh vs Hunter's Admr CR-DC-L/573-88/1807?
Bank of Virginia vs Dunbar SC/H/1819/16-17
Dunbar vs Swan SL-L/583-89/1820
Green vs Strode CR-SC-H/110-0010/1823
Hord vs Richards CR-SC-H/146-4/1825
Clarke vs Finnall LC-H/66-06/1836
Hume vs Hore CR-LC-H/130-1/1835

"Fredericksburg in Revolutionary Days," part 1, *William and Mary College Quarterly Historical Magazine*, vol. 27, no. 2, October 1918, pp. 71-93.

Fredericksburg Political Arena, Feb. 23, 1830, ad for sale of mill stones from Greenbank

Fredericksburg Political Arena, Aug. 6, 1830, notice of a new blast furnace in Fredericksburg

Fredericksburg Political Arena, Jan. 14, 1831, notice of new management at the Fredericksburg blast furnace

Fredericksburg Political Arena, Feb. 5, 1836, ad for the sale of Paoli Mills

Fredericksburg Political Arena, Jan. 6, 1837, notice of James Logan

Fredericksburg Political Arena, July 7, 1837, Ficklen and Conway to build cotton mill

Futhey, J. Smith and Cope, Gilbert. The History of Chester County, Pennsylvania. Philadelphia: Louis H. Everts, 1881.

Gale, W. K. V. Ironworking. Buckinghamshire, England: Shire Publications, 1998.

Gardner, Robert E. Small Arms Makers: a Directory of Fabricators of Firearms, Edged Weapons, Crossbows and Polearms. NY: Crown Publishers, Inc., 1963.

Garnett-Mercer-Hunter families, papers, 1713-1853, Accession #20624d, Richmond, Virginia, Library of Virginia.

Gill, Harold B. The Gunsmith in Colonial Virginia. Williamsburg, VA: Colonial Williamsburg Foundation, 1974.

Gillespie, Charles C., ed. A Diderot Pictorial Encyclopedia of Trade and Industry. NY: Dover Publications, Inc., 1959.

Goolrick, John T. Fredericksburg and the Cavalier Country: America's Most Historic Section, its Homes, its People and Romances. Richmond, VA: Garrett and Massie, 1935.

Gordon, Robert B. American Iron: 1607-1900. Baltimore, MD: Johns Hopkins University Press,

1996.
Governor's Letters Received, June 29, 1776 – Nov. 30, 1784, Library of Virginia, Richmond, Virginia: Richard Call to Governor, Mar. 29, 1781, Image #GLR 01030
Gray, Gertrude E. Virginia Northern Neck Land Grants, 1775-1800, vol. 3, Baltimore: Genealogical Publishing Co., 1993.
Gray, Col. William F. From Virginia to Texas, 1835: Diary of Col. Wm. F. Gray. Houston: Fletcher Young Publishing Co., 1965.
Green, Raleigh Travers. Genealogical and Historical Notes on Culpeper County, Virginia and a History of St. Mark's Parish. Bowie, MD: Heritage Books, Inc., 1995.
Grinnan, Daniel, Papers, Virginia Historical Society, Richmond, Virginia:
Letter from John Strode to Murray, Grinnan, and Mundell, merchants, July 18, 1805, MSS1 G8855 a 27-34.
Bond of John Strode, Apr. 14, 1803, MSS1 G8855 a 22-26.
Gross, Joseph, ed. The Diary of Charles Wood of Cyfarthfa Ironworks, Merthyr Tydfil, 1766-1767. Cardif, Scotland: Merton Priory Press, 2001.
Hammersley, G. "The Charcoal Iron Industry and its Fuel, 1540-1750." *The Economic History Review*, second series, vol. xxvi, no. 4, (November 1973), pp. 593-613.
Hartzler, Daniel D. and Whisker, James B. The Southern Arsenal: a Study of the United States Arsenal at Harper's Ferry. Bedford, PA: Old Bedford Village Press, 1996.
Hayner, Jennie A., Hayner, Maybelle B., Hayner, Florence W. The Descendants of Johannes Haner, 1710-1966. Rochester, NY: Hayner Family Assn., 1966.
Headley, J. T. Washington and his Generals. Westvaco, 1991.
Heite, Edward F. "Historic Site Report: Accokeek Furnace Property, Stafford County, Virginia." Washington, DC: Smithsonian Institution, 1981.
Hening, William W. The Statutes at Large. Charlottesville, VA: University Press of Virginia, 1969.
Herndon, John G. The Herndons of the American Revolution, Part II. Self-published, 1951.
Hinshaw, William W. Encyclopedia of American Quaker Genealogy, vol. 4, Virginia, Baltimore, MD: Genealogical Publishing Co., 1993.
Historic American Buildings Survey, "Strode's Gristmill," HABS No. PA-251, Washington, DC: Library of Congress.
Hopkins, William L. Some Wills from the Burned Counties of Virginia and Other Wills not Listed in Virginia Wills and Administrations 1632-1800. Richmond, VA, 1987.
Hudgins, Dennis. "King George County Rent Rolls, 1771-1772." *Magazine of Virginia Genealogy*, Vol. 32, no. 4, November 1994, pp. 285-292.
Hudgins, Sally R. The Inhabitants of Woodend: a History of the Immigrant Reveley Family of Virginia. Self-published, Martin, TN, 1981.
Hume, Edgar E. "A Colonial Scottish Jacobite Family." *The Virginia Magazine of History and Biography*, vol. 38, Richmond, VA: House of the Society, 1930, pp. 1-37.
Hunter-Garnett Papers, Alderman Library, University of Virginia, Charlottesville, Virginia:
Letter, James Hunter to John Backhouse, Nov. 6, 1761.
Letter, James Hunter to John Baylor, Apr. 13, 1774.
Letter, James Hunter to Col. George Baylor, Oct. 12, 1779.
Letter, James Hunter, Jr. to Archibald Hunter, June 7, 1787.
Letter, James Hunter, Jr. to John Russel Spence, Dec. 17, 1788.
Hunter, James, 1721-1784, Court Papers, 1795-1797, Accession #23762B, Library of Virginia, Richmond, Virginia.
Hunter, Adam to Thomas Ridout of Alexandria, Letter, Dec. 20, 1784, Maryland Archives, MSA SC 910-B13-F14.
Hunter, James to John Backhouse, Letter, Nov. 6, 1761, Alderman Library, University of Virginia.
Hunter, James to John Baylor, Letter, Apr. 13, 1774, Alderman Library, University of Virginia.
Hunter, James to Col. George Baylor, Letter, Oct. 12, 1779, Alderman Library, University of Virginia.
Ince, Lawrence. The Knight Family and the British Iron Industry. Merton, England: Ferric Publications, 1991.
Jackson, Donald, ed. Diaries of George Washington, Charlottesville, VA: University of Virginia Press, 1976.

Jefferson, Thomas. Notes on the State of Virginia. Chapel Hill, NC: University of North Carolina Press, 1955.

Johnson, John Janney. "The Falmouth Canal and its Mills: an Industrial History." *The Journal of Fredericksburg History*, vol. 2, 1997. Fredericksburg, VA: Historic Fredericksburg Foundation, 25-43.

Johnson, John Janney. "Maps, Mills, and Canals—an Industrial History of Falmouth," Central Rappahannock Regional Library, Fredericksburg, VA.

Kaplan, Barbara B. Land and Heritage in the Virginia Tidewater: a History of King and Queen County. Richmond, VA: Cadmus Fine Books, 1993.

Kentucky Court of Appeals Land Records:

DB B2-141—Mar. 13, 1790—deed—William Green to John Strode

DB U-482—Jan. 31, 1810—deed—William Richards to Thomas Strode

DB B2-9—Nov. 3, 1817—deed of trust—John W. Green, Thomas Humphreys, and Thomas Strode

DB B2-129—June 4, 1818—deed—Thomas and Harriet Strode to Thomas H. Mitchell and George K. Lee

Kentucky Land Office Records:

Land Book 1, p. 186, Feb. 1, 1787—John Strode's entry for 300 acres

Land Book 1, p. 186, Feb. 1, 1787—John Strode's entry for 700 acres

Land Book 1, p. 276, Apr. 28, 1792—John Strode's entry for 200 acres

Land Book 1, p. 276, Apr. 28, 1792—John Strode's entry for 300 acres

Land Grants, Book 12, p. 155, Apr. 28, 1792—James Garrard to John Strode, 200 acres

Land Grants, Book 13, p. 583, Nov. 24, 1784—Beverley Randolph to John Strode, 700 acres

Land Grants, Book 14, p. 447, Nov. 30, 1796—James Garrard to John Strode, 166 2/3 acres

Military Warrant No. 3539, Nov. __, 1784, George Slaughter, 1,000 acres

Military Warrant No. 4178, June 21, 1786, William Saunders, 666 2/3 acres

Survey No. 1634, Apr. 30, 1792—John Strode, 300 acres

Survey No. 3959, Sept. 20, 1797—John Strode, 200 acres

Survey No. 5081, Sept. 16, 1797—John Strode, 166 2/3 acres

Survey No. 6609, Mar. 1, 1787—John Strode, 700 acres

Survey No. 6610, Mar. 1, 1787—John Strode, 300 acres

"The King Family of Virginia." Genealogies of Virginia Families from *The William and Mary College Quarterly Historical Magazine*, vol. III, Baltimore: Genealogical Publishing Co., 1982, pp. 283-288.

King, George H. S. "Maryland Coverts Become Virginia Colverts, Colberts and Calverts." *The Virginia Genealogist*, vol. 14, no. 1 (1970), pp. 3-10.

King George County Deeds:

DB 5-646—July 14, 1766—deed—John Dixon to James Hunter

Kinsey, Margaret B. Ball Cousins: Descendants of John and Sarah Ball. Baltimore, MD: Gateway Press, 1981.

Lee, Elizabeth Nuckols. King George County Virginia Marriages, vol. 1, Marriage Bonds, Book 1, 1786-1850. Athens, GA: Iberian Publishing Co., 1995.

Lee, Thomas Ludwell, Letter to James Steptoe, Nov. 23, 1777, MSS1 L51f258, Virginia Historical Society, Richmond, Virginia.

Litchfield, Carter; Finke, Hans-Joachim; Young, Stephen G.; Huetter, Karen Z. The Bethlehem Oil Mill, 1745-1934: German Technology in Early Pennsylvania. Kemblesville, PA: Olearus Editions, 1984.

Madison, James, Papers, Virginia Historical Society, Richmond, Virginia:

Letter, John Strode to ___ Davis, June 11, 1808, Series 1, Reel 10.

Letter, John Strode to James Madison, Oct. 15, 1808, Series 1, Reel 10.

Deed of trust, John Strode to James Madison, July 6, 1810, Series 1, Reel 12.

Malone, Dumas. Jefferson the President: Second Term, 1805-1809. Boston: Little, Brown and Co., 1974.

Mansfield, Dolorus B. The Briggs-Bridge Family. Self-published, 1960, Library of Virginia, Richmond, Virginia.

Maryland Gazette, Apr. 27, 1748, wedding announcement of Benedict Calvert

Maryland Gazette, Aug. 30, 1759, obituary of Thomas Jennings
Maryland Gazette, Sept. 8, 1763, James Hunter's ad for the sale of Angola slaves
Maryland Gazette, Sept. 8, 1763, James Hunter's ad for the sale of Windward and Gold Coast slaves
Maryland Gazette, Oct. 15, 1767, advertisement for finery and chafery workers for Rappahannock Forge
Maryland Gazette, July 3, 1811, obituary of Baker Johnson
Maryland Herald, Oct. 30, 1800, advertisement for gunsmiths for Rappahannock Forge
Maryland High Court of Chancery Records:
Hoffman, John and others vs Baker Johnson, 132-20
Hunter, James, Mortgagee, 1767, B 132-43
Hunter, James, died December 1784, 1784, B 132-22
Hunter, James, Legatee, 1784, B 132-21
Vernon, Abner, Executor of James Hunter, 1784, B 132-22
Hunter, Adam, Legatee, Executor of James Hunter, 1784, B 132-20
Hunter, Adam and Samuel and Robert Purviance agt. Fielder Gaunt, 1786, #18, Folios 346-399
Hunter, Adam and Samuel and Robert Purviance agt. Fielder Gaunt, 1786, #19, Folio 19
Hunter, Adam and others agt. Fielder Gaunt, 1787, #16, Folio 17
Hunter, Adam and others agt. Fielder Gaunt, 1787, #21, Folio 1-43
Hunter, Adam, Complainant, 1787-1790, B 132, Folio 1-43
Hunter, James of Virginia—Creditor of Fielder Gaunt, 1789/90, # 18, Folio 346
Gaunt, Fielder agt. James Hunter and Purviance, 1789-90, #18, Folios 53, 352, 353, 358, 365, 395, 399
Peter, Robert, deposition, February Term 1790, Liber B-21
Letter, James Hunter to Fielder Gant, Apr. 22, 1764, February Term 1790, Liber B-21
Letter, James Hunter to Fielder Gant, Sept. 26, 1764, February Term 1790, Liber B-21
Letter, James Hunter to Fielder Gant, Feb. 5, 1765, February Term 1790, Liber B-21
Letter, James Hunter to Fielder Gant, Feb. 26, 1765, February Term 1790, Liber B-21
Letter, James Hunter to Fielder Gant, Aug. 2, 1765, February Term 1790, Liber B-21
Letter, James Hunter to Fielder Gant, Oct. 16, 1765, February Term 1790, Liber B-21
Hunter, James, Late of Virginia, 1791, B 21-1-29
Vernon, Abner, Grantor (Bond), 1791, B 132-22
Vernon, Abner, died in 1792, 1792, B 132-20
Maryland Journal and Baltimore Advertiser, July 16, 1790, advertisement for sale of Hunter's iron works
Maryland Journal and Baltimore Advertiser, May 6, 1791, advertisement for the sale of 700 acres formerly belonging to Fielder Gantt
McIlwaine, H.R., ed. Journals of the Council of the State of Virginia, Richmond, VA: Virginia State Library, 1931.
McIlwaine, H.R., ed. Official Letters of the Governors of the State of Virginia. Richmond, VA: Virginia State Library, 1926.
Menchinton, Walter; King, Celia; Waite, Peter, eds. Virginia Slave-Trade Statistics, 1698-1775. Richmond: Library of Virginia, 1984.
Miller, Robert L. "Fredericksburg Manufactory Muskets." *Military Collector and Historian*, vol. 3, Sept. 1951, pp. 63-65.
Miller, Robert L. and Peterson, Harold L. "Rappahannock Forge: Its History and Products." *Military Collector and Historian*, vol. 4, no. 4, Dec. 1952, pp. 81-85.
Moller, George D. American Military Shoulder Arms, vol. 1: Colonial and Revolutionary War Arms. University Press of Colorado, 1993.
Morgan, William James, ed. Naval Documents of the American Revolution, vol. 5, 7, 8. Washington, DC: Department of the Navy, 1976.
Moss, Roger W. "Isaac Zane, Jr., a Quaker for the Times." *Virginia Magazine of History and Biography*, vol. 77, pp. 291-306.
Mulkearn, Lois, ed. George Mercer Papers Relating to the Ohio Company of Virginia. Pittsburgh: University of Pittsburgh Press, 1954.
Musselman, Cynthia R. Stafford County, Virginia Cemeteries. Self-published, Stafford, VA, 1983.
Mutual Assurance Society, policy on the Forge Mill, Apr. 16, 1810, R5/V47/376.
Mutual Assurance Society, policy on the Forge Gristmill, Apr. 16, 1810, R5/V47/378.

Mutual Assurance Society, policy on the Falmouth Gristmill, Dec. 16, 1822, R11/V80/4160.
Mutual Assurance Society, policy on the Falmouth Gristmill, Dec. 31, 1829, R12/V86/6029.
National Archives and Records Administration, Washington, DC:
Papers of the Continental Congress. Board of Treasury. Expenditures for Old Accounts, October 1785. M247, r154, v2, p. 393.
Papers of the Continental Congress. Return of Provisions, Sept. 30, 1777. M247, r95, i78, v9, p. 31.
Revolutionary War Manuscripts, Record Group 93, 035511, Letter of Col. George Muter to Charles Dick.
Revolutionary War Manuscripts, Record Group 93, 035512, Letter of Col. George Muter to James Hunter.
Norfleet, Fillmore. Saint-Memin in Virginia: Portraits and Biographies. Richmond, VA: Dietz Press, 1942.
Northrup, A. Judd. The Northrup-Northrop Genealogy. NY: The Grafton Press, c.1908.
Overman, Frederick. The Manufacture of Steel. Philadelphia: A. Hart, 1851.
Overman, Frederick. Mechanics for the Millwright, Machinist, Engineer, Civil Engineer, Architect, and Student. Philadelphia: Lippincott, Grambo, and Co., 1851.
Palmer, William P., ed. Calendar of Virginia State Papers, NY: Kraus Reprint Corp., 1968.
Papenfuse, Edward C., ed. The Biographical Dictionary of the Maryland Legislature, 1635-1789. Baltimore, MD: Johns Hopkins University Press, 1985.
Payne, Brooke. The Paynes of Virginia. Berryville, VA: Virginia Book Co., 1937.
Pearse, John B. A Concise History of the Iron Manufacture of the American Colonies Up to the Revolution. NY: Burt Franklin, 1876.
Peden, Henry C. Revolutionary Patriots of Frederick County, Maryland, 1775-1783. Westminster, MD: Family Line Publications, 1995.
Peterson, Harold L. The American Sword, 1775-1945. Philadelphia: Ray Riling Arms Books Co., 1988.
Peterson, Harold L. "The American Cavalry Saber of the Revolution." *Military Collector and Historian*, vol. 2, no. 3, Sept. 1950.
Peterson, Harold L. Arms and Armor in Colonial America: 1526-1783. Harrisburg, PA: The Stackpole Co., 1956.
Pleasants, J. Hall, ed. Archives of Maryland, Volume 47: Journal and Correspondence of the State Council of Maryland. Baltimore, MD: Maryland Historical Society, 1930.
Presidential Papers Microfilm: James Madison Papers (Virginia Historical Society):
Letter from John Strode to James Madison, June 20, 1789, Series 1, Reel 4.
Letter from John Strode to ___ Davis, editor of the *Virginia Gazette*, June 11, 1808, Series 1, Reel 10.
Letter from John Strode to James Madison, Oct. 15, 1808, Series 1, Reel 10.
Letter from John Strode to James Madison, Feb. 7, 1810, Series 1, Reel 12.
Deed of Trust, John Strode to James Madison, July 6, 1810, Series 1, Reel 12.
Principio Collection, Historical Society of Delaware:
Principio Furnace Receipts, 1739-1741
Principio Furnace Ledger, 1772-1773
Kingsbury Furnace Journal, 1777
Pulliam, David L. The Constitutional Conventions of Virginia from the Founding of the Commonwealth to the Present Time. Richmond, VA: John T. West, Publisher, 1901.
Purviance, Robert. Narrative of the Events Which Occurred in Baltimore Town During the Revolutionary War. MF 185.P97, Manuscript Department, Maryland Historical Society.
Reilly, Robert M. United States Martial Flintlocks: a Comprehensive Illustrated History of the Flintlock in America from the Revolution to the Demise of the System. Lincoln, RI: Andrew Mowbray, Inc., 1997.
Reveley Family Genealogical Notes, Accession #37009, Library of Virginia, Richmond, Virginia.
Revolutionary War Manuscripts, Record Group 93, Letter, Col. George Muter to Charles Dick, Jan. 19, 1781, #035511, National Archives and Record Administration
Revolutionary War Pension Applications, Washington, DC, National Archives and Record

Administration:
S8043, Mordecai Barbour
S 1517, Jarrett Burton
R 3350, John England
W 8939, George Hood
S 36831, Thomas Turnham
S 7960, William Woodside
S 30810, James Yelton

Robbins, Michael W. "Maryland's Iron Industry During the Revolutionary War Era." Report prepared for the Maryland Bicentennial Commission, June 1973.

Scheel, Eugene. Culpeper: a Virginia County's History Through 1920. Culpeper, VA: Culpeper Historical Society, 1982.

Schoepf, Johann David. Travels in the Confederation, 1783-1784. A. J. Morrison, transcriber, Philadelphia, 1911.

Scott County, Kentucky Deeds and Wills:
WB A-49—Jan. 22, 1798—will of John Mulberry
WB C-374, 390 and WB F-187—after June 29, 1822—will of Jacob Mulberry
WB E-69—Sept. 26, 1829—will of John Mulberry

Selby, John E. The Revolution in Virginia, 1775-1783. Williamsburg, VA: Colonial Williamsburg Foundation, 1989.

Shelley, Frederick, ed. "The Journal of Ebenezer Hazard in Virginia, 1777." *The Virginia Magazine of History and Biography*, vol. 62, (1954), pp. 400-423.

Shelton Family Papers, Accession #24593, Library of Virginia, Richmond, Virginia.

Smith, James M. The Republic of Letters: the Correspondence Between Thomas Jefferson and James Madison, 1776-1826, vol. 3. NY: W. W. Norton and Co., 1995, pp. 1725-1726.

Sparks, Jared, ed. Correspondence of the American Revolution: Letters of eminent Men to Washington, 1776-1789. Boston, 1853.

Spotsylvania County Court Records:
WB B-19—Apr. 3, 1750—will of John Allen
WB B-185—Mar. 5, 1754—will of William Hunter
WB B-187—Aug. 2, 1758—Guardian Bond—William Taliaferro guardian of James, William, and Martha Hunter
DB E-666—June 9, 1760—deed—Alexander Wright to James Hunter and George Frazier
DB F-376—Apr. 2, 1764—Thomas Ward of Liverpool granted power of attorney to James Hunter
WB D-191—Apr. 1, 1765—will of George Frazier
DB G-139—Dec. 8, 1767—deed—Thomas Reeves and wife Sarah to James Hunter
WB D-_______--Dec. 2, 1776—Guardian Bond—James Hunter guardian of William Hunter
DB P-407—Sept. 1, 1801—deed—Patrick Home to Mary Sullivan
DB R-357—Dec. 5, 1807—deed—Mordecai Barbour and wife to John Strode
DB R-500—July 5, 1809—deed of trust—John Strode to Murray, Grinnan, and Mundell
DB S-468—June 19, 1811—deed of trust—John Strode to Isaac H. Williams, trustee
DB W-118—Mar. 30, 1819—deed—John W. Green, Commissioner to John Mundell

Springfield Armory Catalog Record 1155, Accession 02, Springfield Armory, Geneseo, Illinois.

"Stafford County," c.1820. Map. BPW 711(33), from sheet 3. Library of Virginia, Richmond, Virginia.

Stafford County Census Records

Stafford County Court Records:
WB M—Sept. 10, 1746—account of the estate of Christopher Brodrick
WB O-88—Mar. 13, 1749—will of William Baxter
DB S-255—June 13, 1785—executors of the estate of James Hunter named attorney to settle debts in Baltimore
DB P-218—Aug. 14, 1759—lease—Peter Daniel and wife Sarah to William and Thomas Spilman
DB S-78—July 14, 1783—petition of Francis Thornton
DB S-356—June 21, 1785—executors of the estate of James Hunter named attorney to settle

debts in Jamaica
DB S-233—Apr. 11, 1785—will of James Hunter
DB S-283—April __, 1785--inventory of the estate of James Hunter
Scheme Book Court Orders, 1790-1793, p. 381—Feb. 11, 1793—Jane Vernon renounced her right of administration of the estate of her husband, Abner
DB S-271--undated—receipt of payment to James Hunter from Virginia for sundries furnished from Sept. 9, 1780 to Apr. 10, 1782DB AA-54—Feb. 17, 1802—deed—Patrick Home to Robert Dunbar
DB AA-3—Mar. 13, 1809—deed—Robert Patton and John Minor, commissioners, to John Seddon
DB-AA-79—Sept. 23, 1809—mortgage—Robert Dunbar to Bank of Virginia
DB AA-110—June 13, 1810—deed—Robert Patton and John Minor, commissioners, to Meshack Massey
DB AA-206—Mar. 14, 1811—deed—Robert Patton and John Minor, commissioners, to Lawrence Sanford
DB AA-208—Mar. 14, 1811—deed—Robert Patton and John Minor, commissioners, to Samuel Marquess
DB AA-296—Sept. 23, 1811—deed—Charles Withers to James Templeman
DB AA-302—Mar. 9, 1812—deed—Montague Duerson and Elizabeth Hudson to George Banks
DB AA-345—Oct. 7, 1812—deed—Robert Patton and John Minor, commissioners, to John Swetnam
DB AA-424—Aug. 14, 1813—deed—Thomas Strode to William Richards
Aug. 22, 1815—deed—Robert Dunbar to Thomas Williamson (from the collection of the Virginia Historical Society)
DB GG-82—Feb. 13, 1826—deed of trust—William Foushee to R. C. L. Moncure
DB GG-186—June __, 1826—deed—R. C. L. Moncure to William (Robert?) Dunbar
DB LL-123—Nov. 23, 1837—deed—Walker P. Conway to Joseph B. Ficklen

Stafford County Land Tax Records

Stafford County Loose Surveys, survey of Rappahannock Forge, 1918, Office of the Commissioner of the Revenue.

Stafford County Personal Property Tax Records

"Stafford County Legislative Petitions, 1776-1781." *Virginia Genealogist*, vol. 31, no. 4, October-December 1987, pp. 306-310.

Stafford County Personal Property Tax Records

Stewart, Catesby W. The Life of Brigadier General William Woodford of the American Revolution. Richmond: Whittet and Shepperson, 1973.

Stewart, Catesby W. Woodford Letter Book, 1723-1737. Verona, Virginia: McClure Printing Co., 1977.

Stewart, Robert A. History of Virginia's Navy of the Revolution. Richmond, VA: Mitchell and Hotchkiss Pub., 1933.

Strode, John. Deposition taken in case of Thornton and Dunbar vs. Stephen Winchester, Sept. 19, 1800.

Strode, John. *William and Mary Quarterly Historical Magazine*, vol. 27, Letter to Gov. Thomas Jefferson. NY: Kraus Reprint Co., 1977, pp. 83-85.

Swank, James M. The History of the Manufacture of Iron in all Ages and Particularly in the United States from Colonial Times to 1891, also a Short History of Early Coal Mining in the United States and a Full Account of the Influences which Long Delayed the Development of all American Manufacturing Industries. Philadelphia: The American Iron and Steel Association, 1892.

Swayze, Nathan L. The Rappahannock Forge. American Society of Arms Collectors, 1976.

Tayloe, W. Randolph. The Tayloes of Virginia and Allied Families. Self-published, Berryville, VA, 1963.

Temin, Peter. Iron and Steel in Nineteenth-Century America; an Economic Inquiry. Cambridge, Massachusetts: MIT Press, 1964.

Thompson, Michael D. The Iron Industry in Western Maryland. Morgantown, WV, 1976.

Thornton and Dunbar vs. Stephen Winchester. Thornton Family Papers, MSS1 T3977 c46-49. Virginia Historical Society, Richmond, Virginia.

Twohig, Dorothy, ed. The Papers of George Washington, Revolutionary War Series, Charlottesville, VA: University Press of Virginia, 1999.
Tyler, Lyon G., ed. "Fredericksburg in Revolutionary Days." *William and Mary College Quarterly Historical Magazine*, vol. 27, Richmond, VA: Whittet and Shepperson, 1920, pp. 71-95, 164-175, 248-257.
United States Circuit Court Records, Library of Virginia, Richmond, Virginia:
Lyde, Svg. Ptr. vs Hunter's Exors, 1794, Miscellaneous Reel #663
Backhouse's Admx vs Hunter's Exors, 1795
Ward's Admr vs Hunter's Exor, 1797
Ward's Admrs vs Patrick Home, 1800
Hunter's Extx vs Homes Admr, 1805
Cooch and Hollingsworth vs Strettle's Admr, 1806
Evans, Oliver vs Dunbar, Robert, 1812
Strode vs Patton & als, 1832, Library of Virginia
United States District Court Records, Library of Virginia, Richmond, Virginia:
Backhouse Admx vs Hunter's Svg Ptr, 1797
Hunter's Exors vs Backhouse's Admx, 1797
Justices of Stafford County vs Strode als, 1810
Oliver Evans vs Robert Swann, 1823
Vernon, Abner, Letters, 1791-1792, Accession 26660, Personal papers collection, Library of Virginia and Archives, Richmond, Virginia.
Virginia Board of Trade, Minute Book, 27 Nov. 1779 - 7 Apr. 1780, Miscellaneous Reel #638, Library of Virginia, Richmond, Virginia.
Virginia Board of War, Journal, 1777-1780, Miscellaneous Reel #302, Library of Virginia, Richmond, Virginia.
Virginia Board of War, Journal, June 30, 1779 - Apr. 17, 1780, Miscellaneous Reel #498, Library of Virginia, Richmond, Virginia.
Virginia Board of War, Journal, Dec. 23, 1779-Mar. 25, 1780, Miscellaneous Reel #632, Library of Virginia, Richmond, Virginia.
Virginia Board of War, Letter Book, June 30, 1779-Apr. 7, 1780, Miscellaneous Reels #302 and #498, Library of Virginia, Richmond, Virginia.
Virginia Board of War, Letter Book, Dec. 23, 1779 - Mar. 25, 1780, Miscellaneous Reel #632, Library of Virginia, Richmond, Virginia.
Virginia Board of War, Letter Book and Index, 1780-1782, Miscellaneous Reel #264, Library of Virginia, Richmond, Virginia.
Virginia Colonial Records Project, Library of Virginia, Richmond, Virginia:
Admiralty - Miscellanea, Register of Passes, 1754-1757, Survey Report #5344, Reel #534.
Court of Session Unextracted Processes, June 1756-March 1779, Letter to James Russell pertaining to John Lavendar and sons, Survey Report #06757, Reel #792.
In-Letter Book, 1764-1767, Letter pertaining to James Hunter's shipment of wood to Scotland, Mar. 24, 1767, Survey Report #06566, Reel #562.
High Court of Admiralty: Prize Papers, 1778, Survey Report # 5474, Reel #809.
House of Lords Papers, MS, Lists, 1766-1774, List of Stamp Officials, Survey Report #1954, Reel # 597.
Letter, Patrick Home to James Home, Berwickshire, Scotland, Mar. 5, 1796, GD 1/384/01/1, Reel #846.
Letter, Patrick Home to James Hunter, Duns, Berwickshire, Scotland, May 11, 1801, GD 1/384/10/2/1, Reel #846.
Letter, Richard Call to Governor, Mar. 29, 1781, Governor's Letters Received, June 29, 1776-Nov. 30, 1784, Image #GLR 01030.
Loyalist Claims - A List of Claims, 1762-1798, Survey Report #2504, Reel # 187.
Papers Relating to the loss of Cornwallis' Army, 1780-1781, Survey Report #02446, Reel #599.
Treasury Papers—In Letters, 1770-1771, Report of John Williams, Survey Report #01313, Reel #119.
Treasury Warrant, July 9, 1765, Adam Hunter named inspector of stamps for Virginia and

Maryland, House of Lords Papers, Bundle A, Survey Report #1954, Reel #597.
Virginia Commercial Agent, Day Book, Mar. 17, 1781 - July 17, 1782, Miscellaneous Reel #634, Library of Virginia, Richmond, Virginia.
Virginia Commercial Agent, Day Book, July 18, 1782 - June 27, 1786, Miscellaneous Reel #634, Library of Virginia, Richmond, Virginia.
Virginia Commercial Agent, Journal, Mar. 16, 1781 - Dec. 18, 1782, Miscellaneous Reel #634, Library of Virginia, Richmond, Virginia.
Virginia Commissary of Stores, Ledger, 1781-1785, Miscellaneous Reel # 636, Library of Virginia, Richmond, Virginia.
Virginia Committee of Safety, Account Ledgers, Oct. 31, 1775 – June 5, 1776, Miscellaneous Reel #301, Library of Virginia, Richmond, Virginia.
Virginia Gazette, June 22, 1776, advertisement for skilled workers at Rappahannock Forge
Virginia Gazette, Sept. 6, 1776, advertisement of Alexander Hanewinkel
Virginia Gazette, Nov. 15, 1776, advertisement for skilled workers at Rappahannock Forge
Virginia Gazette, Dec. 20, 1776, advertisement for workers at Rappahannock Forge
Virginia Gazette, July 4, 1777, advertisement for workers at Rappahannock Forge
Virginia Gazette, July 10, 1778, advertisement for products made by the Buckingham Furnace
Virginia Gazette, Dec. 4, 1779, advertisement for wire makers at Rappahannock Forge
Virginia Gazette, Aug. 16, 1783, James and Adam Hunter advertised wine from Madeira
Virginia Gazette, July 19, 1785, notice of Adam Hunter, copartnership of James and Adam Hunter dissolved due to death of James Hunter
Virginia Gazette, July 19, 1785, notice of Adam Hunter and Abner Vernon to debtors of James Hunter and notice of sale of Negroes
Virginia Gazette, May 25, 1808, notice regarding John Strode's debt to James Madison
Virginia Gazette or American Advertiser, Aug. 31, 1782, notice of salt for sale by James and Adam Hunter
Virginia Gazette or American Advertiser, Oct. 12, 1782, notice of products for sale by James and Adam Hunter
Virginia Gazette or American Advertiser, July 9, 1785, Hunter's executors seeking payment of debts to the estate
Virginia Herald, Apr. 30, 1789, advertisement for the sale of James Hunter's land in Fauquier County
Virginia Herald, Dec. 13, 1792, notice of the death of Abner Vernon
Virginia Herald, Dec. 26, 1792 and Jan. 3, 1793—notice to those in debt to the estate of James Hunter
Virginia Herald, Feb. 21, 1793, obituary of Abner Vernon
Virginia Herald, Mar. 28, 1793, notice of the dissolution of the partnership of Thomas West & Company
Virginia Herald, Apr. 4, 1793, advertisement of Abner Vernon's stallion, Bucephalus
Virginia Herald, Apr. 25, 1793 and May 2, 1793, notice to those with accounts at Rappahannock Forge
Virginia Herald, May 9, 1793, advertisement for the Falmouth Nail Manufactory

Virginia Herald, May 30, 1793, advertisement for milling, blacksmith, wheelwright, and tanning business at Rappahannock Forge
Virginia Herald, July 31, 1794, advertisement for the sale of Stanstead
Virginia Herald, Dec. 11, 1794, sale of the personal property of Abner Vernon
Virginia Herald, May 29, 1795, notice of the postponement of the auction of Stanstead
Virginia Herald, Oct. 16, 1795, advertisement for the sale of slaves and land by John Strode
Virginia Herald, Apr. 21, 1798, advertisement for the sale of Hunter's iron works
Virginia Herald, Apr. 28, 1798, obituary of James Somerville
Virginia Herald, May 2, 1798, advertisement for the sale of Rappahannock Forge
Virginia Herald, May 19, 1798, Joseph Ennever's notice to those indebted to Adam Hunter's estate
Virginia Herald, May 19, 1798, advertisement for the sale of Rappahannock Forge and Fauquier tract
Virginia Herald, July 31, 1798, advertisement for the sale of Wheeler's Gun Manufactory
Virginia Herald, Jan. 25, 1799, notice for the sale of two ship carpenters
Virginia Herald, July 9, 1802, notice of Cooch and Hollingsworth—Rappahannock Forge Mill
Virginia Herald, Dec. 10, 1802, notice of the extension of the road from Spotted Tavern to Rappahannock

Forge
Virginia Herald, Feb. 18, 1803, advertisement for sale of the Marsh Tract
Virginia Herald, Feb. 18, 1803, advertisement for the sale of some of James Hunter's slaves
Virginia Herald, Feb. 22, 1803, advertisement for the sale of Stanstead
Virginia Herald, Mar. 11, 1803, Patrick Home's notice pertaining to the estate of John Glassell
Virginia Herald, May 24, 1803, advertisement for the sale of some of James Hunter's slaves
Virginia Herald, May 24, 1803, notice of Joseph Ennever's appointment to collect debts due James Hunter
Virginia Herald, July 15, 1803, advertisement for sale of the Marsh Tract
Virginia Herald, Sept. 6, 1803, advertisement by Thomas West for the sale of Edward West's gun locks
Virginia Herald, Nov. 8, 1803, notice of the sale of land of Patrick Home
Virginia Herald, Nov. 11, 1803, notice of the sale of all remaining lands of James Hunter
Virginia Herald, Dec. 3, 1803, notice of the sale of James Hunter's Culpeper lands and Stanstead
Virginia Herald, Apr. 19, 1804, advertisement for the auction of Wheeler's Gun Manufactory
Virginia Herald, Aug. 10, 1804, advertisement for the sale of two tracts of land
Virginia Herald, Jan. 11, 1805, advertisement for the sale of lands of James Hunter
Virginia Herald, Jan. 15, 1805, notice of Mordecai Barbour pertaining to a cotton machine
Virginia Herald, May 29, 1805, advertisement for Mordecai Barbour's cut nail manufactory
Virginia Herald, June 21, 1805, notice of Dr. Alexander Vass
Virginia Herald, Aug. 9, 1805, Mordecai Barbour's advertisement for the forge mills
Virginia Herald, July 3, 1807, John Strode's advertisement for men to haul flour
Virginia Herald, Jan. 29, 1808, obituary of Robert Walker
Virginia Herald, Aug. 16, 1808, wedding announcement of Thomas Strode and Harriett Somerville Richards
Virginia Herald, Mar. 14, 1810, advertisement for the sale of John Strode's property
Virginia Herald, Jan. 12, 1812, sale of land by Thomas Strode
Virginia Herald, Oct. 9, 1813, notice of the sale of Thomas Strode's personal property
Virginia Herald, Dec. 2, 1815, notice of the Strode case in the Superior Court of Chancery, Fredericksburg
Virginia Herald, Nov. 28, 1818, notice that Thomas Strode would be in Fredericksburg to settle his affairs in Virginia
Virginia Herald, Jan. 1, 1820, notice of the Strode case in the Superior Court of Chancery, Fredericksburg
Virginia Herald, Oct. 26, 1822, notice of Brooke and Ficklen's rental of the Forge Mills
Virginia Herald, Oct. 26, 1822, notice of John O'Rion's rental of the gristmill at Rappahannock Forge
Virginia Herald, Feb. 2, 1828, obituary of Isaac H. Williams
Virginia Herald, May 6, 1829, obituary of Thomas Strode
Virginia Herald and Fredericksburg Advertiser, Oct. 1, 1789, advertisement for the sale of Hunter's meadow tract in Fauquier County
Virginia Herald and Fredericksburg Advertiser, Jan. 6, 1791, Zephaniah Nooe's advertisement for his "House of Entertainment"
Virginia Herald and Fredericksburg Advertiser, Apr. 8, 1791, advertisement for Thomas West's Nail Manufactory
Virginia Herald and Fredericksburg Advertiser, Apr. 8, 1791, advertisement for the sale of the Fielderia tract
Virginia Independent Chronicle and General Advertiser (Richmond), June 9, 1790, advertisement for the sale of Hunter's iron works
Virginia Journal and Alexandria Advertiser, Jan. 4, 1787, advertisement for the sale of slaves from Rappahannock Forge
Virginia Legislative Petitions, Library of Virginia, Richmond, Virginia:
197-P, Nov. 5, 1776.
198-P, Nov. 5, 1776.
937-P, May 18, 1779.
954-P, May 24, 1779.
Virginia Navy Board, General Correspondence, 1776-1787, Miscellaneous Reels #462 and #635, Library of Virginia, Richmond, Virginia.

Virginia Navy Board, Journals, 1776-1779, Miscellaneous Reel #301, Library of Virginia, Richmond, Virginia.
Virginia Navy Board, Ledger, July 10, 1776-July 10, 1777, Miscellaneous Reel #302, Library of Virginia, Richmond, Virginia.
Virginia Navy Board, Letterbook, July 15, 1776-Oct. 9, 1777, Miscellaneous Reel #302, Library of Virginia, Richmond, Virginia.
Virginia Navy Board, Minute Books, 1777-1778, Miscellaneous Reel #302, Library of Virginia, Richmond, Virginia.
Virginia Navy Board, Order Books, 1776-1779, Miscellaneous Reel #301, Library of Virginia, Richmond, Virginia.
Vogt, John and Kethley, T. William. Culpeper County Marriages 1780-1853. Athens, GA: Iberian Publishing Co., 1986.
Vogt, John and Kethley, T. William. Stafford County, Virginia Tithables: Quit Rents, Personal Property Taxes and Related Lists and Petitions, 1723-1790. Athens, GA: Iberian Publishing Co., 1990.
War Department Collection of Revolutionary War Records, Entry 9—Photographic Copies of State Records, c.1775-1783, Record Group 93 Box 42, National Archives and Records Administration, Washington, DC:
File # 035514—letter from George Muter to W. Nuttall, Jan. 19, 1781
File # 035548—letter from George Muter to William Armistead, Feb. 12, 1781
File # 035676—letter from William Davies to James Hunter, Mar. 26, 1781
File # 035675—letter from William Davies to James Hunter, Mar. 29, 1781
Ward, Harry M. and Greer, Harold E. Richmond During the Revolution, 1775-1783. Charlottesville, VA: University Press of Virginia, 1977.
Watkins, C. Malcolm. Cultural History of Marlborough, Virginia. Washington, DC: Smithsonian Institution Press, 1968.
Watson, John F. Annals of Philadelphia. Philadelphia, PA: Leary Publishing Co., 1909.
Wayland, John W., ed. Hopewell Friends History, 1734-1934, Frederick County, Virginia. Strasburg, VA: Shenandoah Publishing House, Inc., 1936.
Wayland family papers, 1769-1888, Accession #22665, Personal papers collection, Library of Virginia, Richmond, Virginia.
West, Sue Crabtree. The Maury Family Tree: Descendants of Mary Anne Fontaine (1690-1755) and Matthew Maury (1686-1762). Self-published, Birmingham, AL, 1971.
Westham Foundry Ledger Book, 1776-1786, Miscellaneous Reel #638, Library of Virginia, Richmond, Virginia.
Whisker, James B. Arms Makers of Philadelphia: 1660-1890. UK: Edwin Mellen Press, 1990.
Whisker, James B. Arms Makers of Virginia and West Virginia. UK: Edwin Mellen Press, 1990.
Whisker, James B. Gunsmiths of Virginia. Bedford, PA: Old Bedford Village Press, 1992.
"The Williamson Family." Genealogies of Virginia Families from *The Virginia Magazine of History and Biography*, vol. 5, Baltimore, MD: Genealogical Publishing Co., 1981, pp. 698-699.
Works Projects Administration. Spotsylvania County WPA Report. Fredericksburg, VA, 1937.
"The Wormeley Family." Genealogies of Virginia Families from the Virginia Magazine of History and Biography, vol. 5, Baltimore: Genealogical Publishing Co., 1981, pp. 885-904.
Worrall, Jay, Jr. The Friendly Virginians: America's First Quakers. Athens, GA: Iberian Publishing Co., 1994.
Wright, Louis B., ed. The Prose Works of William Byrd of Westover: Narratives of a Colonial Virginian. Cambridge, Mass.: Belknap Press, 1966.
Yates, Robert S. R. A History of James Somerville of Culpeper County, Virginia. Self-published 1988.

Contributors:

Gordon Barlow
Steven T. Bashore
Mr. and Mrs. Clarence Blaisdell
Mr. Preston Blaisdell

Nancy Brantingham
George Bryson
Betty Burgess
Peggy B. Chapman
Robert L. Chew
Ray Cooke
Giles Cromwell
John Daniels
Mark Delcourt
Joshua Duncan
Dr. John L. Eby, Jr.
Mr. and Mrs. Ralph L. England
Robert W. Fisch
Ben Gantt
Tim Garrett
George L. Gordon
Robert J. Hadeen
Mary D. Hayden
Bill Henderson
Sherri Heyse
Alice F. Hills
Robert Jefferson
Roy G. Jinks
Janna Johns
Michael Kallam
Kris G. Leinicke
Myron "Mike" Lyman
Alaric R. MacGregor
Patrick McCurdy
Barry McGhee
Kyle F. McGrogan
Mary Catherine O. Moncure
Thomas M. Moncure
D. P. Newton
Tom Peterson
Dr. and Mrs. Donald Plucknett
Sarah Reveley
Stephen L. Ritchie
Walter B. Roberts
Walter V. "Pete" Roberts
Robert A. Sadler
William G. Scroggins
Bart Sinclair
Robert Strode
Darlene Tallent
Jo Thiessen
Gordon Torrence
Peter Watkinson
Jay Worrall
Ken Wray

Index

Numbers in brackes { } denote pages in the ledger

Index

Numbers in brackes { } denote pages in the ledger

Index

Numbers in brackes { } denote pages in the ledger

Numbers in brackes { } denote pages in the ledger

Index

Numbers in brackes { } denote pages in the ledger

Numbers in brackes { } denote pages in the ledger

Index

Numbers in brackes { } denote pages in the ledger

Index

Numbers in brackes { } denote pages in the ledger

Numbers in brackes { } denote pages in the ledger

Index

Numbers in brackes { } denote pages in the ledger

Numbers in brackes { } denote pages in the ledger

Numbers in brackes { } denote pages in the ledger

Numbers in brackes { } denote pages in the ledger

Numbers in brackes { } denote pages in the ledger

Index

Numbers in brackes { } denote pages in the ledger

Index

Numbers in brackes { } denote pages in the ledger

Index

Numbers in brackes { } denote pages in the ledger

Numbers in brackes { } denote pages in the ledger

Index

Numbers in brackes { } denote pages in the ledger

Numbers in brackes { } denote pages in the ledger

Index

Numbers in brackes { } denote pages in the ledger

Index

Numbers in brackes { } denote pages in the ledger

Numbers in brackes { } denote pages in the ledger

Index

Numbers in brackes { } denote pages in the ledger

Index

Numbers in brackes {} denote pages in the ledger

Numbers in brackes { } denote pages in the ledger

Numbers in brackes { } denote pages in the ledger

Numbers in brackes { } denote pages in the ledger

Index

Numbers in brackes { } denote pages in the ledger

Index

Numbers in brackes { } denote pages in the ledger

Index

Numbers in brackes { } denote pages in the ledger

Index

Numbers in brackes { } denote pages in the ledger

Index

Numbers in brackes { } denote pages in the ledger

Index

Numbers in brackes { } denote pages in the ledger

Index

Numbers in brackes { } denote pages in the ledger

Numbers in brackes { } denote pages in the ledger

Numbers in brackes { } denote pages in the ledger

Index

Numbers in brackes { } denote pages in the ledger

Index

Numbers in brackes { } denote pages in the ledger

Numbers in brackes { } denote pages in the ledger

Index

Numbers in brackes { } denote pages in the ledger

Index

Numbers in brackes { } denote pages in the ledger

Index

Numbers in brackes { } denote pages in the ledger

Numbers in brackes { } denote pages in the ledger

Index

Numbers in brackes { } denote pages in the ledger

Index

Numbers in brackes { } denote pages in the ledger

Index

Numbers in brackes { } denote pages in the ledger

Index

Numbers in brackes { } denote pages in the ledger

Made in the USA